The Body as Object and Instrument of Knowledge

STUDIES IN HISTORY AND PHILOSOPHY OF SCIENCE

VOLUME 25

For other titles published in this series, go to
www.springer.com/series/5671

Charles T. Wolfe · Ofer Gal
Editors

The Body as Object and Instrument of Knowledge

Embodied Empiricism in Early Modern Science

Springer

Editors
Charles T. Wolfe
Unit for History and Philosophy of Science
University of Sydney
Sydney NSW 2006
Australia

Ofer Gal
Unit for History and Philosophy of Science
University of Sydney
Sydney NSW 2006
Australia

ISBN 978-90-481-3685-8 e-ISBN 978-90-481-3686-5
DOI 10.1007/978-90-481-3686-5
Springer Dordrecht Heidelberg London New York

Library of Congress Control Number: 2010924982

Cover image: W. Swanenburg, Leiden anatomical theatre, engraving (after drawing by Woudanus), 1610. Courtesy of the Leiden University Rare Books Collection, shelfmark: p. 315 III n. 19

Printed on acid-free paper

Springer is part of Springer Science+Business Media (www.springer.com)

Contents

Contributors

Peter Anstey is Professor of Early Modern Philosophy at the University of Otago. His research focuses on early modern philosophy with special reference to the writings of John Locke and Robert Boyle. He is currently editing (with Lawrence Principe) Locke's writings on natural philosophy and medicine and has recently commenced a 3 year project on 'Experimental Philosophy and the Origins of Empiricism' which is supported by the Marsden Fund of the Royal Society of New Zealand.
E-mail: peter.anstey@otago.ac.nz

Victor Boantza earned his Ph.D. in history and philosophy of science from the University of Toronto. He is presently a postdoctoral fellow at the Department of History at McGill University. He works on the history and philosophy of early modern science, especially matter theories and (al)chemistry in the seventeenth- and eighteenth-centuries, on which he has published several articles in professional journals; he is currently co-editing a volume on Controversies within the Scientific Revolution.
E-mail: boantza@gmail.com

Raz Chen Morris teaches history of medieval and early modern science at the Science, Technology and Society Graduate Program in Bar Ilan University, Israel. He has published several papers on Kepler's optics and its cultural and intellectual contexts in *The Monist*, *Journal of the History of Ideas*, and *Renaissance Quarterly*; is currently working on the poetics of early modern scientific texts, on Baroque Science (with Ofer Gal), and on the discourse of visibility before Copernicus (with Rivka Feldhay).
E-mail: razdchen@gmail.com

Tobias Cheung is Heisenberg Research Scholar at the Max-Planck-Institute for the History of Science and Privatdozent at the Institute for Cultural Studies of the Humboldt-University in Berlin. He has held research and teaching positions at the Technical University of Munich, the Department of the History of Science of Harvard University, Tokyo University, and at the CNRS unit REHSEIS of the University of Paris VII-Denis Diderot. His research interests reach from history of science and cultural studies to philosophy, anthropology and Japanese studies. He is the author of various articles and books, most recently *Res vivens. Regulatorische Theorien und Agentenmodelle organischer Ordnung 1600–1800* (2008).
E-mail: t.cheung@gmx.net

Harold J. (Hal) Cook is Professor and Director of The Wellcome Trust Centre for the History of Medicine at UCL, London, and has published a great deal on early modern English medicine, including the monograph *The Decline of the Old Medical Regime in Stuart London* (1986); he now is focusing mainly on medicine and natural history in the Dutch Golden Age in an attempt to reassess the relationships between the beginnings of a world-wide trading system and a world-wide exchange of information about nature, with the recent publication of *Matters of Exchange: Commerce, Medicine, and Science in the Dutch Golden Age* (2007). He is also co-editor of the journal *Medical History*, and has been elected to an honorary Fellowship of the Royal College of Physicians.
E-mail: ucgahjc@ucl.ac.uk

Ofer Gal teaches in the Unit for History and Philosophy of Science at the University of Sydney. He has published extensively in the history of early modern science and on various issues in the philosophy of science, as well as the book *Meanest Foundations and Noble Superstructures* (2002). His main current research project is 'Baroque Science': a study of the tensions, anxieties and compromises that shaped the New Science of the seventeenth century and enabled its spectacular success.
E-mail: o.gal@usyd.edu.au

Guido Giglioni is the Cassamarca Lecturer in Neo-Latin Cultural and Intellectual History at the Warburg Institute, University of London. He obtained his Ph.D. in History of Science and Medicine at the Johns Hopkins University, Baltimore in 2002. He has published a book on Jan Baptiste van Helmont (*Immaginazione e malattia*, 2000) and edited a volume of manuscript papers by Francis Glisson (1996). He is currently working on Girolamo Cardano and Tommaso Campanella. His interests lie in history of early-modern medicine and philosophy, with special emphasis on sixteenth- and seventeenth-centuries Italy and England.
E-mail: Guido.Giglioni@sas.ac.uk

Snait Gissis is on the faculty of the Cohn Institute for the History and Philosophy of Science, Tel Aviv University. The main focus of her research and teaching in recent years has been the interactions between the social and the biological from eighteenth to twentieth century; and more particularly, that interaction and the constitution of sociology as an autonomous discipline, and the emergence of 'race' as a scientific category and the uses of the concept of 'race' until the present.
E-mail: gissis@post.tau.ac.il

Cynthia Klestinec is an assistant professor in the English department at Miami University, Ohio, where she teaches courses on early modern drama, the representation of disease, and the rhetoric of science. She completed her doctorate at the University of Chicago and has held fellowships from the Max Planck Institute for the History of Science (Berlin), the Villa i Tatti-Harvard University's Center for Renaissance Studies (Florence), and the American Council of Learned Societies. She has published articles on the history of anatomy in the *Journal of the History of Medicine and Allied Sciences* and *Renaissance Quarterly*. She is currently completing a monograph on the post-Vesalian history of anatomy, which uses the records

of medical students to characterize the dramatic, philosophical and pedagogical aspects of anatomical theaters.
E-mail: klestic@muohio.edu

Alan Salter has just completed a doctoral dissertation on the empirical practices of William Harvey. Publications include an article in *Bulletin de la SHESVIE* (2009) jointly authored with Dr. Charles Wolfe on the experimentalism of Harvey and Locke, a chapter on Harvey's application of touch in his inquiries in a book entitled *Word and Self Estranged in English texts 1500–1700* (forthcoming 2010) and an article on Harvey's borrowings from artisanal practice in a forthcoming edited collection. He is currently working on the importance of myth to scientific discovery.
E-mail: sereniti@bigpond.net.au

Lisa Shapiro is Associate Professor of Philosophy at Simon Fraser University in Vancouver. Her research has focussed on early modern metaphysics and philosophy of mind, with particular attention to import of accounts of the passions to these questions. She is the editor and translator of *The Correspondence of Princess Elisabeth of Bohemia and René Descartes* (2007). Her current project concerns the affective dimension of early modern accounts of sense perception, and she continues to be interested in the work of early modern women philosophers.
E-mail: lshapiro@sfu.ca

Justin E.H. Smith is Associate Professor of Philosophy, Concordia University, Montréal. His principal research focus is on the philosophy of G.W. Leibniz, as in his book entitled *Divine Machines: Leibniz's Philosophy of Biology* (2009), and he is currently preparing a critical edition and translation, with François Duchesneau, of the *Negotium Otiosum* and related texts pertaining to the controversy between Leibniz and Georg Ernst Stahl, for the Yale Leibniz series. He has published widely on the intersection between philosophy and the empirical life sciences in the seventeenth and eighteenth centuries, with a particular focus on the philosophical problems arising in early modern biological taxonomy and in generation theory. Justin Smith also has a developing research interest in philosophical anthropology from sixteenth to the eighteenth centuries, and in particular the philosophical debates concerning human origins, the nature of human diversity, and the place of human beings in the order of nature.
E-mail: justismi@alcor.concordia.ca

John Sutton is Professor of Cognitive Science at Macquarie University in Sydney, Australia, where he was previously Head of the Department of Philosophy. He has held visiting fellowships in the UCLA Center for Seventeenth- and Eighteenth-Century Studies, and at the Institute for Advanced Studies in the Humanities, University of Edinburgh. His research primarily addresses the philosophy, sciences, and history of memory. He is co-editor of *Descartes' Natural Philosophy*, and of the new interdisciplinary journal *Memory Studies*. In 2007, his book *Philosophy and Memory Traces: Descartes to Connectionism* was reissued in paperback. His current collaborative projects address social and shared remembering (with Amanda Barnier); kinesthetic memory and skilled movement (with Doris McIlwain and Wayne Christensen); and distributed cognition in early modern England (with Evelyn Tribble).
E-mail: John.Sutton@scmp.mq.edu.au

Anik Waldow is a Lecturer in the Department fo Philosophy, University of Sydney, and the author of *David Hume and the Problem of Other Minds* (2009). She mainly works in seventeenth- and eighteenth-century philosophy and has a special interest in David Hume, the belief-generating function of sympathy, the problem of other minds, scepticism and personal identity. Her current research focuses on the question of how the emergence of empirically based approaches in early modern natural philosophy gave way to a general shift of interest away from the realm of the metaphysical to the natural world of sentient human beings.
E-mail: anik.waldow@arts.usyd.edu.au

Charles T. Wolfe is an Australian Post-Doctoral Fellow in the Unit for History and Philosophy of Science, University of Sydney, working chiefly in the history and philosophy of the life sciences in the early modern period and in contemporary thought. He has published articles on La Mettrie, Diderot, Maupertuis, Sade, Locke and medicine, monsters, organism, the identity theory, determinism and dreams, and has edited the following volumes: *The Renewal of Materialism* (2000); *Monsters and Philosophy* (2005), a special issue of *Science in Context* entitled *Vitalism without Metaphysics* on eighteenth-century medical vitalism (2008), as well as *Desiderio del Mostro* (co-edited with Antonio Negri, 2001) and a special issue of *History and Philosophy of the Life Sciences* on concepts of organism (with Philippe Huneman, forthcoming 2010). His current projects include works on determinism and vitalism in the Enlightenment.
E-mail: c.wolfe@usyd.edu.au

Richard Yeo is a Professor in the Centre for Public Culture and Ideas, Faculty of Humanities and Social Sciences, Griffith University, Brisbane, and a Fellow of the Australian Academy of the Humanities, who writes on the history of science and ideas. His books include *Defining Science: William Whewell, Natural Knowledge and Public Debate in Early Victorian Britain* (1993); *Encyclopaedic Visions: Scientific Dictionaries and Enlightenment Culture* (2001); and *Science in the Public Sphere: Natural Knowledge in British Culture, 1800–1860* (2001). His current project is a study of note-taking, memory, and information management, focussing on members of the early Royal Society of London such as Robert Boyle, John Locke, and Robert Hooke.
E-mail: r.yeo@griffith.edu.au

Embodied Empiricism

Charles T. Wolfe and Ofer Gal

Introduction

It was in 1660s England, according to the received view, in the meetings of the Royal Society of London, that science acquired the form of empirical enquiry that we recognize as our own: an open, collaborative experimental practice, mediated by specially-designed instruments, supported by civil, critical discourse, stressing accuracy and replicability. Guided by the philosophy of Francis Bacon, by Protestant ideas of this-worldly benevolence, by gentlemanly codes of decorum and integrity and by a dominant interest in mechanics and a conviction in the mechanical structure of the universe, the members of the Royal Society created a novel experimental practice that superseded all former modes of empirical inquiry – from Aristotelian observations to alchemical experimentation.

It is enlightening to consider that this view is imparted by both the gentlemen of the Royal Society, in their official self-presentations, and by much of the most iconoclastic historiography of our time. Lines like "Boyle's example … was mobilized to give legitimacy to the experimental philosophy,"[1] are strongly reminiscent of Bishop Sprat's 1667 eulogy of the "Lord Bacon in whose Books there are everywhere scattered the best arguments for the defence of experimental philosophy; and the best directions, needful to promote it."[2] One reason for the surprising agreement is that this picture of openness, benevolence and civility does capture some of the moral-epistemological mores of the empiricism of the New Science, but this very agreement of historians and apologists also harbors a paradox. In interpreting the emergence and *modi operandi* of early modern empiricism through the writings of its public champions, we are attending to the rhetoric which supported the new empirical practices – practices that aspired and promised to replace rhetoric.

[1] Shapin 1994, 185.

[2] Sprat 1667, 35.

C.T. Wolfe and O. Gal
Unit for History and Philosophy of Science, University of Sydney

C.T. Wolfe and O. Gal (eds.), *The Body as Object and Instrument of Knowledge: Embodied Empiricism in Early Modern Science*, Studies in History and Philosophy of Science 25,
DOI 10.1007/978-90-481-3686-5_1, © Springer Science+Business Media B.V. 2010

This paradox in the way historians of science approached empiricism is compounded by a similar paradox in the way it is studied by historians of philosophy. Here, it was a *theory* that received the title 'empiricism' – a particular speculative account of the way human individuals acquire their knowledge of the surrounding world. It is yet more obvious in the modern interpretation of this theory, which is completely disinterested in empirical practices. This interpretation of empiricism put at its center an ahistorical, disembodied, isolated 'mind' – quite the opposite of what the savants of the New Science were experiencing or advocating.

Recent scholarship has done much to undo these paradoxes. We know much more about the array of practices of producing and marshalling experience that the New Science benefited from and was instrumental in developing: sophisticated experimentation, instrument-supported observation, astronomical navigation, surveying and mapping, collection and taxonomy. We are also much more familiar with the cultural context in which these were developed: commerce and seafaring, court and city, counter-reformation and education reform. Yet we are still far from a comprehensive view of the arena in which practitioners of various empirical traditions were learning from and competing with those of other traditions for epistemological primacy; in which new empirical practices were being formed as reliable ways of creating and validating knowledge; and in which philosophical reflection and public argumentation sought to legitimize and institutionalize new and reformed empirical habits.

This volume is a contribution towards filling this gap. It explores one aspect of the development of empiricism which the traditional use of the term obscured: the keen interest in the body as both an *object* of research and an *instrument* of experience.

The need to re-embody our understanding of empiricism is enforced, to begin with, by empiricism's patent indebtedness to the sciences of the body – medicine, physiology, natural history and chemistry. It is in those traditions that early modern savants could find paradigms of empirical inquiry which did not suffer from the low esteem accorded to artisanship. Indeed, a quick survey of the active members of the Royal Society reveals that many of them were physicians, and a significant number of those – disciples of William Harvey. Through Harvey's tutelage, these physicians-virtuosi were inheritors of the empirical anatomy practices developed in Padua during the sixteenth century. Furthermore, the primary research interests of the early Royal Society were concentrated on the body, human and animal, and its functions – much more so than on the mechanics the Society is usually associated with. Similarly, the Académie des Sciences devoted a significant portion of its *Mémoires* to questions concerning life, reproduction and monsters, consulting empirical botanists, apothecaries and chemists. Directly contradicting its self-imposed mandate to investigate Nature in 'proper' mechanistic fashion, the Académie kept closer to experience than to the Cartesian standards of well-founded knowledge. 'Empiricks', throughout Europe and through the seventeenth century, were primary agents of 'empiricism'.

As reflections on experience and the acquisition of knowledge by embodied, affective agents, meditations on 'first philosophy' and essays on 'human understanding' are closer to treatises on the passions, hysteria, the curing of fevers or vertigo, as well as to tracts on the construction and use of instruments, than they are to critiques of pure reason or proofs of the external world. Empiricism meant a new attention to the senses and their function from a physiological, practical and epistemological point of view, and all those were never far apart. The bold knowledge claims of new techniques and technologies of observation required justification, which was offered by the analysis of natural and instrumental perception and the relation between them. These optical, physiological and practical inquiries comprise much of the writings of early modern thinkers who are commonly read as pure, contemplative 'philosophers'. Conversely, significant reflections on the epistemological ramifications of these inquiries are to be found in the most 'scientific' of early modern texts.

The papers in this volume are divided according to three perspectives on empiricism and the body. Part I comprises studies of the body as an object of inquiry. In these, empirical explorations of the human body are presented as exemplars and harbingers of early modern empirical practices. The opening paper by Harold Cook lays a claim for the power of 'matters of fact' in the advent of medical and scientific empiricism of the seventeenth century. This was not a change of 'method', he argues, advanced by the learned, but a takeover of the medical marketplace by practicing empiricks. Cynthia Klestinec looks at this change from the point of view of the medical student in Padua – the leading medical school at the turn of the century. New forms of experience, she shows, required and implied new forms of manual skills, from dissection to preparation, which called into question old divisions between public and private, learned and practical. Both the Paduan empirical medical tradition and the need to re-define the relations and hierarchy of the senses emerge in Alan Salter's contribution. Salter reveals the experiential empiricism embedded in William Harvey's work as deeply entrenched in contemporary representations of first-person experience, notably the 'discourse of the senses' of English poetry and drama of the period. Victor Boantza looks at the seventeenth century Parisian chemist and academician Samuel Duclos in order to stress how natural history in its chemical manifestations also affects our picture of empiricism: it emerged less metaphysics-free than its ideologists hoped. The role of 'chymistry' at the heart of early modern thought, whether discussions of substance, body or the program of natural philosophy itself, is also stressed in the following contribution. Peter Anstey presents Locke, the penultimate empiricist philosopher, as a chymical physician; an active pursuer of Helmontian chimiatric medicine.

But the body was not just an *object* of particular 'sciences' or 'practices', the examination of which colors our construct of 'empiricism' in new shades. It was also, as discussed in Part II, the primary *instrument* of empirical knowledge. It was not a transparent instrument at all: both the physiological function of the senses and their epistemological status as means of gathering knowledge presented an ongoing practical and intellectual challenge, with some surprisingly conclusions. As Ofer Gal and Raz Chen-Morris show, the advent of Kepler's optics and Galileo's telescope came at the expense of the trust in the human eye. The naturalization of vision implied the poverty of the human sense organ and the estrangement of the human mind from its objects. Bacon's experimental investigations on the appetites of matter, as discussed by Guido Giglioni, did not at all serve the type of empiricism commonly associated with his philosophy. They implied inescapable subjectivity and necessitated ethical and political consideration of the mechanisms mediating knowledge and appetite in human societies. Mediation through memory presented another challenge to the empiricist project, which had to be met both practically and intellectually. The solution could be provided by the body, as Justin Smith shows in his study of John Bulwer's language of signs and gestures, but this kind of language, despite its apparent immediacy and universality, raised again the tension between nature and artifice associated with instruments of observation. Memory was a challenge and a locus of debate for any régime of sensation and self-possession, as Richard Yeo's paper demonstrates. It demanded the arrangement and condensation of material that Boyle's insistence on matter of facts could not allow but other advocates of Baconian natural history, like Beale and Hartlib, found necessary. The anxiety and wonder concerning knowledge by and of the body did not subside with the triumph of the New Science and its empiricism. Snait Gissis analyzes the interconnections between 'sensation', 'subjectivity' and biological science into the end of the Enlightenment with her discussion of Lamarck on sentiment. As her paper demonstrates, the empiricist approach to the senses continued to cast them as a source of unreliable, highly personal data demanding uncertain deciphering, rather than as neutral particulars to be accumulated inductively.

The embodied approach to the interpretation of empiricism does not turn attention away from the mind. As the contributions in Part III show, empiricist thought extended bodily consideration to all aspects of cognition and mental life. John Sutton attends precisely to embodied cognition in his discussion of inattention, 'mind-wandering' and restlessness in the medico-philosophical context of British Empiricism. Traditional history of philosophy but also, and even more emphatically its contemporary descendents, will seize on a 'concept' or 'problem' – be it personal identity, causality or free will – and extract it from its embedded context. Sutton, in contrast, returns to a richer 'local history', a history of mind-wandering, medicine, and moral physiology, of habit and body and brain. Lisa Shapiro's paper ventures farther into the heart of philosophical empiricism with a new analysis of Locke's account of our simple ideas. Essential to Locke's thought, she shows, and thus to that of sensationist thinkers such as Berkeley and Condillac, were his reflections on pleasure and pain, from which emerged an instrumental and immersed model of experience. Tobias Cheung extends this theme into Enlightenment psycho-physiological discourse with a reconstruction of Charles Bonnet's notion of 'embodied stimuli' in the context of organic models. In Cheung's analysis, Bonnet continues and transcends the work of French empiricists like Condillac by providing models of organic complexity which integrate physical, mental and sensory dimensions of experience. Anik Waldow challenges the primarily epistemological understanding of this 'stance' we have inherited from Kant, by pointing to the Galenic roots of empiricism. Empiricism, she claims, cannot be understood apart from its ever-present relation to skepticism. The volume concludes with Charles Wolfe's reflections on medically motivated, indeed 'vitalistic' bases for empiricism in the early modern period as an embodied yet curiously non-experimental practice. There are many faces to empiricism, his contribution shows, and the mechanistic, gentlemanly, detached version is not the most important of them.

Some of the papers collected in this volume were discussed in a workshop on Embodied Empiricism conducted in February 2009 at the University of Sydney. The workshop, as well as the project on Early Modern Empiricism of which it was a part, has been supported by Australian Research Council grant DP0772706: *The Origins of Scientific Experimental Practices*. We would like to warmly thank Mariela Brozky, Antonio Clericuzio, Stephen Gaukroger, Snait Gissis, Dominic Murphy, Jessica Ratcliff, Justin Steinberg, Yi Zheng and especially Jennifer Tomlinson for their indispensable part in the success of the workshop and the collection.

References

Shapin, Steven. 1994. *A Social History of Truth: Civility and Science in Seventeenth-Century England*. Chicago: University of Chicago Press.

Sprat, Thomas. 1667. *The History of the Royal Society of London for the Improving of Natural Knowledge*. London: Printed by T. R. for J. Martyn and J. Allestry.

Part I
The Body as Object

Victories for Empiricism, Failures for Theory: Medicine and Science in the Seventeenth Century

Harold J. Cook

Abstract For millennia, learned physicians tried to develop theoretical principles that would guide their therapeutic actions. The most enduring foundations were built on the discourse of the four elements, four qualities, four humours, six non-naturals, and the ways these combined to yield individual temperaments and constitutions. As these fundamentals came under attack in the seventeenth century, empiricism and medical specifics once again seemed the best method of finding certainty in therapy. This was no simple change in "method" proposed by the learned, however, since the developing medical marketplace gave empirics many new opportunities for promoting their views and forcing the rest to take account of them. Does this transition in medicine also apply to "science" more generally, giving prominence to those "matters of fact" that have gained our attention in recent years? The case is made for answering "yes."

It has long been my view that the real threat to the old natural philosophy from the new lay not in particular concepts such as corpuscularianism or atomism, much less heliocentrism, nor in something contemporaries called the experimental method, but in new ways of actively investigating natural phenomena that arose from a generalized empiricism.[1] Since the late 1970s, many have argued that science is best interpreted as a set of activities and practices rather than concepts;[2] a number of studies have gone on to point to the importance of the mode of inquiry formed natural history and "matters of fact" rather than simple empiricism per se.[3] I would now go further and observe that given its close engagement with things in themselves, science is something distinct from philosophy, although to be sure they

[1] Cook 1986, 1987, 1989a, 1990b; Lloyd 1979.

[2] Latour and Woolgar 1986, first publ. 1979; Figlio 1978; Lloyd 1979; MacDonald 1981; Shapin and Schaffer 1986.

[3] Levine 1983; Shapiro 1983; Eamon 1983; Daston 1988; Dear 1990; Cook 1991, 1993; Findlen 1994.

H.J. Cook
Department of History, Brown University

C.T. Wolfe and O. Gal (eds.), *The Body as Object and Instrument of Knowledge: Embodied Empiricism in Early Modern Science*, Studies in History and Philosophy of Science 25,
DOI 10.1007/978-90-481-3686-5_2,

share some historical roots.[4] One of the most interesting historical problems, then, is how some kinds of knowledge about ways of interacting with the constituents of the world were elevated to the status of learned knowledge, when that was still so often associated with the discourse of universities, which had indeed grown from philosophy and theology.[5] For that kind of explanation, attempts to view developments from the culture of economy and society rather than philosophy, via subjects that required both engagement with natural beings and articulate generalizations about that experience, are illuminating.[6] Seen from the point of view of those engaged in medicine, for example, empiricism was a fundamental interactive means for investigating the world and, consequently, a chief cause of change during the so-called scientific revolution.

A previous generation placed medicine on the periphery of scientific knowledge, as a field more affected by fundamental conceptual changes than constitutive of them, and so more or less a subordinate part of the "experimental" sciences, which were in turn held to be subordinate to the "mathematical" ones.[7] A. Rupert Hall spoke for this generation when he noted the great power of the work of the historical philosopher of science, Alexandre Koyré, who searched for "the special importance of those currents which led to Newton, and (which continued) by ways that were relatively direct to Maxwell, Planck, and Einstein."[8] Hall himself had therefore written that "perhaps it is not stretching imagination to see practical medicine playing somewhat the same rôle in the development of biology as that of technology in the evolution of the physical sciences." But since "the liberty of the scientist to direct his theories in accordance with the scientific evidence alone" is fundamental, "the prime focus of attention" on "Man" that medicine brought to science in the end "tended to hamper [the] later development" of biology.[9] A similar line was taken by Charles C. Gillispie, who argued that "the sciences of life, therefore, find their place in the scheme of the scientific revolution. The impression is hard to avoid, however, that it was a subordinate place. Despite the very evident appeal of Vesalius' subject, or perhaps because of it, his achievements were of a lower intellectual order

[4] I am of course aware of the argument about why the word "science" is anachronistic when applied to the early modern period, but for reasons that will become clear below I do not think that "natural philosophy" is a good enough substitute for a collection of activities that also included medicine, chemistry, and artisanal practices; the term science therefore remains a useful shorthand. On learning and worldly engagements, see for example Eamon 1994; Smith 1994, 2004; Long 2001; Cook 2007; Harkness 2007. For the counter-argument, see Cunningham 1991; Cunningham and Williams 1993.

[5] For the problematic status of medicine in the medieval university, however, see Getz 1992, 1997.

[6] For related arguments, Hadden 1994; Kaye 1998.

[7] The argument for a distinction between experimental and mathematical sciences is captured best in Kuhn 1976.

[8] Hall 1983, 147.

[9] Hall 1954, quotations from 134, 275, 289.

than those of Copernicus or Galileo. His were not the ideas which changed man's conception of the world, or even of himself."[10]

Hall himself nevertheless noted that the number of physicians among the early members of the Royal Society was high (as did Michael Hunter shortly thereafter), while a pre-Koyré line of argument about the importance of medicine and chemistry to the development of the new science was further developed by Walter Pagel, Alan Debus, Owen Hannaway, Jerome J. Bylebyl, Robert G. Frank, Jr., Charles Webster, and many others.[11] Physicians, medical chemists, apothecaries, surgeons, and even empirics were not only a significant presence among groups engaged in the new philosophy, most were well-educated and left a good deal of documentation behind, so that those among them who opposed the new views, as well as those who were in favor of them, stated their reasons clearly. It is therefore possible to identify the issues of the day that concerned contemporaries, thus avoiding the dangers of anachronism. That is, seen through the eyes of medical practitioners and authors, it is possible to gain a view of the prospect and problems of the new science as a self-defined group saw them, rather than relying on views as selected by historians to represent the members of the "scientific" community.

The economic and social processes that lay behind the scientific revolution can be seen clearly if one examines them through the lens of the medical community. As is well known, until at least the first half of the seventeenth century, university-educated physicians of necessity were learned, being grounded in philosophy, while most of the many other kinds of medical practitioners were not. The surgeons, apothecaries, chemists, and "empirics" presented a major threat to the social status, political influence, and economic flourishing of the physicians.[12] So, too, they undermined the intellectual superiority of learned physicians. The non-physicians often aggressively argued that they were far better and more knowing medical practitioners because, unlike those usually legally-superior learned men, they had learned by doing. In other words, what they had gathered from practical experience was, they claimed, a far better kind of knowledge than the wisdom inculcated in the physicians by long years spent in reading and debating texts. In arguing this, they were clearly saying that the measure of value in medical knowledge lay in successfully treating patients.[13] One can find this sentiment expressed in many sixteenth-century medical pamphlets authored by those opposed to the establishment, from mountebanks and empirics like Leonard Fioravanti, to medical chymists like Paracelsus and his followers, to surgeons like the famous Ambroise Paré.[14]

[10] Gillispie 1960, 57. For his own debt to, and comment on, Koyre's program, see 523–524. For further historiographical discussion, see the introductions to Cook 1993, 1990b.

[11] Hall 1974; Hunter 1976, 1982; and for examples: Pagel 1958; Debus 1966; Hannaway 1975; Webster 1976; Bylebyl 1979; Frank 1980.

[12] Eamon 1994; Kassell 2005; Webster 2008. For a fresh look at the historiography of medical relationships, see Jenner and Wallis 2007.

[13] Cook 1994.

[14] Eamon 1993; Webster 2008; Jones 1960.

Even anatomical investigators like Vesalius – whose *De fabrica* was, with Copernicus's *De revolutionibus*, long held to characterize the beginnings of the scientific revolution – held up experience with the material things of the world as the touchstone of knowledge, mocking the learning of physicians who instead wished to grasp the world according to doctrines found in books.[15] The medical businessmen called apothecaries took a similar point of view. They not only dealt with medicinal imports from the New World and Asia, but were among the first to collect exotica and naturalia in cabinets of curiosity that were open to the public.[16]

By the mid-sixteenth century, the felt need of physicians to be in control of the details of nature that were commanded by their rivals was so powerful that the eminent Parisian physician, Jean Fernel, told his peers that empirical knowledge of this kind had to be mastered just as much as learned theory.

> The knowledge, collection, choice, culling, preservation, preparation, correction, and task of mixing of simples all pertain to apothecaries; yet it is especially necessary for the physician to be expert and skilled in these things. If, in fact, he wishes to maintain and safeguard his dignity and authority among the servants of the art, he should teach them these things.[17]

In other words, by the 1570s, even defenders of the dignity of university-inculcated medical learning were being forced to try to take the lead in knowing the material details that came from experience with worldly things.

The reasons for the shift in value from understanding the causes of nature to knowing the details about its particular instances are many. I have argued elsewhere that one of the most important reasons for the shift in the valuing of one kind of knowledge to another was the so-called consumer revolution of the late middle ages (or "Renaissance"). Discourses about "the good" shifted to discourses about "goods," and the cultivation of virtue was better displayed by a tangible appreciation for fine "art," practical devices, and useful information than fascination with philosophical elaborations.[18] As if to maintain the fundamental importance of immaterial powers in this world of goods, the clerics fought back by claiming that the future of humanity depended on the right interpretation of the large implications that might flow from small differences in understanding texts and learned traditions. It created a nightmare world of hunting down witches and heretics, and terrible misfortunes followed when these arguments captured the political stage. Yet in the background, the growth of commerce continued to transform ordinary life, and a view of it would emerge in the later seventeenth century in the form of new political economies that judged government by whether it could improve the material lives of its subjects. The Lord Chancellor of England, Sir Francis Bacon, became a convert to these views and an often-quoted spokesman for furthering their

---

[15] Carlino 1999.

[16] Egmond 2008; Findlen 1998; Cook 2007.

[17] Jean Fernel, *Methodo medendi*, quoted in Reeds 1991, 25–26.

[18] Hanson 2009; Cook 2007, esp. Chapter 1.

advancement, but he was only one of most articulate members of an ill-defined and widespread movement. And the "movement" itself would have been impossible without the regular travels of large proportions of the population, learned men among them, which allowed interlinked discoverses to develop.

Of course learned physicians sometimes objected to the knowledge claims of their rivals just as much as their clerical brethren did. As with cases of uncertified preachers, it turns out that one of the chief arguments of the learned against medical empirics built on the difference between good and bad judgment, which in turn depended on a view of the proper temperament and character-formation of the professional, whose work was by definition that of "professing" a view, or explaining the world so as to keep their listeners within its right order. The physicians argued that their chief task was the preservation and restoration of health rather than treating disease. They – like their fellow professionals in law and church – gave their clients good counsel, not only helping when danger was already present but, even better, helping their patients to avoid danger. For this, they needed to understand the regular course of nature (*phusis*) and to be able to understand how the circumstances of the moment could be explained in terms of those regularities, so that their patients could remain in accord with it. Such advice was fundamental to both retaining and regaining health. Put another way, explaining the causes and consequences of nature and the individual's part in it was not supplementary to what physician did, but their core business. They might also recommend proper therapies in cases where more drastic measures were needed to bring the patient's personal constitution back into balance with his or her surroundings, but these means, too, had to be in accord with their understanding of the operations of nature. To advise in these ways they needed the discipline of learning formed in the lecture- and debating-chambers of the schools.[19]

In other words, learning gave physicians not only insight into the underlying *logos* of the world that taught the good as well as the true, but inculcated in them the ordered inner discipline that allowed good judgment to relate such verities to the individual circumstances of life. As the English physician John Cotta put it in 1612,

> The dignitie and worth of Physicks skill consisteth not (as is imagined commonly) in the excellence and preheminence of remedies, but in their wise and prudent use. It is an ancient saying, that wholesome medicines by the hands of the iudicious dispenser, are as Angels of God sent for the good of men; but in the hands of the unlearned, are messangers of death unto their farther evill. Good medicines are in themselves excellent instruments of health and life, but require a learned workman iudiciously to guide them unto their destined end. It is order and not confusion, that is ever safe and happie; and knowledge (which worketh by election, and true reason, and not rash boldnesse, which doth good by chance and uncertaine event) that is the light and safe guide of understanding mindes. ... If then all enterprises prosper by wise advice, & it is wisedom in matters of meanest moment to consult with a wise and iudicious friend, in cases of health and life certainly every man is not a sufficient counsellor.[20]

---

[19] Cook 1990b, 1994.

[20] Cotta 1612, 1–2.

For men like Cotta, uneducated empirics were a danger not only because of the strong and dangerous remedies they might employ – which were bad enough – but also, even more, because they did not have the proper character formation to allow them to dispense good advice. Indeed, the traits by which their rivals could be known were not so much practical as personal: loquaciousness, braggadoccio, unmannerly gestures, unnecessary boldness, and so on – all the traits of the quack that formed generation after generation of a literature designed to give the professionals the power to police their rivals.

For much of the early modern period, then, the conservative wing among the physicians continued to argue that only a proper form of learning that was essentially literate and philosophical could form a mind capable of good judgment. But another argument had emerged and by the later seventeenth-century had, in many places, come to dominate: good judgment could even better arise from the kind of mental discipline that demanded the acquisition and ordering of as much of the material detail about nature as possible. In other words, it was not the grasping of fundamental axioms and methods of argumentation (or the *logos* as known through right reason) but the acquisition (and memorizing) of correct material detail – of exacting descriptive information – that established a disciplined inner character and allowed the exercise of good judgment. The active pursuit of information could itself lead to virtue.

The point was made well by the eminent London physician, Samuel Garth. When at the beginning of the eighteenth century he addressed his colleagues in honor of the famous William Harvey, and spoke about their common profession of "physick," he said something quite different than his predecessors of a century earlier. He began not with a discussion of the causes of things but with notable examples of new cures physicians could perform, whether they understood the reasons or not. When he came to broaden his discourse to include the places where certainty could be had in the knowledge of nature he spoke not of fundamental principles and axioms or generalizable reasoning but of investigation into particulars: he spoke of botany (the "shapes and tastes of an infinite number of plants"), mineralogy, and all other aspects of the description of nature. Physic "pursues nature through a thousand windings and meanders; the very Center of the Earth escapes it not, neither is there any thing in the Ocean hid from it."[21] This was a godly art. But the seat of the godly art had become not an Olympian height but an Odyssean journey. The physicians of course continued to apply their knowledge to the preservation of health and remediation of disease (although usually by assuming that people could be treated alike rather than only according to their individual differences as expressed by their temperaments), but physic itself came to rest on an active investigation of nature and nature's secrets.

The new science therefore encompassed not only a complete knowledge of nature but the active hunt for new and deeper knowledge of it. Certainty lay in the particulars rather than the generalities. Hans Sloane, later President of both the Royal

[21] Ellis 1963.

Society of London and the London College of Physicians, wrote in his book about the natural history of Jamaica that true knowledge was now based on "Observations of Matters of Fact," which were "more certain than most Others, and in my slender Opinion, less subject to Mistakes than *Reasonings*, *Hypotheses*, and *Deductions* are …. These are things we are sure of, so far as our Senses are not fallible; and which, in probability, have been ever since the Creation, and will remain to the End of the World, in the same Condition we now find them."[22] His contemporary, Georgio Baglivi, agreed entirely: certainty in the knowledge of physic remained the goal, "For the Art is made up of such things as are fully Survey'd, and plainly Understood, and of such perceptions as are not under the controul of Opinion. It gives certain Reasons which are plac'd in due Order, and chalks out certain Paths, to keep its Sons from going astray." But it did this by paying careful attention to matters of fact. As long as "Observation is the Thread to which Reason must point," all would be well. But "'tis manifest, that not only the Original of Medicine, but whatever solid Knowledge 'tis entituled to, is chiefly deriv'd from Experience."[23] Or as the most famous medical professor of the day, Herman Boerhaave, had it, all true physical knowledge was built on sense experience. Physic therefore could account only for those things "which are purely material in the human Body, with mechanical and physical Experiments." First causes "are neither possible, useful, or necessary to be investigated by a Physician." Looking back to his hero Hippocrates, Boerhaave explained that "the Art of Physic was anciently established by a faithful Collection of Facts observed, whose Effects were afterwards explained, and their Causes assigned by the Assistance of Reason; the first carried Conviction along with it, and is indisputable; nothing being more certain than Demonstration from Experience, but the latter is more dubious and uncertain …."[24]

The attentive and well-informed clinician had trumped the well-read and argumentative professor. Both could be good. But the sources of goodness in each arose from different kinds of activities and different kinds of knowledge. The new philosophy could also point to its relationship to the good, but it was a different sort of good than that held dear by its scholastic rivals, stressing the true knowledge based on careful use of the senses rather than what was increasingly referred to as "speculation." This is one of the chief themes in the debates in the 1660s between Joseph Glanvill and Henry Stubbe about the newly-formed Royal Society, for instance, which was contemporary and intermixed with debates between the London College of Physicians and the Society of Apothecaries and Society of Chemical Physicians. Glanvill was accused by some of his clerical colleagues of being atheistic for favoring the new philosophy, to which he replied by writing a book (*Plus Ultra*) in defense of it. He argued that the Royal Society explicitly disavowed philosophical discussion in favor of practical knowledge of benefit to

[22] Sloane 1707, vol 1 sig. Bv.

[23] Baglivi 1704, 5, 9, 15. Original published as Baglivi, *De praxi medica. Ad priscam observandi rationem revocanda* (Rome: Typis Dominici Antonii Herculis, 1696).

[24] Boerhaave 1743, 63, 71, 42.

mankind, among which were improvements in medicine. Indeed, chemistry and anatomy were some of his favorite examples for showing the usefulness of the new philosophy to life. For instance, Glanvill dismissed phlebotomy with contempt as an outmoded practice of physicians while arguing that chemical medicines, discovered by the new experimenters, resulted in pure and effective drugs. He also claimed that anatomy was pioneered by the virtuosi and led to such matters as transfusions, which would also much improve the health of men. He clearly found medicine important only insofar as it could cure diseases, and he also clearly found learned physic's abilities in that department sadly lacking.[25]

According to Stubbe, Glanvill's opinions were repeated by a gentleman over dinner, having been boiled down to something like his published opinion that

> The Modern Experimenters think, that the Philosophers of elder Times, though their wits were excellent, yet the way they took was not like to bring much advantage to Knowledge, or any of the Uses of Humane life …. And the unfruitfulness of those Methods of Science, which so many Centuries never brought the World so much practical, beneficial knowledge, as would help towards the Cure of a Cut Finger, is a palpable Argument, That they were fundamental Mistakes, and that the Way was not right.[26]

Because of the incident at the dinner table, and that "sense of Injury I supposed to be done to me and all other Rational Physicians, by this barbarous Opiniatour," Stubbe tried to set the public straight about Glanvill's claims, especially about the mistakes Glanvill made in his chapter on medicine.[27] The debates were vigorous, often focusing on questions about the extent to which chemical medicines had improved treatment, and sometimes taking up questions about whether anatomy was of any usefulness. But by this time (the 1670s), all the physicians agreed in principle that finding things out by experiments was a good thing, and that one of the heroes of the new and experimental philosophy was William Harvey.[28]

Harvey indeed had discovered the circulation of the blood by masterful methods of investigation, bringing him to conclusions he had not intended. It is sometimes said that his discovery had no implications for medical practice. That may be true with regard to medical therapy (the hope of benefit from transfusion excepted).[29] But the breaking apart of the distinction between venous and arterial blood undercut the rationale for dietetic regimen, which (as mentioned above) was so crucial to arguments for the necessity of advice from learned physicians for the retention and restoration of health. This was a revolutionary finding, simply contradicting the basis of Galenic physiology, and launching further investigations onto a sea of confusion, where the particularities of the findings in anatomy and physiology no longer added up to a whole.[30] It has usually puzzled historians that Harvey seems

---

[25] Cook 1989a.

[26] Glanvill 1668, 7–8.

[27] Stubbe 1670b.

[28] Frank 1979.

[29] Davis 1971.

[30] Frank 1980; Bylebyl 1979.

both a figure of revolutionary new empirical knowledge and yet also enormous professional and political conservatism.[31] But he obviously had no problem in squaring that circle. He may even have been a crypto-Catholic, which also upsets the common formula for the English world in which Puritanism, or at least Protestantism, seems necessary to move people toward the new science.

Let us begin with Harvey's religion. We have nothing explicit about his theological opinions in his own words – which is perhaps not surprising given the deep controversies over such matters in his own day. But we can infer several things from a few bits of evidence. He was of course baptized in the established church, and had his early education at the King's School in Canterbury. He was also awarded the Matthew Parker scholarship, a scholarship founded by the Archbishop to support a boy from the King's School in the study of medicine. Perhaps it is worth remembering, though, that as chief minister of Queen Elizabeth's church, Parker was certainly no Puritan. Instead, he was one of the chief persons to craft the so-called Elisabethan compromise. The several endowments he established late in his life were meant in part to counter the radical Protestants. The Cambridge college in which Harvey resided from 1593 had something of a similarly conservative air: while of course within the Anglican fold, Gonville and Caius had been refounded in 1567 by the then-President of the London College of Physicians, John Caius, who was both a conservative humanist – he lectured against Vesalius – and a Catholic.

Catholicism brings us to the fact that Harvey travelled to Padua in 1598 (at the age of 20) to further his studies in medicine. Of course Padua offered a far superior medical education to any then available in England, but it was after all a Catholic place. Because of the growing threat from counter-reformation Spain, the assassination of the Prince of Orange in 1582 by a Catholic, and the discovery of several plots on Elizabeth's own life in the 1580s, the Parliament had passed a proclamation "for the revocation of sundry of the Queen's Majesty's subjects remaining beyond the seas under colour of study," while sermons were preached against "Oure Italienated Papistes." Not until the peace treaty with Spain in 1604 did traveling to Catholic lands for study bring less suspicion.[32] Harvey was not, of course, the only medical student to defy the law and go abroad in the 1590s: Robert Fludd also left for the Continent in 1598, although he returned to take his M.D. at Oxford in 1605. Nor need we suspect any or all of the English travelers of harboring Catholicism merely because they studied in Italy: the eminent London physician, Simeon Foxe, for instance, was the youngest son of John Fox, the Protestant martyrologist, and himself remained a conformist Puritan.[33] And of course, Venice – in whose territory Padua resided – was from the 1590s battling the Pope's claim to supremacy, leading to a papal Interdict in 1606. But despite toying with alternatives, it remained a Catholic place, and given the times, these intending physicians took some risk of being viewed as Catholic sympathizers upon their return to

[31] Bylebyl 1978.

[32] Quoted from Chaney 1988, 16.

[33] Birken 1977, 306, 328–330.

England. Moreover, as Jonathan Woolfson noted about a decade ago, when Harvey was awarded his doctorate in philosophy and medicine in April 1602 by count palatine, he swore and signed the required oath to the Pope's supremacy and all the articles of the council of Trent, although at his request this was left off the copy of the certificate with which he returned, which is now in the College's Library.[34] It is worth noting here, too, that Harvey's main patron was Thomas Howard, 2nd Earl of Arundel, a scion of one of the greatest Catholic families of the era, although Arundel nominally became a Protestant after 1615. Maybe, then, a whiff of Rome hung about Harvey, which might even account for why Harvey was not appointed a royal physician-in-ordinary until December 1639, by which time the religious policies of Charles I and Archbishop Laud seemed to many to be bent on the restoration of the old religion.

All this is not to argue simply that Harvey was a crypto-Catholic – although that possibility must be entertained – but that most probably, like many other learned men of his day who took an interest in the study of nature, he was not committed to a doctrinal position so much as a broad church. He may be a bit like the great Flemish linguist and botanist Carolus Clusius, who fought the Inquisition but also hated committed Protestants. In Clusius's case, he may even have been a member of the Family of Love, a secretive group that allowed its members to take whatever public oaths were necessary depending on circumstance – an argument for religious doctrine itself being something like "a thing indifferent."[35] Both he and Harvey were well-trained humanist classicists who had learned how to apply the skills of the philologist to the investigation of nature: much reading, a powerful memory, attention to detail, trying things out, and a sense that active investigation alone could help to recover the authentic knowledge of humankind.

As for the last suggestion, Harvey's association with Robert Fludd points in interesting directions that were often associated with this open-ended search for ancient wisdom. As Harvey's best biographer, Sir Geoffrey Keynes, notes, Fludd was not only the first in England to support Harvey's discovery of the circulation in print, but also placed the heart at the center of all animal motion, including the passions. The place of the heart at the center of human experience was then preoccupying many literary and religious authors, and would lead, in the middle of the century, to Catholic worship of the Sacred Heart. Fludd was one of the first so-called Rosicrucians, believing that a true understanding of ancient wisdom could be employed to reform natural knowledge in the present. He was also a pantheistic materialist, and Harvey shows many of the same philosophical attributes, although they have usually been attributed to his sympathy for Aristotle. Another of Harvey's friends was the famous materialist, Thomas Hobbes, of whom Keynes notes that his early treatise *De corpore* shares a theory of the senses with Harvey's notes to his own work *De motu locali animalium*. On his deathbed, Harvey left a ring to Hobbes in order to be remembered by him. And finally we have the several reports that

[34] Woolfson 1998, 21–23.

[35] For Clusius, see Cook 2007; Egmond et al. 2007.

Harvey was a free-thinker who believed that suicide was a proper way out of personal suffering, perhaps even attempting suicide with laudanum first in 1652 and doing it successfully after a debilitating stroke in 1657.[36] All in all, the more one probes, the more one can only agree with those biographers who have found Harvey's personal views to be elusive. But in no sense was he publicly attracted to Puritanism or doctrinal Protestantism, or to the contemporaneous controversies about religious dogma or ceremony.

In the London College of Physicians – it would not call itself the "Royal" College of Physicians until the 1680s – Harvey joined a group of men who generally held similar views: that knowledge of nature was far more important than religion. The private religious views of most Fellows of the College of Physicians are not known. A few can be identified as strong Protestants – or "Puritans" – but they were more or less equalized by several others who were clearly Catholics or Catholic sympathizers. As is well known, the College had been founded by medical humanists, the most famous among them being Thomas Linacre, at a time when both Henry VIII and his wife, Catherine of Aragon, patronized the latest scholarship. Humanism made a powerful return under the Catholic Queen Mary, at which time the College's ambitions for policing the physicians of London were encouraged by her chief adviser, Cardinal Pole, who had been educated in part by Linacre himself. Under Mary, President Caius successfully waged war on empirics in London and even claimed supremacy over Oxford and Cambridge, getting Pole to support the College's demand that medical degrees not be awarded without the students having followed a prescribed course of study.[37] Nor did Caius allow the College's own members to publicly dissent from philologically-inspired medical humanism. From Mary's reign onwards, the sense of priestly order and dignity embodied in the College's own ceremonies would also carry the scent of popery to sensitive Protestant noses. For the first 20 years of Elizabeth's reign, her government had little use for a kind of semi-Catholic fraternity of learned physicians. But by the 1590s, as the policies of the Crown became more conservative, it once again supported more active policies on behalf of the College to police the medical practitioners of London – support that continued under James I.[38]

In the years before Harvey's admission to the Fellowship, the College was especially concerned to bring the surgeons into obedience, portraying themselves as the defenders of learning against the advocates for empiricism. At the same quarterly meeting at which Harvey was elected a Fellow – 5 June 1607 – all members were asked to remember occasions on which one Dr. Bonham had practiced, to help with his prosecution. The resulting "Bonham's case" was truly important for the legal precedents it created. Thomas Bonham was also a Cambridge man, about 15 years Harvey's senior. He likely graduated in medicine in the middle to later 1590s, and certainly before he signed himself "in Medic: Doct." at the end of a laudatory Latin

---

[36] Keynes 1966.

[37] Annals, 1:7b–20a.

[38] Cook 1989b, 1994.

poem in 1602. The poem is prefaced to a treatise on the King's evil by the surgeon William Clowes the elder, and as this suggests, Bonham took the side of the London surgeons in their disputes with the College. In early 1605, he also signed the unsuccessful Parliamentary petition of the Barber-Surgeons' Company asking for the right to administer internal remedies. It may not be surprising, then, that when, just a year after Harvey, Bonham presented himself to the College for examination, the Censors failed him. He underwent examination again on 14 April 1606, but his replies to questions this time were declared "not pertinent."[39]

To summarize the consequences: because Bonham practiced without their permission, the President and Censors sent Bonham to Newgate for contempt, to be held at their pleasure. Bonham's lawyer managed to free him by entering a writ of habeas corpus before the court of Common Pleas, while the officers of the College in turn obtained the support of the royal judges, including Lord Chancellor Ellesmere. Subsequently, they sued Bonham for twelve months' illicit practice (for a total of £60), with their case in King's Bench pleaded for them by Attorney General Hobart – they won this case. Bonham in turn returned to the court of Common Pleas, asking for £100 damages against the College for trespass against his person and wrongful imprisonment, which in 1610 he also won, setting him free, and fining the College £40. The declaration of Chief Justice Edward Coke created a clear distinction between malpractice and illicit practice, and more importantly introduced the view that "the common law will controul Acts of Parliament." He may have intended only to overturn the College's royal patent, but his language allowed eighteenth-century Americans to interpret his words as giving the courts jurisdiction over Parliament, helping to create the idea of a powerful Supreme Court. Lord Chancellor Ellesmere, Sir Francis Bacon, and King James himself were all furious with Coke's decision, and various decisions they took in effect further strengthened the College despite Coke's opinion.[40]

It is worth dwelling on Bonham's case for a moment not to begin a rehearsal of all the politico-legal affairs of the College during Harvey's time in which the physicians battled for supremacy over other kinds of medical practitioners – which they mainly won until the outbreak of civil war in 1641 – but to indicate that Harvey joined a body actively involved with the highest levels of government in efforts to maintain and extend the powers of the medical establishment, and that these were bound up with a defense of a conservative version of learned medicine. He seems to have fully supported his institution. On 16 October 1613, for instance, Harvey brought a letter to the President and Fellows from Viscount Lisle, Chamberlain to the Queen, charging a Mr Talbot (MA) with possible bad practice on his niece the Countess of Rutherford, and requesting they examine him. Afterward, he served several times as a Censor, one of those charged with examining and disciplining medical practitioners. In 1630, he was instrumental in causing the officers of the

[39] Cook 1985.

[40] Ibid.

College to press the City of London for enforcement of the plague orders they had developed, and may have been a go-between with the Privy Council during this period, when the Crown was eager to use the College in its attempts to police the City. And of course he followed King Charles during the civil wars.

But if Harvey was no radical in religion or politics, his pursuit of the study of nature undoubtedly led to important results. My own view of this is, however, that while his discovery of the circulation of the blood was fundamental, it was not based on any radically new approach to either philosophy or natural investigation. He pursued the course of carefully accumulating observations and making trials, something not only common at Padua but equally in Montpellier, Leiden, and other centers of up-to-date medical study. His studies were similar to those of his younger contemporary, Sir Thomas Browne, also meant to clear away the accretions of speculation in favor of matters of fact, and like Browne, and even Descartes, Harvey placed his faith in what the senses can tell us about nature. Like Browne, too, Harvey seems to have collected a cabinet of naturalia. For although it was no doubt severely depleted by the actions of the anti-royalist mob who destroyed so many of his possessions and papers during the civil war, it is sometimes forgotten that Harvey's great gift to the College in 1652 was meant not only to endow a library, but "a Repository for Simples and Rarities" – in other words a collection of botanical and zoological specimens. The contents of his museum deserve further study. As at so many places and, in another decade, at the Royal Society, the possession of a cabinet of natural curiosities was considered to be very important for the study of nature. What some physicians such as Harvey further pursued in addition, of course, was not only the careful visual description of living and dead specimens, but the active inspection by vivisection of bodily processes – which along with alchemy might be considered the first experimental science. The studies Harvey made which are written up in his work *On Generation* (published in Latin in 1651 and English in 1653), are particularly impressive for the range of methods he mentions for the study of nature. Coincident with the publication of his last book, then, he was doing all he could to encourage his College to become the foremost scientific institution in England.[41]

Harvey is therefore a personally fascinating but elusive figure. It was of course his discovery of the circulation of the blood that made him deservedly famous. Without it, historians would find his other published work, as well as his activities on behalf of the College of Physicians, well worth study, but would not think that they were particularly unusual. Nor would his reticence about religion, nor his possible connections with Catholicism and free-thinking, come as a shock. What seemed surprising to twentieth-century eyes is the combination of his activities: his conservatism coupled with innovation. But with Harvey's example in front of us, perhaps it is less surprising that even some of the chemical physicians who tried to establish a society of chemical physicians in the mid-1660s had been royalists during the civil wars (while others had been Parliamentarians), and that they were

[41] Webster 1967, 1979.

equally divided in religious orientation.[42] The point is that despite the strong views of earlier historians that "metaphysical" or ideological assumptions lay behind the work of people like Harvey, people with all kinds of differing fundamental philosophical, religious, and political views could share an interest in empirical investigations, and agree on the results.[43]

In The Netherlands, an equally diverse group of natural investigators were excited about anatomy, too. One of them was a resident Catholic, René Descartes. It was there that Descartes wrote his most famous philosophical works, books deeply affected by local information and conversations: the *Discourse on Method*, including the *Optics*, *Meteorology*, and *Geometry*, *Treatise on Mechanics*, *Meditations*, *Principles of Philosophy*, *The Passions of the Soul*, and the posthumous *Description of the Human Body*. And it was in The Netherlands that he developed his strongest intellectual friendships and most powerful theological enemies. His environment changed him as much as he changed it. And it turned him toward empirical investigations of the physical world.

The general Dutch orientation toward investigating phenomena using physical methods as the foundation for natural philosophy quickly left its mark Descartes – as it did for fellow travelers to the north such as Gassendi and Mersenne. The approach of his close early friend Beeckman, for instance, had been to isolate individual problems and then work toward their solution using the best methods available for each rather to apply a universal set of rules.[44] This hardly compared to the thorough system Descartes dreamed about in his youth, but it worked well. Descartes had seen that everything was interconnected, but he found that starting from first principles was a long way from solving practical problems. Daniel Garber has recently concluded that "it is indisputable that as his system grew, perhaps from the first metaphysics of 1629–1630 onward, method became, first in practice, and then after 1637 in theory, less and less important to Descartes."[45] Put another way, in The Netherlands the predominant method, if that is the right word, was *circumspice*, "look around you." This underpinned, for instance, the eclectic approach of his friend, Constantijn Huygens, secretary to the Prince of Orange and enthusiast for Baconianism, who had been introduced to Descartes in 1632 by Jacobus Golius (the Leiden professor of mathematics and of Oriental languages); Huygens often tried to turn Descartes' attention to practical ends, thinking that his grandest philosophical projects were overly ambitious. As he absorbed more and more of this practical and empirical outlook, Descartes left behind the dreams of his youth while turning to the study of difficult questions about how things are, like so many of the naturalists around him. He took up something he called *expérience* (or experiment).

---

[42] Cook 1987.

[43] I have in mind, for example, the important work of Pagel 1967, 1969, who argued that it was Harvey's "Aristotelianism" that created the possibility for his discovery.

[44] See the fundamental work of van Berkel 1983, a revised and translated version of which is due soon.

[45] Garber 2001, 51, also 85–110; also see Garber 1978.

One of the greatest enthusiasms Descartes developed regarding *expériences* was the exploration of the material structure of animal bodies. These were years of intense public interest about anatomy lessons performed at Leiden, Amsterdam, and Delft, with Rembrandt's famous portrait of Dr. Tulp done in 1632. Descartes, too, became captivated. Almost as soon as he embarked on his new philosophical project, at the end of 1629, he wrote to Mersenne that he did not wish to be distracted with further philosophical inquiries, for "I want to begin to study anatomy." A few months later he wrote that "I am now studying chemistry and anatomy simultaneously; every day I learn something that I cannot find in any book." As he put it in a later summary of his projected *De mundo*, he recognized that "I did not yet have sufficient knowledge to speak of [animal bodies] in the same manner as I did of the other things – that is, by demonstrating effects from causes and showing from what seeds and in what manner nature must produce them." He had come to see that while his first principles could be used to construct the building blocks of creation, when it came to animals and humankind, especially, they could only explain the effects discovered – one could not imagine all that existed merely by proposing first principles. In effect, first principles are far better at *post hoc* explanation than at predicting real things. One of the reasons for not publishing his early *De mundo*, he told the world, was that "every day I am becoming more and more aware of the delay which my project of self-instruction is suffering because of the need for innumerable *expériences* which I cannot possibly make without the help of others."[46]

Accordingly, the section on human physiology prepared for *De mundo* and published posthumously as *De homine*, or *L'Homme* (the "Treatise on Man," also known as "Description of the Human Body"), instructs the reader about human anatomical parts: "I assume that if you do not already have sufficient first-hand knowledge of them, you can get a learned anatomist to show them to you – at any rate, those which are large enough to be seen with the naked eye." Once he had mastered the technique, Descartes dissected diligently on his own from animal material he obtained from butchers. He could not obtain access to human body parts because, as he noted, "I am not a doctor by profession," but he applied himself diligently to looking into the parts of animals.[47] In notes about such dissections, which Leibniz later copied and which take up nearly a hundred pages in the printed edition, Descartes wrote down precise observations about a very large number of studies, often on the fetuses or new-borns of cows: he was clearly interested in the physiology of development, among other matters, for he was still dissecting calves when living in Egmond, in the late 1640s.

Descartes made it very clear, too, that he now considered his earlier work on the foundations of the method to have been undertaken to enable a practical philosophy. The main purpose of his work now was

[46] For Descartes in the Netherlands, see Cook 2007, 227ff.; quotations from Descartes 1985, 3:21, 1:134, 149.

[47] Quotations from Descartes 1985, 1:99, 3:59.

> the maintenance of health, which is undoubtedly the chief good and the foundation of all other goods in this life. For even the mind depends so much on the temperament and disposition of the bodily organs that if it is possible to find some means of making men in general wiser and more skilful than they have been up to now, I believe we must look for it in medicine .... Intending as I did to devote my life to the pursuit of such indispensable knowledge, I discovered a path which would, I thought, inevitably lead one to it.[48]

To accomplish such an ambitious goal, however, he acknowledged the necessity of constant labor: "I also noticed, regarding *expériences*, that the further we advance in our knowledge, the more necessary they become." Yet "I see also that [such *expériences*] are of such a kind and so numerous that neither my dexterity nor my income (were it even a thousand times greater than it is) could suffice for all of them."[49]

By the time he came to write the sixth of his famous *Meditations*, around 1640, he was prepared to be quite different from the Descartes of most undergraduate lectures. Although he never took back his earlier proofs for the mind being distinct from body, "I am not merely present in my body as a sailor is present in a ship, but ... very closely joined and, as it were, intermingled with it, so that I and the body form a unit."[50] He also told the reader that he was especially pleased with the proofs for the existence of the material world. They showed (versus Montaigne and other skeptics) how one could have confidence in the knowledge about the world that came to the mind through the senses. He therefore shifted the apparent reason for publishing his work: while most people interpreted it (and still do) as casting doubt on our knowledge of the world via the senses, so that only pure intellect and God's existence can be known with full clarity, Descartes argued that he only wanted to show that knowledge of the latter was more certain even than knowledge of the world, *which should not be doubted*:

> The great benefit of these arguments is not, in my view, that they prove what they establish – namely that there really is a world, and that human beings have bodies and so on – since no sane person has ever seriously doubted these things.[51]

In other words, his goal in publishing the *Meditations* was to establish the certainty of our knowledge of God and the intellect, not to sow doubts about whether we have bodies intertwined with mind, that we can know about the material world, and so on, matters that he thought were self-evident in his empirical studies.

Turning to the text of the crucial sixth meditation itself, we therefore find him emphasizing how *expériences* are required to know the world, introducing arguments for having confidence in most of what one knows via the body. God is not a deceiver, he says, and he endows us with a variety of faculties by which one can check and correct knowledge that comes via the senses, which "offers me a sure hope that I can attain the truth even in these matters." "Indeed, there is no doubt that

[48] Descartes 1985, 1:143.

[49] Descartes 1985, 1:143, 144.

[50] Descartes 1985, 2:56.

[51] Descartes 1985, 2:11.

everything that I am taught by nature contains some truth." And what does nature teach us in general? "There is nothing that my own nature teaches me more vividly than that I have a body."[52] He went on from these arguments about the body and sensory experience to say something even more surprising in light of his youthful meditations: the original source of his doubts – how could he tell if he were dreaming, or awake, or being deceived by a demon? – had no foundation. "I should not have any further fears about the falsity of what my senses tell me every day; on the contrary, the exaggerated doubts of the last few [meditations] should be dismissed as laughable. This applies especially to the principal reason for doubt, namely my inability to distinguish being asleep and being awake. For now I notice that there is a vast difference between the two …."[53] He had achieved his ambition, to lay to rest the ghost of skepticism, showing not only that the clearest and most distinct ideas one could have were about God and the intellect, but also showing that mind was intermingled with body and that knowledge about the world was dependable.

Descartes' *Meditations* was, then, arguing for our ability to know the world, and for the use of natural reason rather than right reason to do it, even in the maintenance of health, but divided what we know about nature from the more doubtful subject of moral philosophy. It freed the study of nature from the vexed problems of contemporary religion, although it smacked of materialism. If there is any doubt about whether the implications of Descartes' *Meditations* were of medical interest, it should be laid to rest by subsequent events. For following its publication in 1641, Descartes became involved in a grave dispute in Utrecht about where his views were leading, a dispute that arose first not among the philosophers but between the physicians and the theologians.

Moreover, by that time he was also deeply engrossed in analyzing the passions, again arriving at some striking conclusions. He treated the passions as aspects of body that communicate to us how we can be happy and healthful: in other words, he rooted them in his physiological outlook rather than in moral philosophy. He first took the usual line that reason needed to control them, but later declared them all to be good; only a few people alive could manage to regulate them, and the rest of us should not worry about our health or virtue if we could not, so that we should embrace them.

Descartes' close consideration of the problematic relationship between reason and the passions and his changes of view were based on his physiological knowledge coupled with the concerns pressed on him by the young Princess Palatine Elizabeth. Descartes had heard through mutual acquaintances in late 1642 that the young Elizabeth was reading his *Meditations*, and he managed an introduction, which led to a life-long relationship. Elizabeth kept pressing him about his views and forced him to reconsider his too easy assumptions about the superiority of using Reason to govern one's life.[54] Descartes listened carefully, and during the

[52] Descartes 1985, 2:56.
[53] Descartes 1985, 2:61.
[54] Shapiro 2007.

course of their conversation, personal and epistolary, he came to the conclusion that in most cases reason could not control the passions, and even that the passions were all good because they taught us how to live well. Through our bodies they connect us to the world of change, informing us about it for the sake of our own well-being.

The last work published in his lifetime was, then, *Les Passions de l'âme*. For Descartes, the passions mediated between body and soul, and are affected by both. He drew no equivalence between passion and error, as did most classical thinkers. Even more strikingly, in the end he declared the passions to be good. One should not become anxious about one's passions, for they teach us what is necessary for life. Even more powerfully, all the pleasures that are common to both soul and body, such a love, "depend entirely on the passions."[55] This is a large step beyond Aristotle's view that in some cases some of the passions can be good; it is almost unthinkable for Descartes' neostoic predecessors; and it went considerably further than virtually anything else said by his contemporaries toward making the passions into forces for good instead of irrationality and vice. It was a far different view than, say, that of Hobbes, who examined the passions carefully but argued for the necessity of the supremacy of reason over them; I think Descartes' view led more to the monist naturalism of Spinoza, Mandeville, and Hume and other members of the "radical" Enlightenment than to the dualism of Locke, Voltaire, and other moderates.[56] Many scholars are now excitedly pursuing this text and the preliminary correspondence as the key source for understanding the mature views of Descartes. It throws great doubt on the question of Descartes as an advocate for the power of disembodied thought. As a consequence, an empirical and passionate Descartes now stalks the literature.

One last comment about this newly passionate and experimental Descartes: while it is now much doubted whether Descartes ever placed much emphasis on metaphysics as a method for finding truth, it is much admitted that he took a deep interest in medicine. Moreover, many of the first outspoken supporters of Descartes were physicians, most famously Henricus Regius in Utrecht, but soon after François dela Boë Sylvius, Florentius Schuyl, and others in Leiden and elsewhere. Such people had brought into the academic world a tradition of active experiential, even experimental, studies with diseases, anatomy, and chemistry, among other fields: they were not only Cartesians, but Harveians, Helmontians, in due course Newtonians, and other intellectual mixtures. Many of them considered Descartes to be on the edge of outright materialism, or perhaps privately so. But for physicians, what Descartes' writings promised was not so much a new metaphysics – much less one that said the best theories came from deductive reasoning – but a demonstration that the physical investigation of natural bodies on which they had long been engaged was indeed the path to a true understanding of nature. It was also an argument for embodied empiricism,

---

[55] *Les Passions de l'Ame*, in Descartes 1985, I:403.

[56] Cook 1999, and for the "radical enlightenment," see esp. Israel 2001.

in which the passions had far more to do with how we come to know the world than did some sort of disembodied reason.[57]

In conclusion, let me return to the historiographical problems that lie behind the claims I have made above. Medicine is not only an excellent descriptive focus for understanding the causes and effects of the scientific revolution; medicine and closely-related topics of investigative activity helped to cause the changes as well. But because medicine had no single conceptual revolution but countless ones, it fits uncomfortably among the historiographical approaches that flow from philosophical idealism, among which I include many of the cultural studies approaches of the very recent past. The changes indicated above do of course point to changes in cultural values, but they were intimately bound up with changes in both material culture and in bodily activities rather than "mental" events. Certainly the changes were not due to one or another particular party to the religious disputes of the day. Arguing for causal changes arising from careful attention to human entanglements with the material world may seem to some historians to invite the dangers of whiggism and positivism, since investigations into matters of fact were crucial, and a great deal of information that was accepted as true remains so. For almost three decades, questions about "credibility" have been vigorously debated among historians and philosophers of science and medicine. Credibility cannot be reduced to matters of trust, nor can trust be reduced to matters of social status, although of course trust and status must be a part of the complex answer.[58] There is nevertheless also the matter of "discovery" and of the coercive power of how things work, whether those are, in ordinary language, "facts" or "laws."[59] Unless we historians and philosophers continue to address ourselves to the old question of "how did they find that out?" as well as "why did they think that?," we will lose audiences who intuitively understand that science is not the same sort of enterprise as philosophy or other subjects in the humanities. I do not mean by this that we should return to simple-minded accounts of material progress: far from it.[60] But I do mean that we need to pay attention to bodily and material culture, in which the senses and passions come first and something like disinterestedness a far distant second.

For medicine, a fundamental problem remained. If bodily empiricism was the grounding for science, then who was to say whether the empiric, apothecary, or surgeon, or even a peasant woman from South-East Asia, was not better informed about the clinical facts, and therefore a better practitioner, than a well-educated physician? Indeed, many of the historical episodes I have worked with raise the point bluntly, whether it was Jacobus Bontius on Java in about 1630 writing about how deeply impressed he was by indigenous female practitioners,[61] or Robert Boyle speaking of how a knowledge of matters of fact often arose from the illiterate or even

---

[57] Verbeek 1991, 1992, 2000; and, for example, Lennon and Easton 1992.

[58] Most powerfully and provocatively put in Shapin 1994.

[59] Mokyr 2002.

[60] A rather distressing recent example is the paperback edition of Wootton 2006.

[61] Esp. chapter 5 of Cook 2007.

the Chinese. "Where Practitioners of Physick are altogether illiterate, there oftentimes Specificks may be best met with," he wrote in *The Usefulnesse of Experimental Naturall Philosophy*, only to have a physician of the College of Physicians who also wished to advance knowledge retort: "That reasoning [is] equally absurd, which pleads for the Empericks to be countenanced as if their experimentings might very much further this pretended Reformation in Physick."[62] If the questions to which the public wanted answers were about how best to remedy various illnesses, then matters of fact might be all that counted, whereas educated people understood that there was much more to advising about medicine and health than that. But how to make the case? John Aubrey even commented that the public began to shun Harvey's practice after he published on the circulation of the blood, because his ideas had become controversial, and by implication what they wanted was someone known for excellent treatment.[63]

In other words, just as we need to be attentive to the sensory knowledge of matters of fact, and the discovery of the world through the passions, we also need to recognize (as no doubt we all do) that the cultivation of public opinion can indeed lead to quackery. The notorious example of the young Dr. Edward Hannes is therefore a last example. Around 1700, he came to London fresh from his studies in Oxford and tried unsuccessfully to get up a good practice. Finally he ordered his footman to stop the carriages of gentlemen in the streets "and enquire whether they belong'd to Dr. Hannes, as if he was called to a Patient." After failing to find Hannes, the footman was to go to the well-known Garroway's Coffee-House and "inquire upstairs and down." The social elite soon came to think that a medical prodigy had come amongst them, and asked him to call on them as well. Doctor Hannes soon rose to a knighthood and the post of Principal Physician at the Royal Court.[64]

Given the importance of human interests and passions, then, when commerce became the dominant mode of getting on in the world, the knowledge favored by the public was that pertaining to the practical business of improving their material condition – and this can be seen in medicine long before Bacon agreed. But then how to tell good from bad information, or pretenders to knowledge from the authentic voices? That remained the most pressing problem for medical practitioners, as well as for the public. It remains a pressing question for us today.

References

Aubrey, John. 1949. *Aubrey's Brief Lives*, edited by O. L. Dick. London: Secker and Warburg.
Baglivi, Giorgio. 1704. *The Practice of Physick*. London: Printed for Andr. Bell et al.
van Berkel, K. 1983. *Isaac Beeckman (1588–1637) en de mechanisering van het wereldbeeld*. Amsterdam: Rodopi.

[62] Boyle 1663; Hodges 1666.
[63] Aubrey 1949.
[64] Pittis 1715, 40–41.

Birken, William J. 1977. *The Fellows of the Royal College of Physicians of London, 1603–1643: a social study*. PhD dissertation University of North Carolina at Chapel Hill.
Boerhaave, Herman. 1743. *Academical Lectures*. London: W. Innys.
Boyle, Robert. 1663. *Usefulnesse of Experimental Naturall Philosophy*. Oxford: Printed by Hen. Hall.
Bylebyl, Jerome J. 1978. William Harvey: a conventional medical revolutionary. *Journal of the American Medical Association* 239:1295–1298.
Bylebyl, Jerome J. 1979. The medical side of Harvey's discovery: the normal and the abnormal. In *William Harvey and His Age*, edited by J. J. Bylebyl. Baltimore: The Johns Hopkins University Press.
Carlino, Andrea. 1999. *Books of the Body: Anatomical Ritual and Renaissance Learning*, translated by J. Tedeschi and A. C. Tedeschi. Chicago: University of Chicago Press.
Chaney, Edward. 1988. Quo vadis? Travel as education and the impact of Italy in the sixteenth century. In *International Currents in Educational Ideas and Practices*, edited by P. Cunningham and C. Brock, 1–28. Evington: History of Education Society.
Cook, Harold J. 1985. 'Against Common Right and Reason': The College of Physicians against Dr. Thomas Bonham. *American Journal of Legal History* 29:301–322.
Cook, Harold J. 1986. *The Decline of the Old Medical Regime in Stuart London*. Ithaca, NY: Cornell University Press.
Cook, Harold J. 1987. The society of chemical physicians, the new philosophy, and the restoration court. *Bulletin of the History of Medicine* 61:61–77.
Cook, Harold J. 1989a. Physicians and the new philosophy: Henry Stubbe and the virtuosi-physicians. In *Medical Revolution in the Seventeenth Century*, edited by R. French and A. Wear, 246–271. Cambridge: Cambridge University Press.
Cook, Harold J. 1989b. Policing the health of London: The College of Physicians and the early Stuart monarche. *Social History of Medicine* 2:1–33.
Cook, Harold J. 1990a. Sir John Colbatch and Augustan medicine: Experimentalism, character and entrepreneurialism. *Annals of Science* 47:475–505.
Cook, Harold J. 1990b. The new philosophy and medicine in seventeenth-century England. In *Reappraisals of the Scientific Revolution*, edited by D. C. Lindberg and R. S. Westman, 397–436. Cambridge: Cambridge University Press.
Cook, Harold J. 1991. Physick and natural history in seventeenth-century England. In *Revolution and Continuity: Essays in the History and Philosophy of Early Modern Science*, edited by P. Barker and R. Ariew, 63–80. Washington, D.C.: The Catholic University Press of America.
Cook, Harold J. 1993. The cutting edge of a revolution? Medicine and natural history near the shores of the North Sea. In *Renaissance and Revolution: Humanists, Scholars, Craftsmen and Natural Philosophers in Early Modern Europe*, edited by J. V. Field and F. A. J. L. James, 45–61. Cambridge: Cambridge University Press.
Cook, Harold J. 1994. Good advice and little medicine: The professional authority of early modern English physicians. *Journal of British Studies* 33:1–31.
Cook, Harold J. 1999. Bernard Mandeville and the therapy of the 'Clever Politician'. *Journal of the History of Ideas* 60:101–124.
Cook, Harold J. 2007. *Matters of Exchange: Commerce, Medicine and Science in the Dutch Golden Age*. New Haven: Yale University Press.
Cotta, John. 1612. *A Short Discoverie of the Unobserved Dangers of Severall Sorts of Ignorant and Unconsiderate Practisers of Physicke in England*. London: Imprinted for William Iones and Richard Boyle.
Cunningham, Andrew. 1991. How the *Principia* got its name; Or, taking natural philosophy seriously. *History of Science* 29:377–392.
Cunningham, Andrew, and Perry Williams. 1993. De-centring the 'big picture': The Origins of Modern Science and the modern origins of science. *British Journal for the History of Science* 26:407–432.
Daston, Lorraine. 1988. The factual sensibility. *Isis* 79:452–467.
Davis, Audrey. 1971. Some implications of the circulation theory for disease theory and treatment in the seventeenth century. *Journal of the History of Medicine* 26:28–39.

Dear, Peter. 1985. *Totius in verba*: Rhetoric and authority in the early Royal Society. *Isis* 76:145–161.

Dear, Peter. 1990. Miracles, experiments, and the ordinary course of nature. *Isis* 81:663–683.

Debus, Allen G. 1966. *The English Paracelsians*. New York: Franklin Watts.

Descartes, René. 1985–1991. *The Philosophical Writings of Descartes*, edited by J. Cottingham, R. Stoothoff and D. Murdoch. 3 vols. Cambridge: Cambridge University Press.

Eamon, William. 1983. Technology as magic in the late Middle Ages and the Renaissance. *Janus* 70:171–212.

Eamon, William. 1993. 'With the rules of life and an enema': Leonardo Fioravanti's medical primitivism. In *Renaissance and Revolution: Humanists, Scholars, Craftsmen and Natural Philosophers in Early Modern Europe*, edited by J. V. Field and F. A. J. L. James, 29–44. Cambridge: Cambridge University Press.

Eamon, William. 1994. *Science and the Secrets of Nature: Books of Secrets in Medieval and Early Modern Culture*. Princeton: Princeton University Press.

Egmond, Florike. 2008. Apothecaries as experts and brokers in the sixteenth-century network of the naturalist Carolus Clusius. *History of Universities* 23:59–91.

Egmond, Florike, Paul Hoftijzer, and Robert Visser, eds. 2007. *Carolus Clusius: Towards a Cultural History of a Renaissance Naturalist*. Amsterdam: Koninklijke Nederlandse Akademie van Wetenschappen.

Ellis, Frank H. 1963. Garth's Harveian Oration. *Journal of the History of Medicine* 18:8–19.

Figlio, Karl. 1978. Chlorosis and chronic disease in nineteenth-century Britain: The social construction of somatic illness in a capitalistic society. *Social History* 3:167–197.

Findlen, Paula. 1994. *Possessing Nature: Museums, Collecting and Scientific Culture in Early Modern Italy*. Berkeley: University of California Press.

Findlen, Paula. 1998. Possessing the past: The material world of the Italian Renaissance. *American Historical Review* 103:83–114.

Frank, Robert G., Jr. 1979. The image of Harvey in commonwealth and restoration England. In *William Harvey and His Age*, edited by J. J. Bylebyl, 103–143. Baltimore: The Johns Hopkins University Press.

Frank, Robert G., Jr. 1980. *Harvey and the Oxford Physiologists: Scientific Ideas and Social Interaction*. Berkeley and Los Angeles: University of California Press.

Garber, Daniel. 1978. Science and certainty in Descartes. In *Descartes: Critical and Interpretive Essays*, edited by M. Hooker, 114–151. Baltimore: The Johns Hopkins University Press.

Garber, Daniel. 2001. *Descartes Embodied: Reading Cartesian Philosophy Through Cartesian Science*. Cambridge: Cambridge University Press.

Getz, Faye Marie. 1992. The faculty of medicine before 1500. In *The History of the University of Oxford, vol. II: Late Medieval Oxford*, edited by J. I. Catto and R. Evans, 373–405. Oxford: Clarendon Press.

Getz, Faye Marie. 1997. Roger Bacon and medicine: The paradox of the forbidden fruit and the secrets of long life. In *Roger Bacon and the Sciences: Commemorative Essays*, edited by J. Hackett, 337–364. Leiden: E.J. Brill.

Gillispie, Charles C. 1960. *The Edge of Objectivity: An Essay in the History of Scientific Ideas*. Princeton: Princeton University Press.

Glanvill, Joseph. 1668. *Plus Ultra: or the Progress and Advancement of Knowledge Since the Days of Aristotle*. London.

Hadden, Richard. 1994. *On the Shoulders of Merchants: Exchange and the Mathematical Conception of Nature in Early Modern Europe*. Albany, NY: State University of New York Press.

Hall, A. Rupert. 1954. *The Scientific Revolution, 1500–1800: The Formation of the Modern Scientific Attitude*. London: Longmans.

Hall, A. Rupert. 1974. Medicine and the Royal Society. In *Medicine in Seventeenth Century England*, edited by A. G. Debus, 421–452. Berkeley: University of California Press.

Hall, A. Rupert. 1983. *The Revolution in Science 1500–1750*. London: Longman.

Hannaway, Owen. 1975. *The Chemists and the Word: The Didactic Origins of Chemistry*. Baltimore: The Johns Hopkins University Press.

Hanson, Craig Ashley. 2009. *The English Virtuoso: Art, Medicine, and Antiquarianism in the Age of Empiricism*. Chicago: University of Chicago Press.
Harkness, Deborah E. 2007. *The Jewel House: Elizabethan London and the Scientific Revolution*. New Haven; London: Yale University Press.
Hodges, Nathaniel. 1666. *Vindiciae medicinae et medicorum: Or an apology*. London: John Field
Hunter, Michael. 1976. The social basis and changing fortunes of an early scientific institution: An analysis of the membership of the Royal Society, 1660–1685. *Notes and Records of the Royal Society* 31:9–114.
Hunter, Michael. 1982. *The Royal Society and Its Fellows 1660–1700: The Morphology of an Early Scientific Institution*. Chalfont St. Giles: British Society for the History of Science.
Israel, Jonathan I. 2001. *Radical Enlightenment: Philosophy and the Making of Modernity 1650–1750*. Oxford: Oxford University Press.
Jenner, Mark S. R., and Patrick Wallis. 2007. The medical marketplace. In *Medicine and the Market in England and Its Colonies, c. 1450–c. 1850*, edited by M. S. R. Jenner and P. Wallis, 1–23. Houndsmills, Basingstoke: Palgrave Macmillan.
Jones, Ellis. 1960. The life and works of Guilhelmus Fabricius Hildanus (1560–1634). *Medical History* 4:112–134, 196–209.
Kassell, Lauren. 2005. *Medicine and Magic in Elizabethan London: Simon Foreman: Astrologer, Alchemist, and Physician*. Oxford: Oxford University Press.
Kaye, Joel. 1998. *Economy and Nature in the Fourteenth Century: Money, Market Exchange, and the Emergence of Scientific Thought*. Cambridge: Cambridge University Press.
Keynes, Geoffrey. 1966. *The Life of William Harvey*. Oxford: Clarendon Press.
Kuhn, Thomas S. 1976. Mathematical versus experimental traditions in the development of physical science. *Journal of Interdisciplinary History* 7:1–31.
Latour, Bruno, and Steve Woolgar. 1986. *Laboratory Life: the Construction of Scientific Facts*. new ed. Princeton: Princeton University Press.
Lennon, Thomas M., and Patrica Ann Easton. 1992. *The Cartesian Empiricism of François Bayle*. New York and London: Garland.
Levine, Joseph M. 1983. Natural history and the history of the Scientific Revolution. *Clio* 13:57–73.
Lloyd, G. E. R. 1979. *Magic, Reason and Experience: Studies in the Origin and Development of Greek Science*. Cambridge: Cambridge University Press.
Long, Pamela O. 2001. *Openness, Secrecy, Authorship: Technical Arts and the Culture of Knowledge from Antiquity to the Renaissance*. Baltimore: The Johns Hopkins University Press.
MacDonald, Michael. 1981. *Mystical Bedlam: Madness, Anxiety, and Healing in Seventeenth-Century England*. Cambridge: Cambridge University Press.
Mokyr, Joel. 2002. *The Gifts of Athena: Historical Origins of the Knowledge Economy*. Princeton: Princeton University Press.
Pagel, Walter. 1958. *Paracelsus: An Introduction to Philosophical Medicine in the Era of the Renaissance*. New York: S. Karger.
Pagel, Walter. 1967. *William Harvey's Biological Ideas: Selected Aspects and Historical Background*. New York: Hafner.
Pagel, Walter. 1969. William Harvey revisited. *History of Science* 8,9:1–29; 1–41.
Pittis, W. 1715. *Some Memoires of the Life of John Radcliffe, M.D.* London: E. Curll.
Reeds, Karen Meier. 1991. *Botany in Medieval and Renaissance Universities*. Harvard Dissertations in the History of Science. New York: Garland.
Shapin, Steven, and Simon Schaffer. 1986. *Leviathan and the Air Pump: Hobbes, Boyle, and the Experimental Life*. Princeton: Princeton University Press.
Shapin, Steven. 1994. *A Social History of Truth: Civility and Science in Seventeenth-Century England*. Chicago: University of Chicago Press.
Shapiro, Lisa, ed. 2007. *The Correspondence Between Princess Elizabeth of Bohemia and René Descartes*. Chicago: University of Chicago Press.

Shapiro, Barbara J. 1983. *Probability and Certainty in Seventeenth-Century England: A Study of the Relationship Between Natural Science, Religion, History, Law, and Literature*. Princeton: Princeton University Press.

Sloane, Hans. 1707. *A Voyage To the Islands Madera, Barbados, Nieves, S. Christophers and Jamaica*. 2 vols-1725 London: Printed by B.M. for the author.

Smith, Pamela. 1994. *The Business of Alchemy: Science and Culture in the Holy Roman Empire*. Princeton: Princeton University Press.

Smith, Pamela. 2004. *The Body of the Artisan: Art and Experience in the Scientific Revolution*. Chicago: University of Chicago Press.

Stubbe, Henry. 1670a. *Legends No Histories*. London.

Stubbe, Henry. 1670b. *The Plus Ultra Reduced to a Non Plus*. London: Printed for the author.

Verbeek, Theo. 1991. Descartes and the problem of atheism: The Utrecht crisis. *Nederlands Archief voor Kerkgeschiedenis* 71:211–223.

Verbeek, Theo. 1992. *Descartes and the Dutch: Early Reactions to Cartesian Philosophy, 1637–1650*. Carbondale: Southern Illinois University Press.

Verbeek, Theo. 2000. The invention of nature: Descartes and Regius. In *Descartes' natural philosophy*, edited by S. Gaukroger, J. Schuster and J. Sutton, 149–167. London and New York: Routledge.

Webster, Charles. 1967. Harvey's De generatione: Its origin and relevance to the theory of circulation. *British Journal of the History of Science* 3:262–274.

Webster, Charles. 1976. *The Great Instauration: Science, Medicine and Reform 1626–1660*. New York: Holmes and Meier.

Webster, Charles. 1979. William Harvey and the crisis of medicine in Jacobean England. In *William Harvey and His Age: the Professional and Social Context of the Discovery of the Circulation*, edited by J. J. Bylebyl, 1–27. Baltimore: The Johns Hopkins University Press.

Webster, Charles. 2008. *Paracelsus: Medicine, Magic and Mission at the End of Time*. New Haven: Yale University Press.

Woolfson, Jonathan. 1998. *Padua and the Tudors: English Students in Italy 1485–1603*. Cambridge: James Clark & Co.

Wootton, David. 2006. *Bad Medicine: Doctors Doing Harm Since Hippocrates*. Oxford: Oxford University Press.

Practical Experience in Anatomy

Cynthia Klestinec

Abstract Often flagged as an origin of empiricism, experience has a range of meanings in the context of early modern natural philosophy and medicine. It has been aligned with practical knowledge, knowledge of contingent effects, and the un-theorized perception of phenomena accessible to the senses. In the realm of anatomical inquiry, experience joined reason to constitute (according to Galen, Mondino, Berengario da Carpi, Niccolò Massa and many others) the approved anatomical method. For medical students, however, experience also meant manual skill and expertise (*peritia*). Students celebrated the manual expertise – the ability to cut open corpses and by dissecting internal and external structures, reveal them to the audience – of their professors and their peers. In Padua, the home of the famous anatomical theater of 1595, students connected these features of anatomical inquiry with private anatomical exercises rather than public demonstrations, especially those given by Giulio Casseri. This paper queries the private settings in which anatomical knowledge was produced and the roles that private anatomies played in recharging the meaning of experience and embodied knowledge in the fields of anatomy and surgery. Using the exchange between students, professors and local practitioners in Padua and Venice, this paper aims to reconsider the role of practical experience in anatomical training and its connection to learned surgeons and anatomists rather than the rustic ignorance of empirics or the secrets of women.

1 Introduction

In the winter months of 1578, when the weather turned cold, several professors of medicine at the university of Padua found themselves with an opportunity to dissect the bodies of two women. As they were conducting lessons in the hospital of S. Francesco, Albertino Bottoni (?1596) and Marco degli Oddi (1526–1591), both professors of medicine, came upon two women, recently deceased, and decided to

---

C. Klestinec
Department of English, Miami University, Ohio

C.T. Wolfe and O. Gal (eds.), *The Body as Object and Instrument of Knowledge: Embodied Empiricism in Early Modern Science*, Studies in History and Philosophy of Science 25,
DOI 10.1007/978-90-481-3686-5_3,

open their bodies in order "to demonstrate to the listeners the affected places and the kindler [or cause] of the diseases [*morborum fomites*]."[1] Subsequently, they dissected the uterus in each cadaver. Then Emilio Campolongo (1550–1604), professor of theoretical medicine, joined the group and said that in the second cadaver, the body of a woman "consumed by senility [*marasmo*]," he would "penetrate the fistula," which resided beneath her breast, cutting into it in order to show its parts and connections more clearly.[2] At that point, the examination was interrupted by the complaints of a little old woman (*querelis anicularum*), who probably feared at her own death, a similar treatment of exposure and delay of her burial.[3] The woman's complaints were definitely heard because, the administrators threatened degli Oddi and Campolongo with the loss of their salaries. At the beginning of the following year, moreover, they reminded everyone of the statutes governing the use of bodies for dissection: the bodies were to be non-noble (*ignobile*) and more importantly, unknown (*ignoti*). They were not to come from either Padua or Venice, suggesting that at least one of these women was a local.[4] The scene renders dramatically one kind of confrontation that could arise between learned medicine and its patients, where the latter frequently contested and curtailed the authority of the former.[5]

At medieval and Renaissance universities, professors and students studied anatomy by reading and discussing books and by dissecting cadavers and animals. Annually, a public anatomy demonstration was held in the winter months, and it

---

[1] *Atti della nazione germanica artista* [*Acta germanicae artistarum*] (1911–1912), 1578, vol. 1, 143–144: "Circa finem vero Octobris, cum coeli constitutio frigidior aliquanto esset, decreverunt mulierum quae in nosocomio illo morirentur, cadavera aperire et auditoribus locos affectos et morborum fomites demonstrare." Hereafter, this source will be cited as *Acta*, followed by year, volume and page numbers.

[2] *Acta*, 1578, vol. 1, 143–144: "Id consilium feliciter satis ceptum in duobus corporibus subito fuit eversum. Cum enim die sequenti uteros harum mulierum aperire constituissent, et in altera, quae marasmo consumta erat et fistulam sub pectore habebat insignem, incidere et commonstrare quonam penetrasset fistula, Aemylius Campolongus ipsorum aemulus suis eodem die uteri sectionem pollicitus, uteros horum cadaverum abstulerat; unde factum ut re hac, et quaerelis anicularum idem, si morerentur, timentium, ad praefectos nosocomii delatis, interdictum sit et Oddo et Campolongo, sub poena amissionis salarii, ne quod cadaver in posterum aperirent."

[3] Park 1994 explores the relationship between dissection and disrupted burial.

[4] *Statuta Almae Universitatis D. Artistarum et Medicorum Patavini Gymnasii* (1570), bk. 3, ch. 28: "Quod singulis annis fieret anatomia, quodq; pro ea perficienda rectoris urbis, & territoriiq; tenerentur cadaver cuiuscunq; delinquentis, de quo capitis supplicui summeretnr [sic], dare anatomiae deputatis, nisi cadaver esset alicuius civis veneti, aut patavini, quum iam multis dicta anatomia, nisi raro facta fuerit cadaverum defectu. Ideo utilitati non modo scholarium, sed etiam universo mortalium generi consulentes, non in aliquod quorum vili pendium, statuitur quod urbis prasides, ac omnium locorum Patavini districtus praetores teneantur dictis nostris anatomistis dare quodcunque cadaver cuiuscunque delinquentis capitis supplicio puniti, nisi sit venetus, aut Paduanus, civis, veli ex comitatu ex aliqua familia alicuius aestimationis, & nisi consanguinei eius ex eadem familia contradicant. Vel advena nobilis, vel alicuius existimationis, sub poena in dicto statuto contenta. Ad quam videndam omnes scholares nostri matriculati possint intrare, dummodo solverint, non obstante dicto statuto, quae solutio non possit excedere summam marcellorum trium argentorum, ex quibus fiant impensae necessariae, & praefertim exequamur." See also Carlino 1999.

[5] Pomata 1998.

was attended by professors and medical students as well as administrators and a few dignitaries.[6] Throughout the academic year, but most often in the winter months, additional lessons would be held. These were called private and seem to have taken place when bodies were found or made available. The public and private traditions of anatomical inquiry developed concurrently, often with the one shaping the content of the other. In Padua, because the private tradition of anatomy gained more independence by the end of the sixteenth century, the public anatomy demonstration became obsolete as a pedagogical venue. Set within this historical context, degli Oddi and Campolongo's lessons were private, and for medical students, improvised excursions into the practical, medical issues that arose from anatomical inspection. They were medical anatomies, held at the hospital, directed at the causes of disease, and attentive to the particular anatomical traits of these female bodies.[7] As Campolongo indicated, his dissection would penetrate the body and reveal the characteristics and causes of the diseases lodged within. Oriented by medical topics – the nature and causes of diseases in the womb and of a fistula – these private anatomies featured the activities of dissection as a general method of inquiry and one that was more than loosely associated with the masculine, virile authority embodied by the professor of medicine.

For medical students, these lessons were also a stark contrast to the public anatomy demonstrations of Girolamo Fabrici of Aquapendente (Hieronymus Fabricius ab Aquapendente, 1533–1619), who took the chair of anatomy and surgery in 1565, 3 years after the death of Gabriele Falloppio (1523–1562).[8] Fabrici did not combine, as Bottoni, degli Oddi and Campolongo did, the processes of dissection with the assessment of anatomical particulars; nor did he encourage proximity between himself, his students and the specimens. During his long tenure at Padua, which lasted into the seventeenth century, he emphasized the natural philosophical causes of anatomy, developing a lofty, high style for the demonstration of his research.[9] Some students found his eloquence ennobling. As early as 1578, the transalpine students, Samuele Keller and Ioanne Wolfango Rabus, said that his lecture and private colloquium "filled and elevated their souls with many things."[10] Fabrici inspired his students not because he was dissecting corpses and vivisecting animals, but rather because he connected anatomy and natural philosophy. He made the focus of anatomy the coordinated processes or

[6] Ferrari 1987; Carlino 1999.

[7] Park 2006, 39–76 characterizes the medieval origins of this tradition, found in the dissections of nuns, done in search of the anatomical signs of sanctity.

[8] These interim anatomists included Prospero Borgarucci and Nicolò Bucella. See Rinaldi 2009 and Stella 1961–1962 respectively. For the biography of Fabrici, see The Embryological Treatises of Hieronymus Fabricius of Aquapendente (1942), 1–32.

[9] Cunningham 1985, 1997; Klestinec 2004.

[10] *Acta*, 1578, vol. 1, 144: "De Excellentissimo etiam Anatomico compellando tractatum, quod tantum minus necessarium videbatur, cum ipse non tamen in publicis praelectionibus sed etiam in privatis colloquiis anatomen polliceretur et animos multa spe impleret et erigeret."

actions of the organic soul.[11] These included motion, digestion, respiration, sensation and generation. The object of his anatomy lesson was not the many anatomical structures that could be isolated within the venter of the body or even the various connections between this multitude of structures. Rather, Fabrici's object was a coordinated action, traced through both human and animal specimens and the discussion of their structural and morphological characteristics. This approach meant that strange, marvelous, accidental or rare phenomena such as a diseased womb or fistula emerged in Fabrici's teaching and in his publications as a case to explain at the end of a fully philosophical (and normative) account of a process of the soul. They usually emerged as something to explain away. As it evolved over the 1570s and 1580s, Fabrici's program specialized the discourse of anatomy by making it contribute more fully and more clearly to natural philosophical inquiries. He effectively situated the study of anatomy within the theoretical branch of the medical curriculum. It was thus clearly distinguished on its own merits and in the minds of students from the practical, medical anatomies held in the hospital of S. Francesco.

Practical experience meant many things in the early modern worlds of craft production, philosophical speculation, experiment, and medical practice.[12] It has been associated with practical knowledge, knowledge of contingent effects (trial-and-error), and the un-theorized perception of phenomena accessible to the senses. In the medical anatomies discussed above, practical indicated the kind of anatomy lesson underway, a lesson that advanced the understanding of medical topics related to disease and did so by investigating the particular traits of the cadaver. Medical students did not describe these anatomies in terms of their un-theorized perception of the bodies – Campolongo said there was a fistula, and so he would dissect the infected area with the preconception of the fistula (and all the commentaries on fistulas) in mind. While this sense of practical experience has been used to determine early signs of an experimental apparatus, it is not evident here (Dear 1995, 124–38). Nor did students describe these anatomies in terms of the skillful dissections that Bottoni, degli Oddi, or Campolongo made. Handiwork was not the issue. It was a practical anatomy, a study of particular anatomical signs, found and read in the bodies of two female cadavers. Less than two decades later, however, medical

[11] Cunningham 1985; Park 1998.

[12] On experiment, see Agrimi and Crisciani 1990; and Dear 1995, especially 63–92 and 124–50. In medical and related venues, experience came to mean practical knowledge, knowledge of contingent effects, and the un-theorized perception of phenomena accessible to the senses. Experience, for example, is especially prominent in the tradition of 'empirical' medicine, appearing frequently in discussions of recipes and potions and in the marketing strategies of its itinerant practitioners (called empirics). It is also mentioned in the context of household physic or domestic knowledge held by men and women, *rustici e vecchiette*, that derived from trial and error and years and even generations of practice with specific ailments, treatments, and conditions. These kinds of experience have been treated by Eamon 1994, 134–203, Daston and Park 2001, 115–139, and Gentilcore 2006, 200–233; on northern traditions, Smith 2004, 59–94 focuses on experience in the artisanal traditions and trades. For the English context, see Harkness 2006.

students would look to these kinds of events in order to celebrate *peritia* or the technical, manual skills and expertise associated with dissection and with surgery. They would promote *peritia* – manual expertise gained from the accumulation of experiences (often private experiences) – as a virtue, and in doing so, they would help situate it within the academic world of medicine and its study of anatomy and surgery.

In the late sixteenth century, in fact, the tradition of private anatomy was marked by an ebullient reappraisal of *peritia* and its transformation into a virtue of the university-trained practitioner. The expected division of labor between the physician and surgeon and the different access to the body (and treatments) meant that physicians treated with diet and regimen, the internal, humoral balance of the body while surgeons treated external ailments. Despite the fact that medical students were training to be physicians, they talked positively and openly about their knowledge of surgery, surgical operations and instruments. For these students, handiwork was not relegated to a lower order of practitioners such as barbers or barber surgeons. It was pursued. Private rather than public anatomies provided increasingly specialized and exclusive opportunities for medical students to encounter lessons on the techniques of dissection and especially on the learned traditions of surgery.

Students were drawn to these private events initially because they supplemented Fabrici's public and highly philosophical anatomies. Subsequently, however, they gravitated to the private arenas of anatomy because they sought exclusive experiences, that is experiences not open to a wider public. Fabrici had opened his public anatomies, sometimes called 'free' anatomies, to the community. Students were outraged by the presence of *meccanici* and craftsmen of a lower order at public anatomies while they relished the security, familiarity and exclusivity offered by private anatomies. From the concerns about mixing orders in these arenas, *peritia* was secured as an elite, virtue because it was not connected to the experiences of *rustici* or lower-order crafts and trades. Instead, in the late sixteenth century, it was recognized in the technical expertise of professors and learned practitioners, especially learned surgeons.

This transformation can be traced through the extant records of the transalpine students at Padua, the *acta* of their nation. It is estimated that 6,000 foreign students attended the university in the sixteenth century; and the transalpine students constituted the largest nation of foreign students.[13] This made their complaints especially important to administrators, who wished to maintain the number of students at the university. The records of the transalpine nation provide a running commentary on the activities at the university, including the outcomes of voting assemblies, the organization of the nation's library (and the donation of books for it), and the details of pranks, rituals and festivities as well as the content of medical courses and anatomical exercises. With respect to medical and anatomical lessons, students often described the pedagogical styles of their medical professors, the degree to which such lessons satisfied them, and their desire for more frequent lessons on anatomy.

---

[13] Kagan 1986; M. Saibante et al. 1924.

These aspects serve as the basis for this essay, illuminating the ways in which students engaged and responded to the materials of their medical education and to the idea of practical experience. Practical knowledge in these accounts was defined as that (embodied) knowledge issuing from technical expertise, and by the end of the sixteenth century, medical students enthusiastically promoted it as a significant part of their elite education.[14]

2 Natural Philosophical Anatomy, Practical Anatomy and the Public and Private Spheres of Inquiry

By the late 1580s, students attended Fabrici's natural philosophical anatomies as well as the practical anatomies of other professors. They began, however, to judge Fabrici's public demonstration according to the merits of private anatomies. In 1590, a transalpine student described one of Fabrici's demonstrations:

> [Fabrici] has already spent two months on the exposition and description of the bones of the head. Having been brought on to the muscles he has completed three, devoting one hour to each muscle. There are so many muscles that, proceeding in this way, two years will not suffice. When then will he deal with the viscera? In addition, everything is treated confusedly and in a disorderly way: once he discussed the detached arm, going on after many days to discuss the foot. I don't see how anyone can learn the sequence and connection of the whole from looking at these.[15]

Although public demonstrations typically provided an introduction to anatomy by covering the external lineaments and the viscera (or "the sequence and connection of the whole" from the parts), Fabrici here specialized his treatment to deal with the muscles alone.[16] He was working on the coordinated action of motion, a focus that was pedagogically evident but that this student all but rejected. That focus led him to talk about muscles, but not continue his dissection of them. Not only was this student tired of hearing Fabrici talk, he was frustrated because Fabrici spent too much time on specific parts of the body without connecting them to the larger structural framework of the body – the bones and the muscles attached to them, a system that clarified sequences of muscles from origination to insertion point.

The students were subsequently asked, "Why do you desire the private [anatomy], do you [not] find the public one pleasing?" In their records, the transalpine student recorded his nation's full response: "We respond that Aquapendente provided a

[14] On this concept of embodied knowledge, see Smith 2004.

[15] *Acta*, 1590, vol. 1, 286: "Iam ossium capitis enarrationi et descriptioni duos impendit menses: ad musculos devolutus tres absolvit singulis musculis singulas horas tribuens. Tot autem enumerantur musculi, ut hac via incedens biennium non sufficiat. Quando igitur de visceribus? Accedit quod confuse et tumultuarie omnia pertractantur: iam branchium detruncatum attulit, post multos dies pedem allaturus, quando aliquis ex horum inspectione seriem et connexum totius discere posset, non video." Cited by Cunningham 1985, 199.

[16] Berengario da Carpi 1521 refers to public or "common" anatomies (*anatomia communi*), 119v, 489v.

most exact anatomy, but that we ask that what the anatomist delivered profusely and in a casual style of speaking be put before us in the form of a visual synopsis [or chart] that can be remembered, as we do in private anatomy."[17] The student referred to Fabrici's anatomy as exact, which meant topical – a sense that Fabrici himself had used to describe his demonstrations.[18] The student also softened his criticism of Fabrici's teaching, transforming Fabrici's "confused and disorderly" approach into a merely "profuse" and "casual" style. The student also indicated that private anatomies contained few words, were memorable, and profoundly visual experiences – that included not only looking at dissected specimens but also looking at charts that summarized important anatomical information. By 1590, then, students were no longer content with the notion that public and private anatomies belonged to separate but equal traditions. Indeed, this student suggested that public demonstrations could benefit by following the pedagogy and other guidelines that were improvised and practiced in private.

The student wanted more attention to be devoted not only to the structures of the body and their sequences and connections but also to the techniques of dissection. These features emerged prominently in private anatomies. Like public anatomy demonstrations that helped to specialize the study of anatomy as a natural philosophical enterprise, private anatomies promoted the study of particulars, of the medical causes of disease. According to the students, they also emphasized dissecting techniques and treated various parts of the tradition of surgery, thereby providing a fertile ground for a concept of expertise related to technical skill to take root and accrue significance. These private explorations could be undertaken by chance during hospital rounds and be led by professors, but more often they were annual events, lasting for weeks at a time, and were led by both local practitioners and professors, who were all more youthful and engaging than Fabrici.

In the 1570s and 1580s, students became dissatisfied with Fabrici, who promised to give "most complete and highly illuminating" anatomy demonstrations but who, in the end, provided ones that were "very obscure and imperfect."[19]

---

[17] *Acta*, 1590, vol. 1, 288: "Quid, inquit, privatam desideratis, nunquid placuit publica? Exactissimam, respondemus, nobis dedit Aquapendens; verum ut illa quae fusius et licentiori sermonis genere tradidit Anatomicus, nobis iam quasi in synopsi ob oculos proposita memoriae mandemus, de privata agimus."

[18] Fabrici of Aquapendente 1618, 3: "Toto hoc tempore, quo non popularem, sed exactam anatomen administramus, agere in vestram gratiam, auditores, divino favente auxilio constitui de motu, quo totum animal loco movetur: seu de motu locali totius animalis, seu mavis dicas, de motu, quo totum animal locum, sue positionem mutat."

[19] In 1576–1577, Padua witnessed several outbreaks of plague that forced the university to limit and then to close its doors, but activities began to resume their normal pace by 1578. *Acta*, 1580–1581, vol. 1, 170–171: "Promiserat hic egregius vir a principio suarum lectionum auditoribus suis tractatum de tumoribus praeter naturam, de fracturis et ...forte etima de luxationibus. Sed, quod maioris erat momenti, pollicebatur nobis anatomen luculentissimam et absolutissimam; Namque animum ipsius obstinatum neque praeceptorum intercessiones neque nostrorum preces frangere potuerunt, patientia sola nobis reliqua fuit quae et obscurissimam et imperfectissimam anatomiam loco luculentissimae et absolutissimae sustinere potuit." Favaro 1922 explores the conflicts between Fabrici and his students.

As a result, they eagerly sought to arrange private anatomies. They contacted Niccolò Bucella in the winter months of 1572 because he was known to offer anatomies from his house in the quarter of S. Maria dei Servi in Padua.[20] They contacted the surgeon, Michael Aloisius, about holding an anatomy in the church of St. Catherine in January 1582.[21] These events were often delayed or not completed because cadavers could not be found. While the administration of the university did not immediately satisfy the students' wishes for more anatomies or explicitly regulate the proceedings around private anatomies, it initiated a set of reforms that clarified the distinction between public and private studies of anatomy.

In 1583, the administration instituted a series of reforms that encouraged specialization in the fields and inquiries of anatomy and surgery. One reform addressed Fabrici's lessons in particular, mandating that he would not need to interrupt the demonstrations of anatomy with those of surgery because the two would be kept separate (and separate resources would be allocated to them).[22] In practice, Fabrici did not provide surgery demonstrations. He was fully occupied with his anatomical inquiries and his study of natural philosophy. Moreover, the statutes allocated only two bodies a year for the study of anatomy, suggesting that one would be used for the anatomy demonstration and the other for the surgery demonstration. Throughout these years, Fabrici used both cadavers to pursue his anatomical inquiries. Information on surgery was transmitted to students in private

---

[20] *Acta*, 1572, vol. 1, 85: "Verum huc quoque pertinet, quod Excellentissimus Dominus Buccella chirurgus, Nationis Germanicae studiosissimus, ubi constabat brumales ferias absque publica anatome frustra scholaribus abire, interpellatus primum a me ipso, deinde et ab aliis, omnem operam suam ac benevolentiam in negocio anatomico detulit, paratissimum se aiens cadaver aliquod in nostram gratiam secare, dum id ipsi a nobis suggeratur. Mox vero et Itali torpidi illius ocii pertaesi, ac praesertim Massarii duo promptitudine Buccellae perspecta (ut lucro etiam ex mortuis comparando inhiant), in aedibus ipsius sub natalitia festa anatomen apparent, quae sane non exiguum splendorem atque gratiam prae se ferebat, nisi partim incuria et perfidia dictorum ministrorum partim difficultate nanciscendorum cadaverum laboratum fuisse laborious." See Stella, n. 6.

[21] *Ibid.*, "19 Ianuarii Anno 82 habita fuit anatomia privata in collegio Vicentinorum prope templum S. Catharinae, quae cum aliquibus Nationis nostrae non approbatur, semel atque iterum me convenerunt ac de anatomia privata inter nos facienda mecum deliberarunt, Nationemque convocari voluerunt, id quod sequenti die pro ac debui libenter feci. Convenimus omnes in aedibus meis, consilium et propositum omnibus placebat, omnes sumptus necessarios contribuere volebant. Excellentissimus vir Michael Aloisius chirurgus omnem operam et diligentiam nobis pollicebatur; in qua re satis ego quantum potui laboravi, ad quinque parochos ministrum publicum misi, qui omnes promittebant se prima quaequam cadavera oblata communicaturos, at quia propter penuriam demortuorum cadavera nulla acquirere potuimus, deliberatio nostra irrita fuit."

[22] Raccolta Minato: 5 February 1583: "Et perchè ognuno sa di quantoa utilità, et honorevolezza sia al detto studio il far l'Anatomia ogn'anno…L'andera parte, che'l D. Hieronimo Fabritii sia condotto a legger nello studio nostro di Padoa l'Anatomia ordinamente di anno in anno, et la Chirurgia insieme in questo modo però, che tutti li mesi dell'Inverno sia tenuto leggere, tagliare, et mostrare l'Anatomia come lettura ordinaria, passati veramente i mesi d'Inverno non possendosi per li tempi caldi maneggiare i corpi morti, che si putrefanno sia tenuto legger per ordinaria la Chirurgia non intermettendo però quando si possa tagliare, mostrare, e trattare anco le cose di essa Anatomia."

anatomies, taught by a range of local practitioners and professors, including Bucella, Paolo Galeotto, and the very successful Giulio Casseri.[23] The allocation of corpses for these lessons remains obscure. With its decision, the administration distinguished the content of the annual anatomy demonstration from surgery demonstrations. Previously, the same anatomist would move between public and private venues; and in both, he would highlight the admirable form of the body and the way its study disclosed natural knowledge (natural philosophical), and then emphasize the importance of knowing anatomy in order to understand sources of disease (medical) and the ways to set fractures (surgical). The administration's decision helped to distribute these aspects among faculty members: Fabrici gave public anatomy demonstrations that were more philosophical in character; and other professors, especially Giulio Casseri, gave private ones that were oriented around medical anatomy and surgery.

Though detailed evidence of this private tradition is sometimes scarce, private anatomies seemed to have thrived alongside Fabrici's magnificent public anatomies. This parallel development helped focus attention on dissection as a set of procedures and on surgery, which was flagged for its manual operations and its instruments. Whereas Fabrici demonstrated his work with eloquence and enthusiasm (polishing his earlier "disorderly" manner) in the two permanent anatomical theaters, built consecutively in 1583–1584 and 1594–1595, the professors and practitioners who offered private anatomies struck a different note. Casseri called it a 'naked style', a clear, rhetorically 'unflowered' and simple style that addressed the topics of anatomy, the procedures of dissection, the maneuvers of surgery, and the accompanying visual phenomena of charts, instruments and dissected specimens.[24]

These private anatomies were smaller events held in the hospital of S. Francesco as well as local pharmacies, in the church of St. Catherine, and in the homes or basements of these practitioners. These spaces were smaller than the public theater, and therefore, the audiences were also smaller. Though exact numbers cannot be generated from extant records, the transalpine students rarely mention the attendance of other students, suggesting that these private events may have been organized by them and held solely for their benefit. In these private venues, attention was paid to specimens and structures as well as to dissecting techniques. In 1585–1586, Casseri offered a private anatomy, and the transalpine students said that "by dissecting very carefully all the parts of the body – not only the internal parts but also the external (i.e. the muscles and veins) – and in addition by demonstrating the main surgical

[23] Galeotto remains relatively unknown. Sterzi 1909–1910 provides a full account of Casseri's life and works.

[24] In his treatise on speech, he refered to his style of presentation by way of his style of speech, which he described as both "philosophical" and "nude;" see Casseri 1600, 4–5: "Dicam autem, non oratorio dicendi genere (siquidem vocis dignitatas, et maiestas, pro ineffabili rerum, quae in eius celebratione occurrunt, ubertate, sine concinno, ac modulato verborum sono, etiam solis instar elucescet) sed dicam, philosophico more, nudis, ac simplicibus verbis; qualia veritatis propria esse, ait Euripydes."

operations," he "acquitted himself excellently and satisfied us all."[25] Galeotto also gave a private anatomy for which students commended its "special usefulness."[26] In contrast to Fabrici's soul-elevating treatments of anatomy, these events relied on the dissection of the internal and external parts of the body and on the display of surgical operations. These descriptions emphasize the careful techniques of dissection used for the muscles and veins, suggesting that the muscles and veins were the objects that best displayed the anatomist's manual expertise. Moreover, usefulness designated surgical knowledge and often a kind of intervention; it was an active, operative concept rather than the passive or contemplative one suggested by the Galenic category of use. More speculatively, the distinction between internal and external treatment, the former done by physicians and the latter by surgeons, suggests that the attention to the muscles would be related, for the surgeon, to fractures and perhaps also to tumors; and the attention to arteries, veins and nerves would be connected to both bloodletting and the cautery of ulcers, sores, and wounds. The description indicates a broad interest in the study of anatomy and in the uses of anatomy for surgical intervention. It thus runs counter to our expectations about the division of labor between physicians and surgeons, suggesting that medical students in Padua participated in learned medical traditions which included the tradition of learned surgery (not barbering).[27]

The experiences gained at these private anatomies were called "useful" and organized around topics and texts as well as manual activity. Praising Galeotto, this student noted that the "whole matter" of his demonstration "was useful not to him but to his audience," suggesting by contrast that Fabrici's demonstrations tended to be oriented around his own, topical interests rather than the interests of his students.[28] The student explained that "once [the demonstration] had begun and the external lineaments of the body had been explained, the Anatomist [Galeotto] in the account of the cadaver went straight to the eye, and lectured so clearly and so learnedly on its action and structure that he conducted himself as a man most experienced [*peritissimum*] in anatomical matters and in optics."[29]

---

[25] *Acta* 1585–1586, vol. 1, 210: "In qua sane diligentissime omnes non modo internas corporis partes, sed et externas, id est musculos atque vasa dissecando, insuperque praecipuas operationes chirurgicas monstrando optime sese gessit, nobisque omnibus satisfecit."

[26] *Ibid.*, 211: "Rogamus itaque universos et singulos, ut benevolentiam et labores huius clarissimi viri, propriamque utilitatem considerando, frequentes huic interesse velint."

[27] Although physicians in the sixteenth century tried to reinforce their position in the medical community by constructing the idea of a tripartite and hierarchical division of labor with physicians on top, surgeons in the middle, and apothecaries and barbers on the bottom, these boundaries were drawn differently in Venice. Learned surgeons and physicians practiced together, sharing patients and networks of associations. See Palmer 1979 and Cipolla 1978.

[28] *Acta*, 1585–1586, vol. 1, 211: "Potissimum cum sciret nostram prae reliquis Nationem anatomicis ut plurimum delectari studiis, haecque res omnis non in suam sed auditorum suorum utilitatem esset redundatura."

[29] *Ibid.*, 225–226: "Facto autem principio, post declarata externa corporis lineamenta cadaveris ratione protinus ad oculum aggressus Anatomicus, de eius actione et fabrica ita plane ita erudite multis disseruit lectionibus, ut virum sese gesserit anatomicarum rerum et opticae peritissimum."

The student's preference for Galeotto depended first on Galeotto's demonstration of the external lineaments, his ability to show and explain those parts, and secondly on his ability to discuss the eye in relation to optics. Galeotto embodied *peritia*, but the student extended the category of technical expertise to include both the manual activities of dissection and the more theoretical discussion of optics.

Galeotto continued to engage students. He was surely gifted, but his success is also a sign that students recognized manual skill and technical expertise as important. Elaborating on that importance, they began also to assign aesthetic qualities of beauty and pleasure to these demonstrations. Although Fabrici sought to have Galeotto's lessons prohibited, Galeotto gave a private anatomy in 1588–1589. It lasted for three weeks and took place at a pharmacy near Palazzo del Bo.[30] In that practical setting, the students said Galeotto gave "a thorough and complete anatomy, in which not only did he demonstrate most clearly and with amazing and beautiful ease and method the way to dissect bodies, the structure of all the parts and their actions and functions, but he also showed us the ways…through the whole body of the veins, arteries and nerves, to the great delight of us all."[31] In the following year, he gave a notable private demonstration, in which "he urged all things complex [be presented] with brevity because of the late time of year." As the student said, the warmth of the season did not detract from the demonstration in which he "showed to us perfectly the [ways of the] nerves, arteries and veins."[32] While public anatomies in Bologna have been associated with the bawdy spectacles of Carnival and those in Padua with a more solemn ceremony, these students found the experiences of private anatomies to be delightful, a response notable for its consistency and its intensity.[33] In them, they developed an awareness of the process of dissection as potentially beautiful or beautifying. *Peritia* or technical skill was

[30] *Ibid.*, 271.

[31] *Ibid.*, 248: "His itaque peractis, et publica anatome tandem ad finem perducta, Eccellentissimus vir Dominus Paulus Galeoti ne quid a pristina sua de scholaribus bene merendi voluntate recessisse videretur (quam vis iniuriis publici Anatomici theatrum ipsius esset destructum, ipse variis modis ab eodem fuerit lacessitus et in minus commoda carnisprivii tempora reiectus), in officina *al Corallo* per integras tres septimanas luculentissimam et absolutissimam habuit anatomiam, qua non solum mira facilitate ac pulchra methodo, modum secandi corpora, partium omnium structuram ac earundem actiones et usus evidentissime monstravit, sed etiam venarum, arteriarum et nervorum in universo corpore ductus et propagines, non sine maxima omnium nostrorum iucunditate, sine ullis nostris impensis (quamvis ad eas ut aequum erat a spectatoribus exigendas a nobis hortaretur) nobis ostendit."

[32] *Ibid.*, 290: "Ad XII Kal. Aprilis privatae anatomiae tantopere desyderatae administrationem aggreditur Excellentissimus Paullus Galleottus…Quamquam autem tempus anni maturandam suadebat, tamen succincta brevitate omnia complexus, parerga praecidens, et fragrantes rerum succos libans, nihil praeteribat quod ad rem faceret; immo nervorum, arteriarum et venarum perfectam ostensionem nobis exhibuit." In the same year, Johann Jessenio provided a surgery lesson, which the students did not describe with any detail. See *Acta*, 1590, vol. 1, 290: "Aquapendens interim saepius a quibusdam nostrum et praesertim a Domino Ioanne Iessenio excitus, sese ad VI Kal April. Ad administrationes chirurgicas accingit, easque ad finem deducit, sumptus faciente et cadaver administrante inclyta nostra Natione."

[33] Ferrari 1989; Klestinec 2007.

also accompanied by the qualities of industry and generosity and linked to the idea of expertise found in the world of learned surgery. There were other practical lessons at this time – those of Oddi and Mercuriale that dealt with disease and mysteries or *secreti* – but they were not associated with *peritia*, which belonged to another class of practical medicine, that of learned surgery.

3 Learned Surgery Returns from Exile?

Histories of learned medicine have been slow to characterize and incorporate the learned traditions of surgery. Surgery and the surgeon, however, have enjoyed a ubiquitous presence in histories of vernacular health care in the early modern period. In his recent study of the Renaissance hospital, Jon Henderson[34] includes an account from one of the institutions in the city of Florence, which paid 3 florins to a bone doctor, 6 to a physician and 18 to the surgeon. This distribution, Henderson argues, reflects both the expertise and level of activity of each practitioner. The surgeon not only spent more time at the hospital doing procedures, but was also able to combine the roles of barber, surgeon and apothecary.[35] These categories of healers were, as David Gentilcore reminds us, internally unstable in the period: Gentilcore found accounts of barbers practicing physic, and Henderson also notes that at hospitals, both surgeons and barbers "cut the hair of the staff and patients" and "let blood."[36]

Perhaps because the spheres of medical practice overlap, the surgeon has remained an ambivalent, protean figure in the history of vernacular health care. There was, however, a vibrant tradition of learned surgery in Italian cities, especially northern ones, in the Renaissance. In their studies, published nearly three decades ago, Katharine Park[37] and Richard Palmer[38] indicated that the *medico chirurgo* or "graduate surgeon" was closely associated with the Renaissance physician, with colleges of medicine and surgery, with apprenticeships and licensing examinations, and with patient care.[39] At Italian universities, students could take a doctorate in surgery, and though not many did, most medical students had considerable experience with surgery.[40] In Italy, surgery was not limited to manual skill; it also

[34] Henderson 2006, 237.

[35] Henderson 2006, 243.

[36] Gentilcore 1998, 228.

[37] Park 1985, 8.

[38] Palmer 1979.

[39] See also Siraisi 1990, 153–86. The historiographical situation derives from Cipolla's 1976 argument for the radical distinction between physicians and surgeons and O'Malley's 1970 critique. See also Pelling and Webster 1979.

[40] Palmer 1979, 453–54 explains that the statutes of the University of Pisa of 1478 required two years of university education in surgery plus one year of practical experience for the surgical degree, indicating that an academic training was available for Tuscan surgeons at least at that date, and precisely the same requirements were laid down in the statutes of the College of Arts and Medicine of Padua as revised in 1607.

included knowledge of theoretical principles underlying medicine, the practices of textual commentary, and specific operations.[41] In contrast, north of the Alps, where degrees in surgery were not available, the surgeon was consistently aligned with lower trades and the barber.

Renaissance medical students were deeply committed to the topic and practices of surgery. As the foregoing discussion has indicated, they were fascinated by those who exhibited expertise in the hands-on activities of dissection and surgery. That fascination could produce surprisingly fervent rhetoric. In a letter from June 1597, a medical student at Padua praised his fellow, transalpine colleague Johannes Richter for pursuing practical medicine, debating parts of the commentary tradition, and "devoting himself to the laborious work for the administrations of anatomy and surgery."[42] The student elaborated the topos of devotion by suggesting that it was Richter's diligent study and labors that earned him "so much ingenious and erudite honor that it caused the eyes of all to turn to him" and transformed him into "an ornament of the entire transalpine nation."[43] Richter's devotion to the labors of anatomy and surgery was matched by the students' reverence: he became an ornament of the nation while they were converted into his supporters (his followers). Wrapped in a rhetoric of religiosity, the praise suggests that among students, *peritia* was important, a virtue and a sign of virtuosity.

The Renaissance university, however, was not an especially fertile environment for the development of manual expertise. Or so the claim goes. Scholars have pursued other locations for evidence of applied epistemologies: artisans' workshops; venues and books associated with the marketplace, and starting in the seventeenth century, scientific societies, sufficiently detached from the local university.[44] At the university in Padua, a similar argument was made about the study of anatomy. In *De methodo anatomica* (1593), Girolamo Capivacci, then professor of medicine, explained, following Galen, that anatomy is not only "to be considered by way of dissection, but also truly by way of actions and uses."[45] Elsewhere, he employed an

---

[41] Nutton 1985.

[42] *Epistolario della nazione artisti*, 1565–1647, n. 476–477: Anonymous, letter Johannes Richter Oppaviensis, June 1597, 141r–142v: "Is enim Scholar Excellentium Virorum frequentando, praxim medicam sectando, quaeque vel intellectu vel visu dignior occurrerunt diligenter persequendo, de gravioribus artis medicae controversii saepe cum aliis dissertando, demum anatomicis et chirurgias administrationibus operam indefessam navando, tantum ingenii atque eruditionis famam acquisivit ut omnium oculos animosque mi se converteret et singulare nationis Germanicae ornamentum a plurimis haberetur." Bylebyl 1979 noted that medical students, especially foreign students, sought practical training in clinical diagnosis as well as surgery. While his ideas are located loosely throughout the sixteenth century, my argument suggests that they emerged in the last two decades of the sixteenth century and as a result of institutional and professional changes.

[43] *Ibid.*

[44] E.g. Long 2001; Gentilcore 1998; Shapin 1994.

[45] Capivacci 1593, 3: "Non modo circa dissectionem versari, verum etiam circa actiones et usus… Artis anatomicae haec sit definitio. Ars Anatomica est ars naturae hominis sectione, actione, et usu comparata." The treatise was formulated earlier alongside other methodological formulations, which were subsequently published in the 1590s.

Aristotelian terminology of causes, emphasizing the necessity of moving from material to formal and final causes. Pursued in this way, he argued, anatomical inquiry too could produce philosophical knowledge [*scientia*]. This, Capivacci declared, "is to be distinguished from the domain of practica [*peritia*], which is pertinent to dissections and to manual activities [*manuales*]."[46] This was the method that underlay Fabrici's anatomical inquiries; and it situated Richter's handiwork at the bottom of a hierarchy of legitimate claims to knowledge. For Capivacci, the anatomist did not leave behind the manual activities associated with dissection. Instead, knowledge of causes was seen to derive from the description of a part, although the anatomist (along with his students) is celebrated for his ability to explain why a thing exists, rather than 'merely' describe it.[47] But the students seemed to see it in a different way. For them, Richter's handiwork was praiseworthy by itself.

Capivacci constructed an academic medical hierarchy with surgeons at the bottom and philosophers rather than physicians at the top.[48] Although it may seem as if he reproduced the age-old division between learned and popular medicine, between professors and practitioners, and, finally, between intellectual and manual labor, the boundary that he attempted to draw was a response to contemporary events: to the innovative work of Fabrici in the field of anatomy, which would be given a more secure place in the institution if it was accompanied by a clear, humanist method; and to the students who protested at the lack of practical, medical and surgical instruction in anatomy demonstrations.

Moreover, by the 1580s, both physicians and surgeons were publishing widely in Latin and the vernacular; they practiced together, sharing patients as well as thoughts on diagnoses and treatments; they collaborated in examinations and voting assemblies.[49] These men shared networks of associations that ran to the upper echelons of the Venetian Republic, serving on military expeditions and public health boards; their interests and their commitments were part of the medical profession as it took shape in the late sixteenth century. It is in this specific intellectual and

---

[46] Capivacci, 4–5: "Declaro, loco forme et proinde generis dictum fuit, esse scientiam, ut distingueretur a peritia, quae pertinet ad incisiores, ad manuales...Loco autem subiecti, et proinde differentiae dictum fuit. Hoc est non modo sensu, sed etiam ratione comparatum; Quantum ad brevem expositionem, sciendum hanc traditam esse per formam, subiectum, et finem, omittitur autem causa efficiens, quoniam ex necessitate supponitur, ut si proponatur superva canea videatur, quoniam non est in causa, ut hoc definitum a quocunque alio magis distinguatur quam ex particulis propositis."

[47] The *demonstratio propter quid*, which gave knowledge of causes, was seen to develop from the *demonstratio quia*, which was the description of the part.

[48] Although the university curriculum intended a student to pass from logic to natural philosophy, on his way to medical studies, Capivacci seemed to reverse the direction, arguing that anatomy was a philosophical endeavor and that anatomical knowledge as *scientia* could be an end in itself. Capivacci likely had the following dictim in mind: "where the philosopher ends, there begins the physician." Drawn from Aristotle's *De Sensu et Sensato* and parts of *Parva naturalia*, this dictum was quoted by numerous Paduan philosophers and professors of medicine: Jacopo Zabarella (1533–1589), Cesare Cremonini, Giambattista Da Monte, Oddo degli Oddi, Fabrici, Casseri and others. See Schmitt 1985.

[49] Ongaro 1981; Palmer 1983; Bonuzzi 1996.

geographical context that Capivacci sought to distinguish the professor from the practitioner, to make the surgeon subservient to the physician and the philosopher.

This world of surgery provided medical students, who were dissatisfied with Fabrici's philosophical treatment of anatomy or perhaps eager to establish professional relationships and networks with successful practitioners, with a context for applied-learned surgery. By the mid-sixteenth century, the Venetian College of Surgeons hosted annual anatomies as well as disputations and examinations for degrees and licenses. These anatomies attracted both professors and students from nearby Padua.[50] They took place more regularly in the second half of the sixteenth century at the church of S. Paternita, which is no longer extant, and the church of S. Stefano.[51] These public anatomies were not only important because they introduced basic structural anatomy to local practitioners but also because they established professional relationships.[52]

These anatomies did not follow Fabrici's idea of anatomy as a natural philosophical pursuit. Nor did they adhere to Capivacci's program, for Capivacci focused on the philosopher, not the physician, further distancing the academic world from the world of professions and practice. They introduced techniques related to surgical operations, interventions into the body. As one Venetian surgeon, Giovanni Andrea della Croce (discussed more fully below) explained, the surgeon's activities were aimed at uniting and consolidating or mending the broken parts of the human body:

> I say, in the human body (to demonstrate the difference between the art of Surgeons and and that of Horsemasters [*marescalchi*], who work upon inhuman or animal bodies); and I say living to make it understood that Surgery is very different from anatomical actions, which work solely upon dead bodies.[53]

---

[50] Palmer 1983; Bernardi 1993.

[51] BMV, It.VII 2370 (9668), Cap XXXVI, 81, De anotomia. Between 1550 and 1605, the records indicate anatomies took place in 1574, 1585, 1594, 1602, 1603. See BMV, It.VII.2327–2335 (9721–9729) Collegio medico chirurgico di Venezia: Atti (1476–1805), Libro D: 1549–1628 (cc.171).

[52] Palmer 1985, 458 has documented a fascinating case in the 1560s, when the famous surgeon Niccolo Massa testified on behalf of Zuata Francesco da Burano, who was then petitioning the Doge to allow his son Giacomo to practice with him in the service of the Republic; Massa said that Giacomo was not only a doctor in the art of medicine, learned in philosophy and medicine, both physic and surgery, but also "a man of worth." Massa went on to say that this year, when the anatomy was held, "finding myself with him and discussing the anatomy, I saw and heard that Giacomo has a very good understanding of these matters." Massa's father was a doctor of physic, and he testified that Giacomo took his doctorate with the College of Physicians in Venice and that the vote on him was unanimous; he then noted that Giacomo had "observed anatomies and also with his own hands has practiced privately with other doctors." Giacomo was a learned surgeon, a *medico chirurgo*, engaged by these public anatomies and "practiced" at private ones.

[53] Croce 1583, preface: "Et dico, che opera a fine di correggere, over ricuperar la perduta unità di alcun membro; perche questa è l'intentione, il commodo fine, e la utilità, che si aspetta, e si conseguisce da questa arte: essendo tutte le sue attioni, et ogni suo fine à unire, e consolidare quella parte, che è rotta, tagliata, quasta, o altramente divisa; dico, nel corpo humano (per dimostrar la differenza fra l'arte Chirurga de'Medici, e quella de'Marescalchi, che operano ne'corpi inhumani, e bruti) dico vivente per far conoscer la Cirugia esser molto diversa dall'attioni anatomiche, che operano solamente nei corpi morti."

The distinctions, while not overly refined, did become points of contention. These practitioner-based anatomies became a prominent feature of a graduate surgeon's reputation, a prominent feature of his claim to expertise. In *Lo specchio*, the outspoken empiric, Leonardo Fioravanti (1518–1588) criticized surgeons for their faith in anatomy. These surgeons, he claimed, treated their patients like anatomical objects, scraping their bones like "the chops of a pig."[54] Academic anatomists, Fioravanti ranted, "continue in public studiums cutting dead men's bodies, making anatomy in order to teach the students how the human body is composed, so that they would then know how to treat [patients] when they practice the art of surgery."[55] Fioravanti has in mind the figure of the graduate surgeon, trained formally at the university by professorial anatomists. Ending on a strong note, Fioravanti states that unlike the butcher and his art, which "are very necessary to human life," anatomical knowledge is of little importance: dissection is unnatural, for "we see many dogs that never give themselves over to destroying the bodies of other, dead dogs."[56] Fioravanti tapped the concern that anatomical practices verged on sacrilege. This concern derived from the fact that dissection comprised the integrity of the body and often delayed or deferred indefinitely the burial of the body (Park 1994; Carlino 1999). Fioravanti, however, set the concerns about dissection within the context of the professions of academic anatomists, graduate surgeons, and butchers. He deployed the rhetoric typically reserved for the critique of vocations, and importantly, made anatomy the occasion and the graduate surgeon the target.

While anatomy and surgery were not synonymous, it was in the context of these professional debates that different parts of the study of anatomy and of surgery entered into constructions of expertise. Capivacci narrowed the surgeon's role and decreased his significance by defining him according to his manual skill; Fioravanti criticized what must have been a serious competitor, the graduate surgeon, by connecting anatomy to butchery rather than to the traditions of learned medicine (where, for example, Vesalius sought to locate the study of anatomy). In contrast, the published works of surgeons emphasize the theoretical principles of surgery, the ancient sources of the tradition, and the practical benefits of cautery, setting fractures, trepanning the skull and other activities associated with the instruments of the art. The rational surgery of the middle ages had emphasized the theoretical principles of the discipline (McVaugh 2006), and this aspect was given due attention by Venetian surgeons in the sixteenth century.

---

[54] Fioravanti 1572, 49–50: "Ma io per me ho sempre veduto, che i cirurgi, che sono buoni anatomisti, quando medicano pianghe, sempre vogliono fare la loro anatomia coi ferri tagliando le povere carni humane, come se fossero brasuole di porco, vogliono raschiare gli ossi, dare fuoco."

[55] *Ibid.*, 49: "I Cirugici, i quali volgiono sostenare, che loro sono stati gli inventori di questa anatomia, allegando che di continuovo ne'studii publici tagliano huomini morti, facendo notomia di essi, per insegnare alli Scolari, come sta la compositione dei corpi humani, accioche poi sappino medicare, quando eglino pratticaroanno la Cirugia."

[56] *Ibid.*, 53, 51, 51: "La beccaria e arte molto necessaria al vivere humano; e quante vene, nervi, muscoli, e ossa vi sono. Ma, al giuditio mio, questo sapere importa molto poco; noi vediamo i cani, che mai non danno molestia alcuna a i corpi de cani morti."

These aspects of the learned tradition of surgery are reflected in the life and work of Giovanni Andrea della Croce (Bernardi 1826). Son of a surgeon and from the Dorsoduro parish, Croce was licensed by the College of Surgeons in Venice and accepted as a member in 1532. He spent time as a surgeon for the naval fleet and subsequently became prior of the medical college, where he entered a culture of learned, sophisticated and urbane practitioners.[57] A sign of that sophistication was his connection to Francesco Sansovino with whom he worked on an Italian translation of Giovanni di Vico's *Practical Surgery* (1560).[58] In addition to treatises on military medicine and the treatment of wounds (1560), Croce published *The Universal and Perfect Surgery (of all the parts necessary for the optimal surgeon)* in Latin in 1573 and in Italian in 1574. In that monumental work, he identified himself as "medico venetiano" and offered "the theory and the practical aspects necessary in surgery," emphasizing his debt (as a humanist surgeon) to Hippocrates, Galen, Celsus, and Avicenna.[59] Croce's concept of surgery was a learned one. He explained that surgery required the "work of the hands" and was guided by the practical intellect[60]:

> Surgery is the oldest and most certain part of medicine, and it is a habit of the practical intellect, acquired with many rules and by experience [isperimenti], that is, with artful operations of the hands and proper instruments, joining, separating, and tying together the many wounds that run through the parts of the body; I say that one does this quickly, with certainty and with little pain, the artificial operation is done by the hands of the Medico, which is different from the other operations done by him with the intellect, such as seeing, composing, risolving, defining, demonstrating or other various actions completed with the parts of the soul...[Surgery] is the artificial operation, done with order, with art, with prudence, and not without a light touch, and it is structured by anatomical learning and extended practice.[61]

---

[57] *Ibid.*

[58] Bareggi 1988.

[59] Croce 1583, title page. Nutton 1985 has shown that classical texts on surgery were slowly made available in new editions mostly of the sixteenth century: while Celsus was available as early as 1478; Galen's *Method of Healing* appeared in Latin in 1519; the main surgical sections of Avicenna's *Canon*, Books XV and XVI, were not available in Latin until 1549; Book VI of Paul of Aegina's *Compendium* of medicine appeared in 1528; and Oribasius's surgical books were translated and made available by Vidius in 1544.

[60] According to Aristotle, the practical intellect had two parts *ars* and *prudentia*: *ars* or artificial operation was concerned with changeable things in nature, things which can be manipulated by manual, mechanical operations; and prudence, by contrast, was the faculty responsible for making moral decisions. Croce wished to include both in his definition of learned surgery. On the practical intellect in early medicine, see Ottosson 1984, p. 90.

[61] Croce, bk. 1, preface: "La cirugia è la più vecchia e la più certa parte di tutte la medicina, e è un'habito dell'intelletto prattico, acquistato con molte regole, e isperimenti, accioche con artificiosa operatione delle mani, e stromenti accommodati, uniendo, separando, e togliendo via molti affetti nel continuo delle parti del corpo humano: presto, sicuramente, e con poco dolore danar possi dico, che è artificiosa operatione fatta con le mani del Medico, a differenza di molte altre operationi fatte da lui con l'intelletto, di videndo, componendo, risolvendo, diffiniendo, dimostrando, o altramente operando con le parti dell'anima...Et è artificiosa operatione, cioè fatta con ordine, con arte, con prudenza, e non senza leggiadria, e è regolata dall'anatomia, e da una lunga prattica."

Croce's work on surgery is part of a larger development on the part of these surgeons to publish often and widely.[62] Their works treated the subjects of medicine and public health. In Latin and the vernacular, they demonstrated their knowledge of natural philosophy and of the ancients, their ease with traditions of commentary and the practices of medical humanism, and their knowledge of current treatments (especially for syphilis and plague). They also published in the fields of literature and art. Michelangelo Biondo (1497–1565), for example, published Latin treatises on physiognomy, memory, and medical topics – both practical and theoretical – and in the vernacular, a plague tract, pastoral poem, and study of the *paragone* of the arts.[63] He moved easily between languages, disciplines and genres. Similarly, at the end of the century, in addition to two, short tales, Fabio Glissenti (ca. 1550–1620) produced his enormous study on death, *Discorsi morali* (1596), which breached generic boundaries in order to probe the literary, philosophical and medical significance of his subject (McClure, 1998, 2004). Venice brought medical students into contact with these men, with these models of intellectual and professional success. Culturally sophisticated, these surgeons could not be reduced to the manual skill of a 'mere' dissector. Indeed, Croce argued, anatomy was instrumental to surgery – not the other way around.[64]

4 'Class-Consciousness' and the Venues of Anatomical Inquiry

Back at the university, various aspects of this tradition of surgery, which was at once practical, theoretical, humanist, and cosmopolitan, intersected with the study of anatomy. Recalling Richter's experience with the study and labor of anatomy and surgery, we must consider for a moment that this experience was particular not only because it depended on particular features of corpses, but also because it indicated something exclusive, an experience not available to all medical students. It did not derive from the public tradition of anatomical demonstrations, which was open to students who paid three "marcelli d'argenti" so long as they had matriculated into the university at least 1 year before and were studying medicine.[65]

Public and private anatomical traditions were never entirely separate, however. They continued to respond to each other, reshaping the significance of both, as the

---

[62] On publishing strategies and professional development, see Carlino 2009 and Harkness 2006, introduction.

[63] Biondo 1999, introduction. This is a modern edition of Biondo's treatise on physiognomy, originally published, in Rome, by Antonio Blado Asolano in 1544.

[64] See n. 38.

[65] *Statuta Almae Universitatis D. Artistarum et Medicorum Patavini Gymnasii*, bk. 2, ch. 27: "D. rector cum uno socio omnes doctores legentes, & omnes doctores de collegio, & ipsi duo massarii: ac etiam duo ali scholares pauperes de quorum paurpertate saltem per eorum iuramentum constet si fuerint per rectore, & consiliarios electi admittatur fine ulla solutione. Reliqui omnes repellantur. Nec rector aut consiliarii, aut ipsi massarii habeant potestatem aliquem admittendi non matriculatum, & qui non studuerit in medicina per annum, & quis non solverit."

following example will demonstrate. In 1595, Fabrici had written to the Venetian Senate to request permission to host a "free anatomy" in the new anatomical theater (1594–1595); and on 12 September 1596, the Senate granted his wish, accepting financial responsibility for the annual demonstration and the maintenance of the theater.[66] He was likely motivated by the large and diverse crowd that attended the premier demonstration in the new theater, a crowd that included (according to the student describing the event) "tailors, shoemakers, sandal-makers, butchers, fishmongers, and, lower than these, porters [perhaps funereal] and basket-bearers."[67] Fabrici's publicity stunt was not appreciated by all medical students. In the records of the transalpine nation, the conciliators, Matthia Jacobaeo Ripensi (a Dane) and Georgio Rumbaum Uratislaviensi (a Silesian) wrote passionately against the idea of a "free anatomy [*liberam anatomiam*]:"

> I counsel that there should be no credence given to the opinion of those who try to introduce anatomy for free in this academic institution, a sort of disease of the humanities, and the most open window of sedition and murder, contrary to our ancient traditions that have been preserved all the way to the present age. Indeed, what would agitate more vehemently or excite to arms more quickly those noble and learned members of the most famous school, those enjoying the extraordinary and unassailable privileges, than to witness with a hostile eye or to allow a crowd of the worst sort of gaping craftsmen and vulgar men to occupy the lower benches – nay! The whole theater – to trample, diminish and impede that sweetest fruit of anatomy?[68]

---

[66] *Acta*, 1595, vol. 2, 57: "Sed volo ut sit libera, omnes ut videant absolutissimam; et oportet ut Universitas Venetiis scribat ut ex pecuniis Studii sumtus habeat, et ego etiam intercedam meis literis; et concedo ut eligatis secundum statuta Massarios quibus singulis septimanis dabo florenos quatuor." On free admission, see also Riccoboni 1598, xviii: De ingressu in theatrum ad spectandam anatomen et honore anatomici. See also *Senato Terra Registro* 66, Archivio di Stato di Venezia (12 September 1596): "L'andera parte, che per l'avvenir l'Anatomia nel predetto studio nostro di Padoa sia libera, si che cadauno possa entrar nel theatro à vederla senza pagar cosa alcuna, et per le spese, che occorrono farsi siano delli danari delli danari [sic] della cassa del studio ogn'anno che si farà detta Anatomia Ducati XXV. Da esser dati a parte a parte, secondo che farà bisogno, il qual bisogno sia conoscuito da lettor di essa alli Massari over Anatomisti da esser eletti dalla Università al principio del studio per provinar le cose necessarie d'essa Anatomia non potendosi dar più di fiorini quattro alla settimana; per quelle settimane però, che l'Anatomico taglierà, et amminstererà le parti del corpo, et non quanto solamente leggerà, et detti sfiorini quattro hanno contati in mano dell'Anatomico, che taglierà, et administrerà et per lui siano dati a detti Massari secondo che conoscerà esser bisogno."

[67] *Ibid.*, 58: "Confluxerat eo tota quasi civitas, et extremae etiam farinae homines tanquam ad forum cupedinis: subsellia occuparunt hebraei, sedentarii magistri, sartores, calceolarii, solearii, carnarii, salsamentarii et his inferiores baiuti et corbuli illi."

[68] *Ibid.*, 60–61: "Neque eorum standum esse sententiae unquam et consulo et suadeo, qui liberam anatomiam quasi rei literariae pestem in hoc statu academico, atque fenestram seditionis ac caedis latissimam contra morem antiquum et ad nostram hanc usqu aetatem servatum, introducere conantur. Etenim quid nobilissimos Gymnasii celeberrimi studiosos, privilegiis singularibus praeceteris et immunitatibus gaudentes, commoveret vehementius aut ad arma excitaret celerius, quam invido conspicere oculo aut pati oscitantium etiam mechanicorum et proletariorum hominum extremae sortis catervam, subsellia immo universum theatrum occupare, fructum illum anatomiae iucundissimum conculcare quasi, minuere, impedire?"

The transalpine student experienced Fabrici's free anatomy as contaminating. The presence of craftsmen was part of the "disease" afflicting the noble institution of the university and particularly, the liberal arts. These craftsmen were not local, respected medical practitioners, who were welcomed into the theater and eagerly sought out by transalpine students. Instead they came from 'lower' trades, those belonging to the production of food and clothing.[69] Moreover, the student flagged the ancient traditions of learning, insisting on the humanist foundation of his education in anatomy and on an image of the academy that was exclusive, an institution, a pedigree, a set of habits reserved for the noble orders and higher trades. The presence of craftsmen in the anatomical theater threatened to diminish his humanist education.

The description indicates that the craftsmen inhibited the students' ability to see the anatomy. The presence of the former obstructed sight lines and the educational agenda of the event. Students had become used to listening to Fabrici during the public anatomy demonstration; the anatomical theater, newly built, splendid and extravagant, was an auditory space. Fabrici even hired musicians on two occasions to join the annual staging of the anatomy.[70] This time, however, something was different. The theater was crowded, as usual, but when it mixed orders – noble and non-noble – the result was a searching claim for the visual and not only the auditory apprehension of anatomical knowledge. In the language of the description, one hears the discourse of civility: the craftsmen looked the wrong way, they "gaped" at the spectacle, which prevented the students from *properly* seeing its "fruits."

The students' anger at the integrated audience in the anatomical theater culminated (on the pages of the nation's official records) in a rhetorically rich, violent, and futile outburst. The students concluded: "Let it remain hated, if we are wise; let this incitement to idleness remain hated, since it would be better in the present circumstances to use even a slippery knife, however feeble, than it would be to enjoy fully an uncertain hope in future circumstances [for our education]."[71] The students had been told to maintain peace (incited to idleness), and thus could only defend their honor and education with words – violent words rather than deeds though they wished to take up weapons, even feeble ones. The metaphor of "ansa lubrica" or slippery handle emphasizes the students' connection to the scene. It was an anatomy demonstration, but they felt a connection to the more active students at the event, namely the student-assistants (*massarii*), who by election to the position

---

[69] During the discussion of this paper, Alan Salter pointed out that these lower trades were predominantly ones that worked with animals – bleeding and skinning animals, making their parts available to eat, and tanning the hides for shoes and clothing. Perhaps these tradesmen thought they could learn technical skills from the anatomist, or perhaps they wished to compare their technical skills against those of the anatomist. The applicability of anatomical knowledge, however, was not the only reason that spectators came to the theater. On the wide appeal of the anatomy demonstration in Bologna, see Ferrari 1987.

[70] *Acta*, 1594–1595 and 1597.

[71] *Acta*, 1595, vol. 2, 60–61: "Valeat, si sapiamus, valeat detestandus desidiorum fomes, et ansa lubrica, cum melius sit praesentibus exigua uti, quam frui incerta sperandis spe."

were responsible for the details of the event and the preparatory dissections. These student-assistants took up knives, even slippery ones. This was the basis of the metaphor and however obscure, the clue to this demonstration. Future anatomies were uncertain – would bodies be made available; would space be found to demonstrate; would Fabrici uphold his responsibilities as a teacher? – but the students' remarks extend, with the help of a hyperbaton, that uncertainty to humanist education, which at the end of the sixteenth century, looked to them like it was in a state of decline.[72]

This episode should be seen in light of both public and private traditions of staging and studying anatomy. Fabrici's public anatomy made students intensely aware of the nobility of their education. In concrete terms, they criticized the public proceedings according to the merits of private anatomies: based, first, on the ability to see particulars; secondly, on the ability to participate or follow carefully the manual techniques of dissection and surgery; and finally, on the ability to engage in anatomical study and inquiry that was exclusive (private and intimate rather than public and open). Importantly, these features were laden with ideas of honor, masculinity and the uncertainties about life after graduation (how diminished would their degree be if the university was too often 'open' to everyone?).

5 Conclusion

By the end of the sixteenth century, anatomy and surgery were separate fields at the university of Padua, a splitting introduced formally during a series of seemingly innocuous reforms in the early 1580s. The study of anatomy was recast as a natural philosophical enterprise, capable of producing *scientia*, and Fabrici worked to transform his research in this area into a visible, public anatomical tradition. Fabrici promoted this tradition and further refined it in the permanent anatomical theater. Not neglected or stunted, the study of surgery also prospered under these new institutional conditions. Local surgeons and practitioners as well as professors provided regular instruction on dissection and surgery in private venues. Students sought out these events that were simultaneously useful, participatory, and exclusive. They learned about dissecting techniques and surgery; and in accounts of those experiences, they reflect on the nature and importance of technical expertise. Accruing significance in these private anatomies, technical expertise (*peritia*) became a virtue, one that students identified as present or absent in their teachers and in their colleagues. Students were fascinated by expressions of this expertise and devoted to those who embodied it.

While technical expertise (*peritia*) has been a feature of recent studies on artisanal traditions and trades, it appeared in Padua when academic life intersected with

[72] Favaro 1922 describes the many ways that the transalpine students were disappointed by Fabrici's anatomy lessons. Fabrici delayed his lessons, refused to offer them, and shortened them.

the cosmopolitan ethos of nearby Venice. Graduate surgeons were visible examples of successful practitioners and of a new cultural, urban elite – humanist in pedigree and entrepreneurial in spirit. The reputation and allure of graduate surgeons also influenced medical students, giving them reasons to make technical expertise an important part in their academic studies. They considered craftsmen, not surgeons, to be beneath them. As ideas about professions were taking shape against an increasingly refined and expansive marketplace, students sometimes adopted the language of professions, characterizing their teachers and their peers in terms of labor, industry and devotion. Technical expertise found a place in this broadly conceived reassessment.

Several questions might be asked in the wake of this development: how did the various parts of this learned tradition of surgery develop and gain traction in the university (Fabrici, at the end of his career, published an encyclopedic study of surgery); what kind of relationship was obtained between *peritia*, an embodied technical skill, and instruments; and how did questions about the vernacular or the *questione della lingua* provide new legitimacy to vernacular publications, practitioners, and work (is this what Casseri hoped to promote with his desire to use a "nude" or "naked" style for anatomy)?

In a review essay on the nature of natural philosophy, Adrian Johns (1999) situates the critical interest in experiment as the response on the one hand, to the many different systems and theories of nature on offer in the early modern period, and on the other, to the realization of implicit character assumptions about investigators being credible, reliable, honest gentleman. These experiments, Johns explains, "entertained, and in entertaining they exemplified a polite way of learning that could underpin a more stable society. In an age of all but unlimited contention, experiments alone proved persuasive." He then cites Roger French's evocative description of William Harvey's spectacular vivisection of the chambers and structures of the heart: "There is a conviction arising from being covered in blood that recognizes no philosophical opposition."[73] Johns ends his review by noting that the sources of that conviction are difficult to decipher. Private anatomies, while not experiments per se, elicited conviction but not necessarily because they covered the participants with blood or gore. Rather (perhaps occasionally, in addition) they offered experiences that were private, exclusive and not available to lower social orders and thereby reaffirmed the students' membership in the ruling elite. The students' conviction derived from conceptions of social identity. The conviction had an aesthetic character as well. Galeotto's dissections displayed his technical skills and his expertise (*peritia*); but his dissections were also 'exquisite' and 'beautiful,' an aesthetic judgment that emphasized the accumulation of experience – seeing many dissections in addition to doing them – and the clarification and disclosure of the body to sight. While craftsmen endorsed *peritia* in the production of goods and could make qualitative judgments about the use-value of goods, the anatomist-surgeon-student attained expertise in dissection, a moment identified by its aesthetic character.

[73] Johns 1999, 1128–1131.

Expert dissections not only revealed the body, they transformed the body, via dissection, into an aesthetic object. Far from the contaminating presence of lower craftsmen and tradesmen, private anatomies were thrilling because they were exclusive; they secured manual expertise as an academic privilege and perhaps also as an aesthetic accomplishment.

References

Agrimi, Jole and Chiara Crisciani. 1990. Per una Ricerca su *Experimentum-Experimenta*: Riflessione Epistemologica e Tradizione Medica (secoli XIII–DV). *Presenza del Lessico Greco e Latino nelle Lingue Contemporanee*, 9–49. Macerata: Università degli Studi di Macerata.

Atti della Nazione Germanica Artista [*Acta germanicae artistarum*], ed. Antonio Favaro (1911–1912). Padua: Libreria Emiliana.

Bareggi, Claudia di Filippo. 1988. *Il Mestiere di Scrivere: Lavoro Intellettuale e Mercato Librario a Venezia nel cinquecento*. Rome: Bulzoni.

Bernardi, Francesco. 1826. *Elogio di Gio. Andrea Dalla Croce Medico Chirurgo ed Anatomico Veneziano del secolo XVI*. Venice: Giuseppe Picotti.

Bernardi, Francesco. 1993. *Prospetto Storico-Critico. Dell'Origine, Facoltà, Diversi Stati, Progressi, e Vicende del Collegio Medico-Chirurgico, e dell'Arte Chirurgica in Venezia* (1797). Reprint. *Dalla scienza medica alla pratica dei corpi: fonti e manoscritti marciani per la Storia della Sanità*. Venice.

Biondo, Michelangelo. 1999. *Conoscenza dell'Uomo dall'Aspetto Esteriore* (Rome: Antonio Blado Asolano, 1544). Reprint. Rome: Beniamino Vignola.

Bonuzzi, Luciano. 1996. Medicina e Sanità. *Storia di Venezia: Dalle Origini alla Caduta della Serenissima*, 407–440. Istituto della Enciclopedia Italiana, Rome

Bylebyl, Jerome. 1979. Padua and Humanist Medicine. *Health, Medicine and Mortality in the Sixteenth Century*, 335–70. Cambridge: Cambridge University Press.

Capivacci, Girolamo. 1593. *De Methodo Anatomica*. Venice: Io. Baptistam Ciottum Senensem.

Carlino, Andrea. 1999. *Books of the Body*: *Anatomical Ritual and Renaissance Learning*. Chicago: University of Chicago Press. Originally published as *La fabbrica del corpo: Libri e dissezione nel Rinascimento*. 1994. Turin: Einaudi.

Carlino, Andrea. 2009. Introduction. *Vulgariser la Médecine: Du Style Médical en France et en Italie* (*XVIe et XVIIe siècles*), 1–34. Geneva: Droz

Carpi, Berengario da. 1521. *Commentaria cum amplissimis additionibus super anatomiam Mundini una cum textu eiusdem in pristinum et verum nitorem redacto*. Bologna: Hieronymum de Benedictis.

Casseri, Giulio. 1600. *De Vocis Auditusque Organis*. Ferrara: Victorius Baldinus, Cameralis.

Cipolla, Carlo. 1976. *Public Health and the Medical Profession in the Renaissance*. Cambridge: Cambridge University Press.

Croce, Giovanni Andrea della. 1583. *Cirugia Universale e Perfetta di Tutte le Parti Pertinenti all'ottimo Chirurgo*. Venice: Giordano Ziletti.

Cunningham, Andrew. 1985. Fabricius and the 'Aristotle Project' in Anatomical Teaching and Research at Padua. *The Medical Renaissance of the Sixteenth Century*, 195–222. Cambridge: Cambridge University Press.

Cunningham, Andrew. 1997. *The Anatomical Renaissance: The Resurrection of the Anatomical Projects of the Ancients*. Aldershot: Ashgate.

Daston, Lorraine and Katharine Park. 2001. *Wonders and the Order of Nature*. New York: Zone Books.

Dear, Peter. 1995. *Discipline and Experience: the Mathematical Way in the Scientific Revolution*. Chicago: University of Chicago Press.

Eamon, William. 1994. *Science and the Secrets of Nature: Books of Secrets in Medieval and Early Modern Culture*. New Jersey. Princeton University Press.
The Embryological Treatises of Hieronymus Fabricius of Aquapendente.1942. ed. Howard Adelman. New York: Cornell University Press.
Fabrici of Aquapendente, Girolamo. 1618. *De Motu Locali*. Padua: Io. Baptistam de Martinis.
Favaro, Giuseppe. 1922. L'insegnamento Anatomico di Girolamo Fabrici d'Acquapendente. *Monografie storiche sullo studio di padova, contributo del R. Instituto veneto di Scienze, Lettere ed Arti alla Celebrazione del VII Centenario della Università*. Venice. 107–136.
Ferrari, Giovanna. 1987. Public Anatomy Lessons and the Carnival: The Anatomy Theatre of Bologna. *Past and Present* 117: 50–106.
Fioravanti, Leonardo. 1572. *Dello Specchio di Scientia Universale dell'eccell. Dottore, et Cavalier M. Leonardo Fioravanti bolognese, tre libri*. Venice: Marchia Sessa.
Gentilcore, David. 1998. *Healers and Healing in Early Modern Italy*. Manchester: Manchester University Press.
Gentilcore, David. 2006. *Medical Charlatanism in Early Modern Italy*. Oxford: Oxford University Press.
Glissenti, Fabio. 1596. *Discorsi Morali Contra il Dispiacer del Morire*. Venice: Domenico Farri.
Harkness, Deborah. 2007. *The Jewel House: Elizabethan London and the Scientific Revolution*. New Haven: Yale University Press.
Henderson, John. 2006. *The Renaissance Hospital: Healing the Body and Healing the Soul*. New Haven: Yale University Press.
Johns, Adrian. 1999. Identity, Practice, and Trust in Early Modern Natural Philosophy. *The Historical Journal* 42(4): 1125–1145.
Kagan, Richard. 1986. Universities in Italy, 1500–1700. *Les Universitiés Européennes du xvi au xvii siècle: Histoire Sociale des Populations Etudiantes*, 2 vols. Paris. I: 153–86.
Klestinec, Cynthia. 2004. A History of Anatomy Theaters in Sixteenth-Century Padua. *Journal of the History of Medicine and Allied Sciences* 59: 375–412.
Klestinec, Cynthia. 2007. Civility, Comportment, and the Anatomy Theater: Girolamo Fabrici and His Medical Students in Renaissance Padua. *Renaissance Quarterly* 60 (2): 434–463.
Long, Pamela. 2001. *Openness, Secrecy, Authorship: Technical Arts and the Culture of Knowledge from Antiquity to the Renaissance*. Baltimore: Johns Hopkins University Press.
McClure, George. 1998. The *Artes* and the *Ars Moriendi* in Late Renaissance Venice: The Professions in Fabio Glissenti's *Discorsi morali contra il dispiacer del morire, detto Athanatophilia* (1596). *Renaissance Quarterly* 51: 92–129.
McClure, George. 2004. *The Culture of Profession in Late Renaissance Italy*. Toronto: University of Toronto Press.
McVaugh, Michael. 2006. *The Rational Surgery of the Middle Ages*. Florence. Sismel, Edizioni del Galluzzo.
Nutton, Vivian. 1985. Humanist Surgery. *The Medical Renaissance of the Sixteenth Century*, 75–99. Cambridge: Cambridge University Press.
O'Malley, Charles, ed. 1970. *The History of Medical Education*. Berkeley: University of California Press.
Ongaro, Giuseppe. 1981. La Medicina nello Studio di Padova e nel Veneto. *Storia della Cultura Veneta*, Vicenza, 3: 75–134.
Ottosson, Gunnar. 1984. *Scholastic Medicine and Philosophy: A Study of Commentaries on Galen's Tegni* (*ca. 1300–1450*). Naples: Bibliopolis.
Palmer, Richard. 1979. Physicians and Surgeons in Sixteenth-Century Venice. *Medical History* 23: 451–460.
Palmer, Richard. 1983. *The Studio of Venice and Its Graduates in the Sixteenth Century*. Trieste: Lint.
Park, Katharine. 1985. *Doctors and Medicine in Early Renaissance Florence*. Princeton: Princeton University Press.
Park, Katharine. 1988. The Organic Soul. *Cambridge History of Renaissance Philosophy*, ed. by Q. Skinner and E. Kessler, 464–484. Cambridge: Cambridge University Press.

Park, Katharine. 1994. The Criminal and the Saintly Body: Autopsy and Dissection in Renaissance Italy. *Renaissance Quarterly* 47: 1–33.

Park, Katharine. 2006. *Secrets of Women: Gender, Generation and the Origins of Human Dissection*. New York: Zone Books.

Pelling, Margaret and Charles Webster. 1979. Medical Practitioners. *Health, Medicine and Mortality in the Sixteenth Century*, ed. Webster. Cambridge: Cambridge University Press.

Pomata, Gianna. 1998. *Contracting a Cure: Patients, Healers, and the Law in Early Modern Bologna*. Baltimore: Johns Hopkins University Press.

Raccolta Minato. n. 56. (n.d.). Archivio Antico dell'Università di Padova (AAUP).

Riccoboni, Antonio. 1598. *De gymnasio patavino*. Padua: Apud Franciscum Bolzetam.

Rinaldi, Massimo. 2009. Modèles de vulgarisation dans l'anatomie du XVIe siècle: notes sur la *Contemplatione anatomica* de Prospero Borgarucci (1564). *Vulgariser la médecine: Du style médical en France et en Italie (XVIe et DVIIe siècles)*, 35–54. Geneva: Droz.

Saibante, Mario Vivarini, Carlo and Voghera, Gilberto 1924. Gli Studenti dell'Università di Padova dalla fine del 500 ai Nostri Giorni (Studio Statistico). *Metron* 5: 163–223.

Schmitt, Charles. 1985. Aristotle among the Physicians. *The Medical Renaissance of the Sixteenth Century*, 1–15. Cambridge: Cambridge University Press.

Shapin, Steven. 1994. *A Social History of Truth: Civility and Science in Seventeenth-Century England*. Chicago: University of Chicago Press.

Siraisi, Nancy. 1990. *Late Medieval and Early Renaissance Medicine: An Introduction to Knowledge and Practice*. Chicago: University of Chicago Press.

Smith, Pamela. 2004. *The Body of the Artisan: Art and Experience in the Scientific Revolution*. Chicago: University of Chicago Press.

Statuta Almae Universitatis D. Artistarum et Medicorum Patavini Gymnasii. 1570. Padua: Apud Innocentium Ulmum.

Stella, Aldo. 1961–1962. Intorno al medico padovano Nicolò Buccella, anabaptista del '500. *Atti e Memorie dell'Accademia Patavina di Scienze, Lettere ed Arti* 74, pt. 3: 333–361.

Sterzi, Giuseppe. 1909–1910. Giulio Casserio, anatomico e chirurgo (c. 1552–1616). *Nuovo Archivio Veneto*, ser. 3, 18–19: 207–278, 25–111.

Early Modern Empiricism and the Discourse of the Senses

Alan Salter

Abstract The relationship between science and literary texts in the early seventeenth century has only rarely been examined by scholars yet it is of immense importance in explaining the achievement of scientists in the period. The emergence of a language of empiricism and its usage in genres as eclectic as cosmography and drama shaped the practice of science by providing expressions and concepts that could be applied by investigators to their inquiries. But it was not just the language that took effect. Empiricism and the senses became topics in their own right and the works they appeared in displayed energy and intensity and an excitement at the possibilities of using new narrative ideas to explain the world. It is my contention that without this discourse of the senses the empirical practices that enabled the physiologist William Harvey to discover the circulation of the blood and the generation of the animal could not have been devised.

1 Introduction

During the second half of the sixteenth century the key words of empiricism took on their modern meanings. Observation came to signify scrutiny, or careful regard or painstaking attention to a thing; its earlier usages included custom, or practice or the performance of a devotional rite. Experience now meant personal affect as a source of knowledge; it had hitherto meant a trial or an event that affected an individual in some unspecified way. And to its prior usages of meaning, as in the meaning of scripture or a statement, or faculty, organ or instrument the word sense now served to convey concepts such as inward imagination, or the perception of an external object, or as something in opposition to intellect or reason. The frequency with which these words appeared increased greatly; citations of the word sense, its cognates and variants jumped by a factor of four, experience by seven and observation thirteen. Prior usages did not disappear but it was the new empirical usages that accounted for these big increases. The number of texts in which these words appeared increased by similar proportions, well

A. Salter
Unit for History and Philosophy of Science, University of Sydney

C.T. Wolfe and O. Gal (eds.), *The Body as Object and Instrument of Knowledge: Embodied Empiricism in Early Modern Science*, Studies in History and Philosophy of Science 25,
DOI 10.1007/978-90-481-3686-5_4,

above the increase in the number of all books published. So too did the frequency of affiliated words such as diligent or industrious, often used to qualify observation.

These new significations and the ideas that they gave expression to were not limited to scientific texts. They appeared in genres such as poetry, drama, travel and pageantry whose audiences extended beyond philosophers and physicians to merchants, artisans and the common man. Authors took the physiology of the senses as a foundation on which to construct imaginative treatments of epistemology which in turn shaped the new philosophy of empiricism. A discourse of sense emerged that infiltrated public imagination and helped to shape the activity and the texts of scientific inquiry. It is the spread and confidence of this discourse – from obscure epic to popular entertainment – that makes it possible to argue that a work such as William Harvey's *de Motu Cordis* published in 1628 which displays unmistakeable evidence of the language and concepts of popular empiricism could not have been written prior to the early 1600s and that Harvey could not have made his breakthrough discoveries without an intimate familiarity with this public discourse. It is this claim that I discuss in my paper, by an examination of five texts representative of this "genre empiricism" followed by an analysis of Harvey's debt to this discourse in his two major works.

2 Five Texts

2.1 *On Experience: Eden's* A Treatise of the New India (1553)

Eden's text is a translation from the original Latin of Sebastian Munster's work of "universall cosmographie." The book describes strange and wonderful things: of events on the voyage south via Aden where the king was hanged, of elephants "very docible and apt to be taught," of the marital customs of Calicut, of the practice of suttee, of the sale of elderly parents by their children to cannibals, of timber sawn by stone tools and of "the people with great hanging eares."

But it is Eden's own preface that is of greater interest for in it he claims that experience is the necessary and sufficient source of knowledge. Things experienced, he writes, however strange exist: "Whereas in this Booke (wellbeloued Reader) y[u] mayest reade many straunge thinges, and in maner incredible, except the same were proued most certayn by dayly experience & approyued auctorie"; he marvels that Magellan sailing westward from Spain met the Portuguese sailing eastward in the Molucca islands, "(A) thing surely most wonderful, and in maner incredible, but that the same is proued most certayne by experience."[1]

[1] This may be a comment on the possibility of a flat earth, an idea still current in England and referred to in Eden's later passage on the errors of St. Augustine. Under the Treaty of Tordesillas of 1494 the non-European world was divided into two. Portugal was allotted the Eastern hemisphere, which was defined so as to include Brazil which Portugal already claimed, and Spain the Western. The meeting of Magellan in the service of Spain and the Portuguese in the Molucca Islands south of modern Borneo would therefore demonstrate by experience the existence of a spherical earth. I am grateful to Captain Iain Kirk of the Australian Maritime Safety Authority for suggesting the meaning of this passage to me.

When in discussing the existence of the Antipodes he writes that "no man knoweth further hereof then is tryed and founde by experience" he marks out things experienced as the boundary of things known (Eden 1553, 2, 4, 6).

Eden justifies his confidence in experience in part by reference to its success. "Wherfore," he writes, that "whereas men of great knowledge and experience, are to great affaires, theyr attemptes haue for the most parte good successe." Augustine failed in the absence of experience; of his errors in astronomy, Eden writes:

> and to declare my opinion in fewe words, I thinke it no greate marueyle that Saincte Augustyne shoulde fall into erroure in the science of Astronomie in whiche he trauayled but as a straunger... And I beleue playnely that, that excellent witte of hys, could not haue remayned longe in y^{t} erruor yf he had been wel exercysed in Astronomy, or had knowen any suche experyence as is spoken of here beefore.[2]

Eden's claim of the sufficiency of experience is more radical. He denies the separateness of reason. Reason means knowledge, and knowledge is "what we commonly call learning, whether from books or from discourse." But Castiglione showed, Eden writes, that the knowledge found in books is valueless; the wise man relies on his own experiences and the experiences of other men and on the contents of their minds, themselves reliant on their own experiences and those of other men. It follows then that knowledge, or reason, is reducible by infinite regress to experience. Reason is not a class of rules independent of experience; it is an accumulation of individual men's experiences and thus reducible to it.

2.2 *On Observation: William Cuningham's The Cosmographical Glasse (1559)*

The Cosmographical Glasse begins with a flamboyant front piece portraying the great heroes of cosmography, such as Strabo, Ptolomey and Marinus (probably the Phoenician cartographer of the first–second century) as well as classically attired women representing arithmetic, astronomy and so on. In the upper centre of the frame is a motto "Viscerit Vulnere Veritas," literally "Truth becomes green from injury" or more easily "Truth ripens in pain," signalling the immense struggle involved in the acquisition of knowledge. It is a point Cuningham expands on in a short introductory verse:

> In this Glasse if you will beholde
> The Sterry Skie, and Yearth so wide,
> The Seas also, with windes so colde,
> Yea and thy selfe all these to guide:
> What this Type meane first learn a right
> So shall the gayne thy trauill quight[3]

[2] Eden 1553, 5.

[3] Cunningham 1559, 1.

"Type" means a representation or figure so that those who would understand the image in the cosmographical glass and would benefit from it must first learn how to observe the figure properly, through hard work, practice and habit.

The work is part dialogue and part catechism between Spoudaeus, a plodding student, and Philonicus his teacher on the necessity of diligent observation in acquiring knowledge of the world. Cuningham had a close personal as well as professional interest in the topic; though an educated and practicing physician he published several almanacs and was a keen astrologer. Of cosmography (a comprehensive discipline of geography, astronomy, geology, anthropology and so on) he is lavish in his praise:

> If ever there wer Art for all mens vse inuented, Science set forth wherein consisteth Sapience, or Treasure worthy to be had in estimation: no doughte (loungye Reader) either Cosmographie is the same, or els it is not to be founde vppon th'Earth.[4]

And of his glass, the metaphor of his instrument of inquiry, he writes that:

> I... haue deuised this mirrour...(in) which men may behold not one or two personages, but the heauens with herplanets and starres, th'Earth with her beautifull Regions, and the Seas with her meruellous increse.[5]

The importance of the text lies in its treatment of observation. Observation, Cuningham writes, is necessary for knowledge since man is "in all things ignorant, except in such as ther senses and custome teach them" (Cuningham 1559, 1). Language and rhetoric are not knowledge and words prove nothing; only observation, correctly performed, can discover truth. But observation is not some vague procedure or elusive concept; it is an exact and specific technique for the arrangement of our sensory perceptions. Observation can be divided into two parts: the observation of a thing and the observation of a sequence of things. The former is necessary to establish precision, the meaning of which is explained by Philonicus in an example concerning the measurement of geographical dimensions:

> I will in fewe words open up the whole labour. Take a quadrant (...which you shal see among the other instruments) and set it directly vpright vpon some playne in the meridian lyne...and is here marked A.B. as also C...then rayse vp and downe the ruler...vnto the sonne, whan as he is in the meridian line: and obserue diligently that height, in the circle of degrees noted in your quadrant from B. to C. the .xi. or .xii. day of December which is B.D. vntill you find he goeth no lower: and againe in like case the .xi. or .xij. of lune vntyll you perceiue he increaseth nothinge in height.[6]

These instructions, detailed and precise, are followed in the text by a table of longitude, latitude and declination. But this table has more than an informational

---

[4] Cunningham 1559, 2.

[5] Cunningham 1559, 1.

[6] Cunningham 1559, 15.

purpose; by its detail and density it demonstrates the precision and diligence in the technique of observation that Philonicus expounds. Philonicus reveres proper observation; it is attentive and careful, more like a measurement than a description.

The observation of a sequence of things protects the precision of the single observation from the possibility of error and establishes its stability over the course of time. In a passage from Book 2, Philonicus instructs Spoudaeus on the observation of the constellations. Spoudaeus is concerned that there is a diversity in the declination of stars and planets, perhaps caused by instruments "not exactly made." Philonicus refutes this, saying that it is due to the moving of the heavens, and though "it vary not in a mans lyfe any thing sensible" it is important to know "the obseuring of it exactly and also teache other that hereafter shal lerne as you are now instructed your selfe" in order that future measurements can be made in the same way as hitherto.[7]

This notion of diligence pervades the work. Diligent observation guards against the possibility of error, corrects the mistakes of others and protects the inquirer against flaws in instruments, for these flaws must be detected and cannot be held responsible for faulty observations. Spoudaeus is in awe of Philonicus, often referring to the methods he has been taught for drawing maps and measuring and calculating declinations, latitudes and so on; he marvels at the techniques necessary to gather the dimensions required for these maps, of diligence, close and painstaking attention to the thing and the multiplicity and frequency of single inspections that constitute proper observation.

2.3 *On the Senses: George Chapman's* Ovids Banquet of Sence *(1595)*

George Chapman moved to London in his early maturity where he mixed in a group centred on Sir Walter Raleigh, and which was mocked by Shakespeare in *Love's Labours Lost* as the "School of Night." Other members of the group were Christopher Marlowe, Thomas Harriot and Matthew Roydon, a mathematician and one of Raleigh's associates. Chapman dedicated Book 18 of his translation of *The Iliad* to Harriot, completed Marlowe's *Hero and Leander* following the playwright's death and argued in *De Guiana* that Elizabeth should recognize Raleigh's discovery of the territory and fund its colonization. The activities of this so called School of Night provide the background for Chapman's epic. It was, in one editor's words, "a group of men...interested in pushing science and philosophy beyond the bounds of contemporary belief and decorum...bound together by a common curiosity" and for whom *The Shadow of Night*, Chapman's first published work, was its manifesto (Bartlett 1941, 5). Although we can read *Ovids Banquet of Sence* in the banquet tradition in which the hero meets a sequence of provocative

[7] Cunningham 1559, 15.

challenges to be overcome, we can also read it in an empirical context in which the senses are portrayed as discrete instruments of the human body with physiological capacities and imaginative powers.

Early in the poem Chapman deals with the reliability of sight. He introduces the story of Niobe whose statue was brought from Mount Sipylus in Lydia to Augustus' garden in which the action takes place. There Niobe turned into a rock and shed everlasting tears after her children were killed by Apollo and Artemis. The myth of Niobe is a poetic opportunity for Chapman to treat of sight's deception:

So cunningly (is the statue) to optick reason wrought,
That a farre of, it shewd a womans face,
Heauie, and weeping; but more neerely viewed,
Nor weeping, heauy, nor a woman shewed[8]

Chapman dwells on the opposition inherent in sight when he writes of the physics of vision and the eye's superficial appreciation of its object:

Betwixt mine Eye and obiect, certayne lynes,
Moue in the figure of a Pyramis,
Whose chapter in mine eyes gray apple shines,
The base within my sacred obiect is:
On this I will inscribe in golden verse,
The meruailes raigning in my soueraigns blisse,
The arcks of sight, and how her arrowes pierse
(Chapman 1595, 69)

Chapman means there is a pyramid drawn by rays of light that emanate from Corynna, the object of Ovid's affection, and the peak of this pyramid shines into his eyes. But her form, the base of the pyramid, exists only in his mind. He will therefore inscribe a poem on his mind to explain how her beauty pierces his eye but is not captured by it.

The closing stanzas of the poem deal with touch but Chapman's touch is not the base sense of antiquity and the Renaissance but is the divine sense, "(t)he sence wherewith he feeles him deified" that we must use with "feare and reverence" (Chapman 1595, 80, 81). Touch overcomes the blindness of our other senses; it is the goal of all our human actions. It renders the body spiritual and is "the sences Emperor," the "King of the King of Sences," the individual private sense and the sense that disputes against reason (Chapman 1595, 79, 80). It is the true expression of love, grounds all other senses, feels and has feeling and creates eternal memories.

Chapman's ideas on sensory materiality occur throughout the work and argue for the superiority of the senses over reason. Corynna bestows a kiss on Ovid that throws him into a delirium, a swirling of his body and a confusion of his mind; the tunes that were "shooke from her braine (and) first conceiued in her mentall wombe" were "nourisht" by her animal womb and changed in their material form. Ovid's unwelcome entreaty that Corynna grant him a kiss may show him to be weak in character, but his weakness is the result of a sensual strength that claims his reason. It is the senses that allow Ovid's eventual triumph; they satisfy all his exigencies, spiritual and material.

[8] Chapman 1595, 54.

2.4 *On the Senses: Middleton's* The Triumphs of Truth *(1613)*

Thomas Middleton's pageant for the inauguration of Sir Thomas Middleton[9] of the Grocers' Company as Lord Mayor of London in October 1613 was the most magnificent of the era, surpassed only by the 1604 coronation entry of James I into the city. It cost 1,300 pounds, at a time when the average annual wage of an unskilled worker in south-east England was 5 pounds, and consisted of five ships, movable islands, small hillocks and elaborate transportation mechanisms.[10] Its subject is the contest between truth and error and in it the five senses play the central role.

After a song addressed to London, a "Graue Foeminine Shape," mother to many honourable Lords Mayor, the pageant moves to the river, where the first thing it sets its eyes upon is a display of five islands "art-fully garnished with all manner of Indian Fruite-Trees, Drugges, Spiceries."[11] The company returns to Barnard's Castle, where his Lordship is attended by Truths Angel, sent by truth to guard and guide the mayor, accompanied by Zeale, "Truths Champion…the scourge of Sin."[12] It continues on to Pauls-chaine close by St Paul's where Error waits and implores Middleton on the promise of great riches to forego Truth. But Zeale forces Error and Envy, Error's champion, to retreat and make way for truth, and demands of the mayor that he reject Error. Truth is personified as a woman dressed in flowing white silk; she recalls her lesson from the past, that there is but one true way to truth:

> let not thy Heart be led
> In ignorant waies of insolence and pride
> From Her, that to this day hath bene thy guide;
> I neuer showed thee yet more Paths then one,
> And thou hast found sufficient That alone[13]

The company moves to St Paul's Churchyard, the largest public space in the City. At the head of the procession is the chariot of Truth accompanied by Truth's celestial handmaids, the three graces and the four virtues. Close behind are Zeale and the Angel of Truth keeping guard over the Lord Mayor in his chariot, pursued by Error. The narrator describes the scene in the churchyard:

---

[9] There is not thought to have been any familial relationship between the playwright and the mayor.

[10] The Russian ambassador Alexis Ziuzin described Middleton's pageant as 'the most expensive and elaborate lord mayor's pageant ever produced.' See Taylor, 2008. Sir Thomas Smith, of the Haberdashers and Skinners and Governor of the East India Company explained to the Russian entourage that "except for the King's coronation there is no other such great ceremony in England." See Taylor, 2007, 977 fn. Bergeron notes that the coronation pageant cost 4100 pounds. See Bergeron, 73.

[11] Middleton 1613, A3, B. It was a customary for pageants to display the products of the Livery Company to which the new Mayor belonged. The fruit trees, drugs and spices point to the Grocers Company's foreign trade.

[12] Middleton 1613, B2.

[13] Middleton 1613, B2.

> where stand ready the fiue Ilands, those dumbe Glories that I spake of before vpon the water, vpon the heighth of these fiue Ilands sit fiue persons, representing the fiue Sences, *Visus, Auditus, Tactus, Gustus, Olfactus*….; at their feete their proper Emblemes, *Aquila, Ceruus, Araneus, Simia, Canis,* an *Eagle,* a *Hart,* a *Spider,* an *Ape,* a *Dogge*[14]

In what can only be an intentional conjunction of vistas, the islands of the senses, first seen on the river, are now the first pageants to be observed by the company after Truth's claim that there is only one pathway to truth. They are arranged in a line or a triangle, with sight and smell at the vertices of the base line, hearing and taste midway along the inclinations and at the apex touch, which probably occupied the largest island. A ship appears, suddenly, with neither pilot nor sailor, but on a white streamer "these two words set in letters of gold, Veritate Gubernor, I am Steer'd by Truth" and on the ship the King of the Moors and his Queen. On the island of touch there stands a castle, where the king's followers are to be found, two or three appearing on its ramparts.[15]

At this point the procession is led by the five islands who guide the ship, Truth, Zeale, Error and the Lord Mayor along Cheapside into "the little Conduite,"[16] where stands a mount, bearing the title "Londons Triumphant Mount," which is the "chief Grace and Luster of the whole Triumph." The mount has been taken over by Error and is symbolically shrouded in fog and mist. At each corner stands a monster, each one a disciple of Error: barbarism, ignorance, falsehood and impudence. But Truth sends Error on its way so that the mount is returned to glory by the lasting power of truth.

With this glorious "mount triumphant" the figuration of the pageant changes, from the allegory of the power of the senses that point the one true way to truth to a celebration of the city of London and its beneficence. At the foot of the mount sits mother London; above her Religion and at her sides Liberality and Perfect Love, all personified and dressed in manner fitting to their proper character. At the rear are four pageants depicting chastity, fame, simplicity and meekness; and to the sides displayed the city's charitable and religious works. Knowledge and Modesty sit on the heights of the mount. Finally to the Guildhall, the site of the city's government, where London and Truth praise the celebration of the day and its hero, the mayor, though not before Time, who offering to cut off the glories of the day with his scythe, is prevented by his daughter truth. The day ends with the death of Error and "all the Beasts that are joynde to it," set on fire by a flame shot from the head of Zeale.

The Triumphs of Truth describes the victories attained by truth over error during the day, culminating in the final victory, the destruction of error and its beasts. Truth's victory was grounded on the five senses, which show the one true path to

[14] Middleton 1613, B2.

[15] A Moor was an Indian, possibly a reference to the Grocers' Company's interests in India.

[16] Little Conduit Street, which according to present day maps lies at the end of Cheapside.

truth. The senses dominate the action of the pageant. They appear first, atop the islands on the Thames, and next in the long and stately scene in St Paul's Churchyard, where it is most likely that, given the size of the yard, the majority of the audience would be found. This scene has a central allegorical role too, for the senses "stand ready" to receive Truth and protect her during the next stage of her journey. In the previous scene Truth has affirmed that there is one true path to truth and in this next scene she immediately comes upon the senses, which signify this one true path. This same scene portrays Zeale forcing Error to retire by threats of fire and words alone. But these cannot endure for fire is soon extinguished and words are forgotten, so the senses must intervene.[17] The islands are physically dominating too, five chariots on one of which, the middle one, touch, is a castle in which are to be found the followers of the King of the Moors. Though the mount triumphant seems to take the head of the procession on the four journeys from Cheapside to Leadenhall the allegory has by this stage altered to a more gentle form, which tells of the goodness of the City, its religious and charitable works. But the threat from Error does not diminish and the five islands though hereafter unmentioned must I think remain part of the company, for the mount triumphant poses no material threat to Error and Envy.

Middleton has no view on the priority of the senses. It is the case that he probably places touch at the head of the five islands, perhaps in a triangular format borrowed from the Nova foelix Arabia arch erected for James I's coronation entry to the City in 1604, and it is also the case that he accords to touch the responsibility of hosting the King of the Moor's castle, but there is nothing else to suggest that he wished to draw attention to the senses' hierarchy. There is the familiar treatment of the senses in the plot and in the language: an air sung at the beginning of the pageant is called sweet, a description also found in Chapman,[18] and the narrator says that the opening ceremony will give his Lordship's "eare a taste of the dayes succeeding glory."[19] But there is no explanation for the presence of the islands or of the emblematic representation of the senses by animals. There is no comment on the senses' leadership of the procession, nor of the central role they have in the pageant (why did Middleton not call the pageant the triumph of the senses?). All this suggests that the treatment of the senses, their natural leadership, their personification and even the gesture at the primacy of touch was unsurprising and expected. The day is patterned like the passage of life; the Lord Mayor enters into it as a child, borne of Mother London, to be handed over to the city at its end by the triumphs which have throughout been guided by the senses and the circumstances and values of the city so that Middleton elevates the five senses to a position of eminence in the life and society of man.

---

[17] It is possible that fire here is the fire of the four elements of antiquity but it is I think more likely to be flame, since it is flame that destroys Error, shot from the head of Zeale in the closing spectacle.

[18] Corynna has a "sweete voice." See Chapman 1596, 57.

[19] Middleton 1613, A3.

2.5 *On Language, Reason and the Senses: Thomas Tomkis'* Lingua *(1607)*

Lingua: Or "The combat of the tongue and the five senses for superiority" was first performed in 1607. The plot concerns a fierce, angry and sometimes violent contest between Lingua, symbolizing language and reason, and the five senses. The primary characters are Lingua, Auditus, Tactus, Olfactus, Visus, Gustus and Communis Sensus, but there are important secondary characters too, such as Veritas, Mendacio, Lingua's page, Phantastes and his page Heuresis, and Memoria and his page Anamnestes. The action takes place in a grove called Microcosmus.

The matter of the play is introduced in the first scene where Lingua and Auditus debate her contention that she should be granted the status of a sense. Lingua says she will petition Communis Sensus with her demand to be included with the senses but Auditus instantly rejects her argument. Language is deceitful, he claims; she counters that it is the senses that tyrannize for they pass to mind only what they wish the mind to know. I shall take my revenge, she announces, for I know how to deceive the senses and how to imitate them: their weakness is their pride and I shall be satisfied. In a brief end to the scene she instructs her page Mendacio to lock up Veritas and sets about to devise a scheme that will make the senses compete one against the other.

After a short intervening scene Common Sense enters in grave and sombre dress introducing himself as Psyche's sovereign and proceeds to conduct a strange disjointed conversation with Phantastes. Common Sense then instructs Lingua to command the senses to attend him, for the contest is about to begin. The format is to consist of Lingua and each sense showing the objects by which they are to be judged. Lingua begins her advocacy and looks to all her friends to stand with her: rhetoricians, lawyers, women and so forth. Her speech is flamboyant and contains much gibberish, "Gallemaufry" Common Sense calls it, but she makes the apposite point that the senses can only be represented by polished language, for their knowledge is only of things present. But the tongue can recount things past and things to be; it can charm the animals and the plants and prevent wars when all the senses have been defeated; the world would fragment were it not for speech. It is an eloquent speech says Common Sense, but it is not enough for these attributes of rhetoric, logic and narration are not in the nature of a sense.

Appetite, a minor character, then delivers a formal statement signed by the senses listing ten charges against Lingua. They assert that she has imprisoned Veritas; that she maintains a train of artless Empiricks; has made rhetoric wanton, logic to babble and astronomy to lie. To present their own cases the senses arrive in high fashion, attired in grand heraldic costume. Visus defends the validity of sight. Sight reveals all; light, sight's page, discovers falsehood; the eye is the soul's mirror; and wisdom is bred by experience which is seen by the eye and conveyed to the brain. Auditus, Olfactus and Gustus each chime in. Finally Tactus speaks. The hand, he says, is the instrument of all instruments, the root of life, like a spider casting her net over every part of the body. Touch has great skill and all other senses are beholden to it; touch can command them all. [20]

Common Sense gives his judgment. The senses must be classed into two: those that are pleasurable, which deserve the crown, and those that are necessary, which deserve the robe. Pleasurable senses such as sight and hearing serve the queen (that is, the soul), and the necessary senses serve the body. Since sight is the sense of invention and hearing the sense of increase it is right that sight takes the crown. Tactus is to receive the robe because he is the necessary sense. Auditus, Olfactus and Gustus are given minor roles. Lingua however is not a sense for there are only five great bodies in the Universe, the four elements and the pure heavens, and so there can be only five senses. But since Lingua is a woman, Common Sense decrees that henceforth all women shall have six senses, the five known and a sixth, the sense of speaking. Lingua is not pleased at this slight and vows revenge. But her revenge is easily and quickly defeated and she is punished, to be imprisoned and guarded by Gustus until she is 80 years old. As for Mendacio, her page, his punishment is never to be believed.

Tales of deceit and trickery pervade the play. Language and reason deceive. Auditus accuses Lingua "confesse the truth, th'art wont to lie"[21] and Mendacio concurs "My Ladie loues me exceedingly; she's alwayes kissing mee, so that ... Mendacios never from betwixt her lippes."[22] The senses also deceive but with this important point of difference: they deceive themselves. Visus is deceived by Tactus, who pretends he has the plague; Tactus is deceived by his own imagination to think he is a prince, a Caesar, a great Alexander, "transform'd, unto the sacred temper of a King" when he puts on the robe and crown, laid down by Lingua and Mendacio for him to discover. These episodes are intended to amuse but they conceal an acute insight into the character of sense perception, that it is arrogant and mistaken. Visus has no skill in the recognition of the plague, no prior encounter and no experience as a physician; sight can only see. Tactus cannot comprehend language and so is tricked by the words of Mendacio; and Olfactus smells but cannot imagine or reason. Lingua is right when she observes that there is no independent mediation by the brain of the senses' input, for they conspire to influence the brain's judgment:

> O how these senses muzzle common sense:
> And more and more with pleasing objects strive,
> To dull his judgment and preuert (sic) his will
> To their behests[23]

Nor are the senses objective; they are deceived by context and memory. They cannot transcend their own experience; they see what they wish to see and smell what they wish to smell.

---

[20] The hand is the "King of the King of Sences." See Chapman 1595, 80.

[21] Tomkis 1607, Act I, scene I.

[22] Tomkis 1607, Act 3, scene I.

[23] Tomkis 1607, Act I, scene I.

While there are plenty of clues in the play as to contemporary discourse on scientific matters there is only one explicit reference to the increasingly confident expression of empiricism in the science and popular culture of this period when Common Sense invites Lingua "to speake what you can for your selfe." She proceeds as follows:

> My Lord, though the Imbecillitas of my feeble
> sexe, might drawe mee backe, from this Tribunall,
> with the habenis to wit Timoris and the Catenis
> Pudoris, notwithstanding beeing so fairely led on
> with the gratious (επιογχεια) of your iustissime
> (διχαιοσονηζ) Especially so asprimente spurd'
> con gli sproni di necessita mia pungente, I will
> without the helpe of Orators, commit the totam salutem
> of my action to the Volutabilitati (τον γ ϖαιχιλοσ λογοσ),
> which (avec votre bonne playseur) I will finish with
> more then Laconica Brevitate.[24]

The purpose of the passage is to introduce her speech, first apologizing for her weakness as a woman and promising to proceed quickly without assistance and to conclude briefly. But Common Sense is right when he calls it gibberish. It is not so much that the passage uses five languages but rather that in an effort to impress and captivate it succeeds only in being affectatious and foppish. Memoria steps in and highlights the problem: "I remember about the year 1602, many vsed this skew kind of language."[25] This is how people used to talk he says, but no more. We no longer speak this half language, with its Latin, Greek, French and Italian but plain English, the language of experience and the senses. Common Sense likens this mixture of language to a congealing of English tin and Greek gold drawing attention to the contrast between the two. Tin is a hard metal, older than gold, utilitarian, that when alloyed to copper and bronze makes tools and weapons; gold is soft and superficial, for decoration and jewellery. In this short exchange of no more than a few dozen words Tomkis conveys the essence and import of his drama: the legitimation of English empiricism, of the victory of the senses over the false reason represented by Lingua and the break with the past that, though not necessarily caused by the change in monarch, had occurred alongside it.[26] Lingua is no knockabout farce but a serious work whose language and depiction of the intensity of the contest between empiricism and reason provides us with a vivid example of the vindictive reality of this epistemological

---

[24] Tomkis 1607, Act 3, scene 5. I have placed the Greek text in parentheses because the characters are unclear in the EEBO (Early English Books On-Line) version.

[25] Tomkis 1607, Act 3, scene 5.

[26] James I ascended the throne of England in 1603. James' ascension was widely regarded as new era for the Kingdom, attested in part by the appearance of the phoenix at the Nova Arabia Foelix arch in the Coronation pageant and Dekker's comment in the accompanying text that "vpon this day, began a new *Creation*." See Bowers 1955, vol. II, 258–259.

discourse. The work was astonishingly popular, going into five reprints, the last in 1657, some 40 years after Tomkis' death. Its format of naïve comedy, learned allusion and close-textured dialogue suggests that it was written for a mixed audience and that empiricism, reason and learning were topics of intense interest at all levels of society and were not restricted to recondite debates in the universities, philosophical clubs or court.

3 The Discourse of the Senses and William Harvey's Idea of Empiricism

The association between language and thought is not new. It has already been studied in the premodern period by Lucien Febvre who demonstrated that the availability of words was critical to the articulation of new beliefs. More recently, David Rosen has examined the relationship between the scientific and poetic communities in the early modern period, finding a shared discourse and its necessary presence in scientific discovery. To Rosen, this "shared discourse" lived in the shadows of learned England, solitary, subtle and independent of other cultural currents. But there was another discourse, illustrated by the work of Middleton, Tomkis and others. This discourse was public, assertive and self-confident; it gave birth to new genres, not in rejection of earlier notions of the senses and their place in discourse but in acknowledgment of their capacity to amplify and enrich it. In this discourse the senses offered explanatory power, provided a greater linguistic range with which to describe contemporary celebrations and anxieties and introduced a wholly new subject, the senses themselves. The combative action and the flamboyant victories of the senses over language and reason in works such as *Lingua* and *The Triumphs of Truth* introduce an empiricism that is vital and energetic. Sense literature in the early modern period was disputational and densely descriptive, exemplified especially in the work of Chapman; it was as though there was a fervent desire to understand the senses and that writing about them would accelerate the attainment of that desire. New narrative threads emerged and the senses were now enlisted to describe, imagine and explain the world.

Harvey drew on these new threads in his work on circulation and generation. Two fifths of *de Motu Cordis* consists of remarks, reflections and propositions on empiricism. Most of these forms make use of its essential words -sense, observation, experience and their cognates and variants- in meanings that first appeared in the second half of the sixteenth century. *De Motu Cordis* is as much a text about empiricism as it is about physiology and Harvey's discoveries were only possible because of a literature and language that made empirical concepts available to investigations of the animal body.

Consider Harvey's frequent use of the words sense and experience. Except when he uses it in its ordinary meanings of perception and the organs of perception, he identifies the word sense *(sensus)* with knowledge, truth and observation.[27] In the Epistle Dedicatory to Dr Argent at the beginning of *de Motu Cordis* he praises philosophers who are never "so abundantly satisfied in their own knowledge" that they do not welcome the truth from elsewhere,[28] and later in chapter 9 refutes erroneous opinions of the motion of the heart as being "contrary to observation *(sensui)*."[29] The ideas expressed here are exactly what we see in Tomkis when Visus claims that all knowledge and experience is revealed by the sense of sight and in Middleton who puts the senses in dominion over truth. Like Chapman, Harvey endows touch with great imaginative power. His language suggests an acute cognitive capacity, a completeness of information at the moment of perception and a remembrance more firmly embodied, more enduring than sight or hearing. In *de Motu Cordis*, he frequently uses the verb to sense *(sentire)* to mean feel, and throughout the *Prelectiones* and much of *De Generatione Animalium* he ranks touch as the privileged sense, natural as well as mediated, and material rather than intellectual.

As with Eden, for whom experience is the necessary and sufficient source of knowledge, so too with Harvey. Following an assertion as to the cause of breathlessness in a tertian fever he comments that "I say this as one who has had experience *(expertus)* in the dissection of those who have died at the beginning of an attack."[30] In writing of the contamination of the body by poison, the wound itself being unaffected, he writes "the wound made by the bite of a mad dog being healed, yet I have seen fever *(experti sumus)* and other terrifying symptoms ensue."[31] In chapter 17 he remarks on the reasons for the large vessels in the lungs and why they "are filled so full of blood as we know from experience and our own eyesight."[32]

The relative importance of sense or experience compared to reason appears infrequently as a topic of discussion in *de Motu Cordis* but is a central theme in

---

[27] In assessing Harvey's original Latin I have been mindful of the possibility of modern usages creeping into recent translations. I have therefore used the Whitteridge translation which is based on the 1653 English edition. I have also consulted Leake's earlier translation. With very few exceptions, Harvey's Latin was classical rather than Renaissance. In particular, the key words of his empiricism - *observatio, observo, experientia, experimentum, experior, sensus, sentire,* their cognates and variants – are classical not Renaissance. See Giglioni 1994. And as Hoven points out, Renaissance Latin was not really a language but more a supplement to classical Latin. See Hoven 2006, xix. Other than a brief comment by Whitteridge on the inadequacies of dictionaries of classical Latin in translating Harvey's zoological terms neither she nor Leake offer any firm views on the sources of his Latin. Accordingly I have assessed both translations in the context of classical Latin, relying on the Oxford Latin Dictionary and the classical sources cited therein.

[28] Harvey 1628a, 6.

[29] Harvey 1628a, 6, 80.

[30] Harvey 1628a, 113.

[31] Harvey 1628a, 113. The 1653 English translation has "we have notwithstanding observed that a *feaver* and other horrible Symptoms have ensued." See Harvey 1653, 110.

[32] Harvey 1628a, 132.

de Generatione Animalium.[33] Harvey's preface to this work deals in part with the reliability of the senses and the mediation of the intellect, both cardinal themes of *Lingua* and *Ovids Banquet of Sence*. Consider, writes Harvey, a painter about to paint a portrait. He makes a thousand sketches of the person's face, yet when he stores each drawing in his mind it no longer represents a face but something different, an abstraction, an imagined thing, obscured and confused by the act of representation. This mental image differs from the thing itself and from every other mental representation; the class of all such observations is not a single mental object, for our intellect constructs different images of the same object. There is a distinct scepticism here that is characteristic of the work of Chapman and Tomkis. For "those who see foreign countries and towns or the inward parts of the human body only in drawings or paintings, and make for themselves a false representation of the reality" he warns that they "will never attain to a solid and certain knowledge."[34]

Harvey's practice of observation is distinguished by the same qualities of differentiation, frequency and diligence that we find in Cuningham. Writing of his custom of making notes to be later assessed one against the other he writes of "collecting and comparing many observations"; of the motion of the blood in the right auricle of the heart close to death he comments that he had "observed (it) at different times"; and of the lungs in *de Motu Cordis* he notes that "I have discovered many things from the countless observations I have made."[35] When he writes that "what has been rightly spoken may be confirmed, and what is false corrected in the light of anatomical dissection, personal experience many times repeated and diligent and precise observation," of "using daily more search and diligence," of the indistinguishability of systole and diastole that "I have observed these things for hours together" and of "(H)aving observed these things with much caution and circumspection in a great number of eggs" we can be sure that he considers proper observation – diligent, precise and assiduous - as something that lies at the core of his investigative practice.[36] It is a technique vindicated by its empirical success. "(A)t last" he says "using daily more diligence, by often looking into many and different sorts of living creatures, I believed I had hit the nail on the head ... and had gained the knowledge I so much desired, that of the movement and use of the heart and arteries."[37]

[33] Harvey's concerns for the questionable reliability of reason date from his early inquiries. Though not published until 1651, his research into generation was well underway by the time he finished his inquiries into the heart and the blood. In *de Motu Cordis* he writes that '(T)hese things concerning the formation of the foetus will be proved elsewhere by a great many observations'. See Harvey 1628, 128–129. On the possibility that *de Generatione Animalium* was available in manuscript form in 1638, a claim made by Sir Thomas Browne, see Whitteridge 1971, 210, citing Webster, 262–274.

[34] Harvey 1651, 11, 13.

[35] Harvey 1628, 29, 44, 62.

[36] Harvey 1616, 265; 1628, 10, 29.

[37] Harvey 1628, 29–30.

References

Bartlett, Phyllis Brooks. 1941. Introduction. In *The Poems of George Chapman*, ed. Phyllis Brooks Bartlett, 1–16. Oxford: Oxford University Press.

Bergeron, David. 1971. *English Civic Pageantry 1558–1642*. London: Edward Arnold.

Bowers, Fredson. 1955. *The Dramatic Works of Thomas Dekker.* Cambridge: Cambridge University Press.

Chapman, George. 1595. Ovids Banquet of Sence. In *The Poems of George Chapman*, ed. Phyllis Brooks Bartlett, 53–82. 1941. Oxford: Oxford University Press.

Cuningham, William. 1559. *The Cosmographical Glasse*. London.

Eden, Richard. 1553. *A treatyse of the Newe India*. London.

Febvre, Lucien. 1982. *The Problem of Unbelief in the 16th Century, the Religion of Rabelais*. Tr. Beatrice Gottlieb. Cambridge: Harvard University Press.

Giglioni, Guido. 1994. *Indici Delle Opere Di William Harvey*. Macerata: Universita di Macerata.

Harvey, William. 1616. *Prelectiones Anatomie Universalis (The Anatomical Lectures of William Harvey)*, unpublished manuscript. Tr. & ed. Gweneth Whitteridge. 1963. Edinburgh & London: E. & S. Livingstone.

Harvey, William. 1628a. *Exercitatio Anatomica de Motu Cordis et Sanguinis in Animalibus*. Frankfurt. Tr. Gweneth Whitteridge. 1976. Oxford: Blackwell Scientific Publications.

Harvey, William. 1628b. *Exercitatio Anatomica de Motu Cordis et Sanguinis in Animalibus*. Frankfurt. Tr. Chauncey D. Leake. 1928. Springfield & Baltimore: Charles C. Thomas.

Harvey, William. 1651. *Exercitationes de Generatione Animalium*. London. Tr. Gweneth Whitteridge. 1981. Oxford: Blackwell Scientific Publications.

Harvey, William. 1653. *The Anatomical Exercises of Dr. William Harvey Professor of Physick, and Physician to the Kings Majesty, Concerning the Motion of the Heart and Blood*. London.

Hoven, Rene, assisted by Laurent Grailet. 2006. *Dictionary of Renaissance Latin from Prose Sources*, 2nd edition. Leiden & Boston: Brill.

Middleton, Thomas. 1613. *The Triumphs of Truth*. London.

Rosen, David. 1995. Anatomical Comparisons, Metaphysical Conceits. *Caduceus*, 11, 185–206.

Taylor, Gary. 2008. Middleton, Thomas. In *Oxford Dictionary of National Biography*. Online edition. Oxford: Oxford University Press.

Taylor, Gary and John Lavagnino. 2007. General Editors. *Thomas Middleton: the Collected Works*. Oxford: Oxford University Press.

Tomkis, Thomas. 1607. *Lingua*. London.

Whitteridge, Gweneth. 1971. *William Harvey and the Circulation of the Blood*. London: Macdonald.

Alkahest and Fire: Debating Matter, Chymistry, and Natural History at the Early Parisian Academy of Sciences

Victor D. Boantza*

Abstract Samuel Cottereau Duclos (1598–1685) established the chymical research program at the newly inaugurated Parisian Academy of Sciences (est. 1666). While in the years following his election Duclos enjoyed an unmatched level of activity and influence among academicians, during the 1670s, his institutional status and authority had dwindled considerably. The origins of this decline are examined in light of Duclos' preference of solution chymistry over traditional distillation analysis, highlighted by his research into alkahest. The assessment reveals metaphysical contentions within the Academy concerning the nature of matter and the scope chymical analysis. A crucial turning point in these disputes is signaled by Duclos' confrontation with Denis Dodart, defender of distillation, over the direction of the 'Natural History of Plants' project. In epistemological and philosophical contexts, an examination of the debate illustrates the interplay between conflicting perceptions employed in negotiating distinctions such as physical/chymical and organic/mechanic. The different approaches to plants' constitution, or 'inner' natures, depict the interrelationship between empiricism and natural history.

1 Introduction

In 1666, in a letter to Jean-Baptiste Colbert, Louis XIV's minister of finance and the first protector of the French Academy of Science, Christiaan Huygens referred to what would soon become the assembly of members comprising the Academy, established at the end of that year.[1] Huygens referred specifically to

*A version of this paper was presented at The Three Societies Meeting in Oxford, July 2008. I thank Alice Stroup, Trevor Levere, and Janis Langins for useful comments on an earlier version.

[1]For background on the early academy see Saunders 1980; Mallon 1983. Published studies discussing the Academy during the seventeenth-century include: Maindron 1888; Brown 1934; Hahn 1971; Taton 1966; Hirschfield 1981; Stroup 1990; Sturdy 1995.

V.D. Boantza
Department of History
McGill University, Montreal

C.T. Wolfe and O. Gal (eds.), *The Body as Object and Instrument of Knowledge: Embodied Empiricism in Early Modern Science*, Studies in History and Philosophy of Science 25,
DOI 10.1007/978-90-481-3686-5_5, © Springer Science+Business Media B.V. 2010

the physics or natural philosophy faction (as opposed to the mathematical group[2]), suggesting that,

> the most useful occupation for such an assembly would be to work on a natural history project, modeled on Baconian precepts; a history that would consist of experiments and remarks as a supreme way for attaining knowledge of the causes of all that can be seen in nature; for knowing the causes of gravity (heaviness), heat, cold, attraction, magnetism, light, colors, the composition of air, of water, of fire and of all other bodies; that which ascertains animal respiration, the ways metals, stones and plants grow, investigating all things unknown or poorly understood… the method must be one of proceeding from effects to causes… the descriptions should be numerous and detailed.[3]

Despite the broadly defined research program, Huygens specified that chymistry[4] and the dissection of animals should be part of such a project, which should primarily treat "matters judged good, beneficial and useful."[5] In his vision of a "useful occupation" for the Academy, Huygens linked natural history with "knowledge of the causes" seamlessly, while referring to controversial themes such the causes of gravity, attraction, and the composition of bodies. Huygens' proposal represents a way by which natural investigations, informed by Baconian natural historical methods, could touch upon contentious subjects in a relatively inconspicuous manner. The reference to Bacon and to natural history implied a meticulous, extended, and varied collection of natural phenomena and effects, "the descriptions [of which] should be numerous and detailed." Despite this emphasis, however, Huygens spoke clearly of "proceeding from effects to causes."

Huygens' formal proposal helped not only to convince Colbert and his advisers to found the Academy but also played an influential role in shaping the early Academy's research agenda. When the Academy was established in 1666 two chemists had been appointed: the senior Samuel Cottereau Duclos (1598–1685), who soon became the

---

[2]Among the founding members of the Academy were seven mathematicians, responsible for research into geometry and astronomy; and seven philosophers, in charge of physics, zoology, chymistry, anatomy, medicine and botany. Both groups met on a weekly basis: the former on Saturdays and the latter on Wednesdays. See Sturdy 1995; Stroup 1990.

[3]Huygens 1888–1950, VI, 95–96 (letter undated): "La principale occupation de cette Assemblée et la plus utile doibt estre, à mon avis, de travailler à l'histoire naturelle à peu pres suivant le dessein de Verulaminus. Cette histoire consiste en expériences et en remarques et est l'unique moyen pour parvenir à la connoissance des causes de tout ce qu'on voit dans la nature. Comme pour sçavoir ce que c'est que la pesanteur, la chaud, le froid, l'attraction de l'aimant, la lumière, le couleurs, de quelles parties est compose l'air, l'eau' le feu et tout les autres corps, à quoy sert la respiration des animaux, de quelle façon croissent les métaux, les pierres et les herbes, de toutes lesquelles choses on ne sçait encore rien ou très peu… L'utilité d'une telle histoire faite avec fidélité s'estend a tout le genre humain et dans tout les siècles à venir, parce qu'outre le profit qu'on peut tirer des expériences particulières pour bastir une philosophie naturelle, dans laquelle il faut nécessairement procéder de la connaissance des effets à celle des causes."

[4]I use 'chymistry' to denote early modern chemistry, with particular reference to the transitional phase between alchemy and modern chemistry (especially sixteenth- and seventeenth-century), bridging classical and medieval alchemy with modern chemistry. See Newman and Principe 1998.

[5]Huygens 1888–1950, VI, 95–96.

most active and influential of all founder-members; the younger Bourdelin (1621–1699), practicing apothecary and skillful experimenter, was hired to help Duclos equip, build up and manage the intended laboratories, and to help him develop the program for chymical analysis.[6] Duclos led the Academy's inaugural meeting of December 1666, presenting a memoir entitled "recherche des principes de mixtes naturels." Discussing ways of analyzing *mixtes*, Duclos pointed to the importance of understanding the *generative* capacities of water, since "all natural mixts are produced with water, and that without it none could be produced." Duclos mentioned four examples: meteors, which originated from the union of exhalations and water; simple minerals, "which are engendered by water, from earth"; plants, which germinate and grow only in the presence of water; and living bodies, which originate from an "aqueous humor."[7] Duclos' latter two references have been grafted onto Huygens' general proposal to give rise to the early Academy's two major natural history projects: the natural history of plants and the natural history of animals.

Shortly after the establishment of the Academy, its members became increasingly involved in a prolonged series of animal dissections, brought together in Claude Perrault's 'Comparative Anatomy of Animals' project, which was published in 1671 as *Mémoires pour servir à l'histoire naturelle des animaux*. Chymical research was even more prominent and was mainly represented by Duclos and Bourdelin, his assistant. Duclos was a highly capable experimenter who performed numerous lecture-demonstrations for the assembly.[8] Yet even the experimental aspects of his work carried a pronounced theoretical mark.[9] Bourdelin, in contrast, remained solely within the practical realm. Between 1666 and his death in 1699 he performed and systematically recorded vast numbers of distillations, mostly of plant matter. Some of Bourdelin's distillations were commissioned by Duclos for various research purposes, but for the most part they were carried out in the context of the second natural history project of the early Academy – the 'Natural History of Plants' – initially drafted and proposed by Perrault, Huygens' friend, in January 1667.

Perrault distinguished between two types of research required for a comprehensive study of plants: either by collecting plant material and studying its external features and medical applications (natural history; *l'histoire*); or, by examining the causes of the medical properties of plants and of vegetable reproduction and

[6] Duclos is discussed briefly in Metzger 1969, 266–272. He is mentioned in Partington 1961–1970, III, 11–13; Debus 1991, 151; Principe 1998, 40; all these authors erroneously state his date of death as 1715. Tedoriciu 1974 clarifies basic biographical details. Duclos is recently receiving increasing scholarly attention, especially by Stroup 2002; Clericuzio 2000, 178–180; Kim 2003, esp. 48–52; Holmes 2003; Jacob 2006, esp. 52–65; Boantza 2007; Franckowiak 2008, 2009. On Bourdelin see Dorveaux 1929; Partington 1961–1970, III, 13; Stroup 1991, *passim*; Sturdy 1995, esp. 110–115.

[7] AdS, PV, 1: 9–10.

[8] Duclos examined various themes in this context. One of his most elaborate discussion-examinations is of Boyle's works, especially the *Certain Physiological Essays*, during the winter 1667–1668.

[9] Duclos published only two works. On behalf of the Academy, he published a natural history of the mineral waters of France (Duclos 1675). His only private treatise is a cosmological treatise on the principles of natural mixts, which was denied publication by the Academy and which he ultimately published in Holland with Elzevier, Van Helmont's publisher (Duclos 1682).

nutrition mechanisms (natural philosophy; *la physique*[10]). The latter way, Perrault thought, would require a broad application of chymical analysis, alongside microscopic observations of seeds and shoots, evaluation of theories of propagation and generation as well as studies of sap circulation, mostly related to the question whether sap circulated like blood.[11] Having been proposed only a short while after the establishment of the Academy, the Natural History of Plants started slowly and on a conservative note, consisting mostly of critical assessments of previously published botanical works (such as Gaspard Bauhin's *Pinax theatri botanici*, which academician–botanist Nicholas Marchant had already started to revise).[12]

Academicians, however, keen to study nature rather than texts, soon undermined Perrault's bookish natural history and focused on empirical work, incorporating chymical analyses as means to provide causal explanations to their descriptive matter. This was Duclos' main contribution to the project. By June 1668 he delivered a memoir delineating the method to be applied in the natural history of plants. Duclos found Perrault's plan, which focused on illustrations, incomplete. He subsequently added the requirement for further various textual descriptive details such as whether a plant was tall or rested its branches on the ground; whether it sent out roots from these branches; he also demanded precise descriptions of the root, trunk, leaves, flowers, seeds, fruit and other natural products such as resins, gums or liquids.[13]

2 Duclos' Chymical Natural History of Plants

Duclos' main contribution was the addition of chymical analysis to the work plan. The combination of a natural historical investigation and chymical analyses indeed formed a "most treacherous intellectual problem," in Alice Stroup's words. Between the late 1660s and the early 1680s the project suffered from various problems such as inconsistent funding, tensions between collaborating academicians, and editorial rivalries.[14] Yet in a sense, most of these difficulties – especially the institutional funding problems, which reached a peak with Louvois' 1686 memoir, as well as the editorial rivalry between Duclos and Denis Dodart (1634–1707) – can be attributed to the introduction of chymical analysis. The most controversial aspects of the project stemmed from a clash between chymical analysis as a causal (natural philosophical) means of inquiry and explanation, and an enduring insistence on a descriptive natural historical methodology. Chymical analysis encompassed both dimensions.

[10] It is important to note that the term 'physique' had many meanings in the early modern period, denoting various pursuits ranging from the metaphysical to the empirical, experimental, and medical.

[11] AdS, PV, 1: 30–38; Stroup 1990, 70.

[12] Stroup 1990, 70.

[13] AdS, PV, 4: 48r–48v: "Il a dict que pour procéder avec méthode en cette histoire, il juge a propos d'expliquer premièrement en peu de mots le portrait de la plante faict par le graveur, on en doibt faire la description exacte claire et succincte. Exacte pour instruire le lecteur de tout ce qui concerne ce subiect. Claire pour ne pas embrasser son esprit de termes ambigus, et succinct pour ne le point fatiguer d'une lecture superflue."

[14] Stroup 1990, 69.

As such, under the hands of academicians of various philosophical inclinations, it increasingly came to be both the central object and subject of controversy within the project.

The general framework Duclos suggested consisted of seven major points concerning the study of plants: "species; differences; denominations; place [provenance]; time [season of growth and maturation]; culture; and uses [medical or other]." Under the heading "differences," Duclos included "differences which can be gleaned from the plant's size, its appearance ["port," carriage], its parts and their products." As to parts, Duclos specified that each of the following parts – "stem or trunk, leaves and flowers, fruits or seeds, etc" – must be examined according to "size, number, figure, consistence, color, odor and taste," followed by an examination of "their *constitution*."[15] Duclos' emphasis on a close empirical observation of the plants is evident. The "place and culture" of a plant, he specified, "must be reported by way of observation and practice instead of [drawing upon] the traditional writers, most of whom do nothing but copy things already written."[16] By constitution Duclos meant chymical composition and prescribed the employment of color indicators.[17] The other means of analysis would be distillation, followed by a study of the crystals obtained from the coagulated (dried up) received juices. Duclos remarked that "one can finally know the constitution of plants by the qualities of their separated constituent parts, which are their distilled juices (*eau distillées*), their spirits, both acrid and sulfurous as well as acidic and mercurial, their oils and their fixed or volatile salts."[18]

Duclos' consideration of water, acid, oil and salt, squares well with contemporary views regarding the received products of distillatory fractions, usually consisting of the Paracelsian triad of salt, sulfur, and mercury (*Tria Prima*) with the common addition of water and earth; the former three were regarded as active constituents and the latter two as passive. By referring to these distillants or constituent parts as extractions or the end products of *extractive* processes, however, Duclos hinted at their status as products of a non-radical separation or partial decomposition. This is further evidenced by his recommendation that "after such extractions, we must proceed to examine each extracted substance."[19] The examinations implied revolved around the content and characteristics of various salts, typically considered, in traditional chymical philosophy, as vital-generative constitutive elements.[20]

[15]AdS, PV, 4: 48v–49r (italics added).

[16]AdS, PV, 4: 54r: "Les temps et la culture doibvent estre rapportez sur l'observation et la practique plustot que sur la tradition des ecrivains dont la pluspart n'a faict que copier ce qui en estoit desia escript."

[17]AdS, PV, 4: 49r–v. Especially *noix de galle* (oak marble galls, used in the production of dyes; contain large amounts of tannic acid).

[18]AdS, PV, 4: 51r.

[19]AdS, PV, 4: 51v: "Apres ces extractions il fault passer a l'examen de chaque matiere extraicte."

[20]For a detailed discussion on the history of salts in French chymistry see Franckowiak 2003; for a recent and informative study of salts in chymical and medical contexts in Britain see Roos 2007; see also Newman and Principe 2002, esp. 275–281. I use 'chymical philosophy', paraphrasing Debus' well-known term 'chemical philosophy' (Debus 1977). For further details see also Debus 1991, 2001.

For instance, Duclos noted that, "the distilled water of plants that are humid and cold, such as lettuce, purslane or chicory, carries with it some portion of a sulfurous salt." This salt could be traced when the corresponding distilled water was mixed with a solution of salt of lead, which rendered the solution milky and turbulent. In contrast, the same could not be observed in distilled waters proceeding from dry and earthy plants since their salts were less volatile. Dryness, Duclos argued, impeded volatility, change, and action while fermentation and humidity encouraged vitality, motion, and growth. Accordingly, the particular combination of constitutive salts indicated the chymical composition of the plant in question. Similarly, the presence of fixed or alkali salts – typically found in burnt plant matter – was ascertained by the dissolution of vitriol of iron in common water; upon mixing the two liquids, if fixed salts were present, the iron was precipitated, as indicated by the apparent yellowish-reddish color of the solution.[21]

The significance of the salts as vital-constitutive entities could not be overstated. Duclos went as far as to propose that,

> the various salts of all the constituent parts of a plant can be reunited into one single salt, which will contain all the virtues of that plant. This salt can then be finally resolved into an insipid watery liquid and a pure and dead earth, devoid of all virtue, [and that,] without any notable diminution of weight. This is the extreme [radical] analysis of plants, which serves in acquiring the most exact knowledge of the constitution of a subject.[22]

Duclos' view of what constitutes a radical analysis of plants, or other mixts, echoes his vision of matter. In line with Van Helmont, Duclos held that a radical chymical resolution or decomposition of matter should yield "an insipid watery liquid." Over the following two weeks, in two consecutive memoirs, dated 16 and 23 June 1668, Duclos communicated to the assembly his views concerning chymical analysis, pitting distillation against solution analytical chymistry and physical mechanism against organic chymical vitalism, in the context of particulate theories. These memoirs elucidate how a traditional chymist like Duclos charted the physical–chymical territory against the backdrop of the emergent mechanical philosophy.

Duclos' decision to dedicate two consecutive meetings to the theoretical aspects of chymical practice is telling, especially in light of the early Academy's pronounced empirical-Baconian agenda, upheld in attempt to minimize debate among members of a new and varied collective research enterprise. The informal reason behind Duclos decision to discuss these topics was clearly linked to his research plans for the Natural History of Plants project, which by 1668, given the significance of chymical analysis and research, Duclos presided over. The official excuse, however, is found at the beginning of the first memoir, dedicated to fire analysis, or distillation:

[21] AdS, PV, 4: 51v–52v.

[22] AdS, PV, 4: 53r–54v: "le divers sels de toutes les parties constitutives d'une plante peuvent estre réunis en un sel, qui contiendra toute la vertu de la plante. Et ce sel peut finalement estre reduict en eau insipide et en terre pure et morte, sans diminution notable de son poids. Et cela est l'extreme analyse de la plante, qui ne sert qu'a avoir une connoissance plus exacte de la constitution du subiect."

> Mr. Duclos has said that, lacking a laboratory for conducting chymical analyses, and wishing to avoid idleness, while all other members of the company are busy working, he had made up his mind to propose to the assembly to render advice as to the methods for performing chymical analyses, which will be useful once a laboratory is established; having received the company's approval, he went on to claim that the principal means of analysis are fire, air and dissolutive liquors.[23]

The last statement in this passage sets the background to the ensuing discussion. Fire, Duclos noted, acts to separate (decompose) the parts of a mixt in two ways: either by the action of its heat alone, in which case no inflammation occurs, or by way of combustion and inflammation.[24] He explained the mechanism underlying fire analysis, or distillation, in the following manner:

> The heat of the fire excites a motion in the mobile parts [of the mixt], according to their degree of mobility, to the effect that those that share the same degree of mobility cannot separate at the same degree of heat, which in agitating them equally, makes them rise together, and [hence] they separate only from the [relatively] less mobile parts. And those that are unequally mobile separate from each other, because the easiest to move [most prone to motion and excitability], being most agitated and most rapidly rarified by the heat, rise first and detach from the less mobile, that can follow when chased by a stronger fire.[25]

This explained at once how distillation works to separate between the constituents of mixts and signaled the limitations of fire as an analytical tool; the underlying principle being that two constituents can be *different* (in essence, nature, or constitution) yet by virtue of sharing the same degree of mobility (excitability), will not be separated during distillation, since the fire will cause them to "rise together."

Duclos underscored the relation between heat and motion by challenging "those who say that it is a property of heat to bring together things of the same nature and to separate those of various natures." This conveys Duclos' critique of traditional Scholastic and Paracelsian views according to which like acts upon like, as applied to distillation practices. In effect, it is not the heat that actively either separates or unites the constituents of bodies, since its *sole action* is to impart motion, according to the given constituents' degrees of mobility. The subsequent separations or unions are but consequences of the *motion* – which depends upon the constituents' propensity to move

---

[23]AdS, PV, 4: 58r: "Mr du Clos a dict que n'ayant point encore de laboratoire pour travailler aux analyses chimiques, et ne voulant pa d'ailleurs demeurer oisif, pendant que toute le monde de la compaignie travailloit chacun de son coste, il avoit jugé a propos de proposer a l'assemblée la metode qu'il seroit d'advis qu'on observant pour procéder a ces analises lors qu'on aura un laboratoire ; et la compagnie ayant approuvée sa proposition, il dit qu les pricipaux moyens des analyses chimiques sont le feu, l'air, et les liqueurs dissolutives."

[24]AdS, PV, 4: 58r–v: "Pour ce qui est du feu il agit dans la séparation des parties des mixtes ou par sa chaleur seule sans embrassement ou par embrassement et combustion."

[25]AdS, PV, 4: 58v: "La chaleur du feu excite du mouvement dans les matières mobiles, selon le degré de leur mobilité, de sorte que celle qui sont mobiles en mesme degré ne se séparent point les unes des autres par un mesme degré de chaleur, qui les agitant également les fait monter ensemble, et les sépare seulement de celles qui sont moins mobiles. Et celles qui sont inégalement mobiles se séparent les unes des autres, car les plus faciles a se mouvoir estant plus agitées et plus tost raréfiées par la chaleur s'eslevent les premières et quittent celles qui sont moins moniles mais qui les peuvent suivre estant pressées d'une chaleur plus forte."

– and not of the *heat* itself. Duclos' discussion of possible *unions* suggests yet another sense in which he was critical of the analytical capacity of fire, since some constituents will "only liquefy and attach to those which are fixed, producing a new composition of parts."[26] Significantly, Duclos avoided interpreting such unions by recourse to affinities, resemblances, or correspondences (whether understood as occult qualities or not) acting between two entities that bear essential similarities. Duclos allowed for only the physico-mechanical principle of motion in explaining the separation as well as creation of new compounds during distillation. Echoing Boyle's critique of fire analysis (especially as seen in his *Sceptical Chymist*), Duclos suggested that fire may alter the components – not only did fire fail to decompose mixts into their elementary constituents, but it created new ones that were not part of the initial mixt.[27]

Closely acquainted with chymical experimental reality, Duclos admitted that, "fire can occasion, by the power of its heat, not only separation, but also union." Herein, then, lies the explanation of volatility or fixity of chymical substances, since fire

> separates the volatile parts from the fixed ones by the motion occasioned by its heat … and it unites, into a new compound, the parts which are less fixed with those that are more fixed, melting and liquefying the humid parts which penetrate into the dry ones, combining together, as in the case of carbon and glass.[28]

Heat, however, can produce some unions that will resist all degrees of heat; such unions can be resolved by means of "inflammation and combustion" alone. The latter necessitate the presence of air and hence point to the limited capacities of distillation, which customarily takes place "in closed vessels," joined and "sealed carefully." This also supports the claim that air is a "means of analysis." By means of its heat and inflammation, fire can resolve "mixts that [were] composed of volatile or combustible parts." For the analysis of the most fixed (i.e. immobile, non-volatile) mixts – those that are usually earthy and dry, since fixity is related to lack of humidity – Duclos prescribed the use of "resolutive menstrua."[29]

At the beginning of the following memoir, dedicated to solution analysis and "dissolutive liquors" Duclos provided a crucial clue regarding the action mechanism of such solvents:

> since most mixts that do not have a strong compaction between their parts, are independently resolved [resolve themselves] by way of putrefaction in their own humidity, the chymists have taken the opportunity to conduct the resolution of less humid substances by way of putrefaction, by the addition of some regulative liquor.[30]

---

[26]AdS, PV: 4, 59r–v.

[27]For details on the controversial nature of fire analysis in the early modern period see: Debus 1967; Holmes 1971. See also the important study by Kim 2001.

[28]AdS, PV, 4: 60r: "fait séparer les parties volatiles des fixes par la mouvement que sa chaleur… et il fait coniondre en un composé nouveau les matières qui sont moins fixes avec les plus fixes, fondant et liquefiant les humides qui pénètrent dans les seiches et sentient avec elles, comme il se fait au charbon et au verre."

[29]AdS, PV, 4: 60r–61v.

[30]AdS, PV, 4: 63v–64r: "Parce que la plus part des mixtes qui n'ont pas une forte componction de leus parties, se résolvent d'eux mesmes par la putréfaction en leur propre humidité ; Les chymistes ont prins de la occasion de faire des résolutions des metieres moins humides par putréfaction a l'aide de quelque liqueur adioustée [ajustée]."

The common function of the "regulative" liquors is to "facilitate putrefaction" or fermentation. Duclos distinguished between three kinds of menstrua, solvents, or regulative liquors: corrosive, extractive, and resolutive. The corrosive, when applied to solid mixts, brings about the discontinuation of their mass and breaks them down into "integrant, highly subtle, particles"; the extractive is employed for the extraction of a certain part of the mixt (usually acts by precipitation); and the resolutive is used for occasioning a "radical resolution."[31] Duclos referred to the latter in his research proposal for plant analysis as the "extreme analysis of plants, which serves for acquiring the most exact knowledge of the constitution of a subject."[32] This type of resolution, moreover, was highly regarded and sought after in analytical chymistry; hence Duclos' suggestion that resolutive menstrua and their actions should be studied most closely. The corrosive resolution, on the other hand, is considered as a preparatory step to the extreme resolution in that it rarifies the parts of the mixt and endows them with a kind of heightened activity and enhanced mobility, making them less compact.

All these menstrua, Duclos proclaimed, consist of "salts, resolved and spiritualized," that is, "reduced into highly penetrating liquors." These salts are either mercurial, sulfurous, or mixt.[33] The mercurial liquors are acidic and merely corrosive (like *aqua fortis*) while the liquors of sulfurous salts are acrid and merely extractive (like spirit of wine or alcohol). The mixt spirits, containing both mercurial and sulfurous salts joined together, are solely capable of occasioning a radical resolution: "these mixt menstrua are the true solvents, intended for real chymical analyses... they are useful in the research of the principles of natural mixts as well as in performing analytical observations, which facilitate our knowledge of the nature and qualities of mixts."[34] The radical resolutive analysis, then, is an organic-vital process. Its activity generated by the combination of mercurial and sulfurous salts, it is analogous, like other types of resolution, to putrefaction and fermentative processes.

By 1668, when Duclos expounded his views on chymical solution analysis and argued for its relevance to plant analysis – a matter that had been recognized by the assembly – a basic framework for the natural history of plants existed. The plan was multi-authored to some extent, drawing on an accumulation of ideas and proposals by Huygens, Perrault, and Duclos. For carrying out the actual research, however, the Academy depended almost entirely on Bourdelin to analyze plants in the laboratory and on the botanists Nicolas and Jean Marchant to cultivate and describe them. Bourdelin refined chemical techniques, especially for analyzing oils, and

[31]AdS, PV, 4: 64r–v.

[32]AdS, PV, 4: 53v.

[33]Duclos' reference is reminiscent of the traditional view of composition, comprising the two principles of sulfur and mercury.

[34]AdS, PV, 4: 64v–65v.

kept detailed records of his experiments and laboratory expenses.[35] The Marchants, father and son, cultivated rare plants for the Academy's use and were in charge of sections of the Royal Gardens, nurseries and the *orangerie*; they cultivated seeds from all over the world, collected by friends and colleagues. After cultivating a plant, the Marchants described it, gave it to the illustrators and supplied the rest to Bourdelin for analysis.[36]

3 Dodart Enters the Arena: Natural History by Fire

The ambitious Natural History of Plants collective project went down in the chronicles of the Academy as an overall failure. Despite efforts to minimize controversy, it was plagued by numerous disputes between academicians over theory as well as practice, and was never completed or published as intended. After Colbert's death in 1682, Louvois became the Academy's protector and it was under his inconsistent financial support that the project suffered and deteriorated.[37] Almost a decade after it was first proposed by Perrault and subsequently expanded by Duclos, in 1676, two publications appeared, drawing on the vast work carried out under the banner of this project: Marchant's *Descriptions de quelques plantes nouvelles* and Dodart's *Mémoires pour servir à l'histoire des plantes*. True to its title, the *Descriptions* consisted of a descriptive history of plants, featuring illustrations but lacking reference to plant analysis. Dodart's *Mémoires*, in contrast, discussed chymical analysis at great length. Dodart, however, advanced a different view of the subject than the one envisioned by Duclos, the initial director of the project.

Since 1671, the year Dodart joined the Academy, Duclos' power over the project as well as his general status within the Academy dwindled manifestly.[38] Dodart was about 35 years younger than Duclos, who was 68 years old when he had initially joined the Academy in 1666. Dodart was ambitious and energetic and soon took

---

[35]MS. n. a. f. 5147. *Registre des depenses faites pour laboratoire (1667–1699).* Part (XV) of collection MS. n. a. f. 5133–5149 entitled *Proces-verbaux des analyses et experiences faites au laboratoire de l'Academie des sciences, etabli dans le batiments de la Bibliotheque du roi, depuis 1667 jusq'en 1699.*

[36]On Nicolas and Jean Marchant see Laissus and Monseigny 1969; Sturdy 1995, *passim*; Stroup 1990, *passim.*

[37]Three years into his office, Louvois published an (in)famous memoir, in which he attacked the chymical research in the Academy both implicitly and explicitly. The memoir singled out Duclos and asked that all speculative, non-pragmatic, research be abandoned. In particular, Louvois criticized all work that could be associated with alchemy. See Stroup 1990, 107–112 on Louvois "ministerial intervention." For the full text as well as a translation into English of the memoir (and other details) see Meynell 2002.

[38]On Duclos' decline see Stroup 1990, Kim 2003, esp. 51–52. MS. n. a. f. 5147 charts Duclos' decline by showing how during the 1670s and 1680s he had placed increasingly fewer laboratory orders while Dodart had taken over, his name being increasingly associated with the laboratory working transactions, requests, and expenses (all recorded and organized by Bourdelin).

over the Natural History of Plants, marginalizing Duclos.[39] Duclos' institutional power decline cannot be attributed to one single reason but the most prominent factor in his demise arose from his research agenda, and particularly his promotion of solution analysis alongside – but mostly over and above – distillation. The minutes of the Academy chart this decline. During the late 1660s, Duclos was by far the most active academician; between 1667 and 1669 his memoirs fill roughly 500 pages of the *procès-verbaux*, discussing chymical analysis, coagulation and cohesion, as well as providing a detailed scrutiny of Boyle's *Certain Physiological Essays* and *The Origin of Forms and Qualities*; from 1675 to 1683 his memoirs fill some 20 odd pages.[40]

Dodart was Perrault's *protégé* and both subscribed, in different fashions, to both traditional as well as mechanistic views. Inspired by the greater intellectual climate and by Louvois' public demand for practical, useful, and non-speculative research, Dodart increasingly privileged natural historical over natural philosophical pursuits. Preceding the brief preface, the "avertissement" to the *Mémoires* is instructive in this respect; of particular interest is Dodart's depiction of the division of labor within the project and the general characteristics of the research collaboration. "This book," he began, "is the work of all the Academy" and it "is the result of propositions, experiments, and reflections." Accordingly, Dodart proceeded, whereas nearly all chymical subjects ("chymie") were handled by Duclos (and some by Pierre Borel), Bourdelin performed virtually all the "operations chymiques." The reflections, drawing on the former two, are attributed to most academicians involved in the project. Although they would seem to pertain most closely to the author of the book, Dodart singled out Bourdelin as having "provided many advices, made many remarks, and handled most the records, from which I have drawn the chymical experiments discussed in this book."[41] The emphasis on the practical, non-speculative, and natural historical aspects of the project is evident.

As seen, Duclos did not reject mechanistic principles; he even upheld a version of corpuscularianism, mentioning particles under the influence of motion. Yet he drew a clear and distinct line between the *physical* (mechanical) and the *chymical* (vital), arguing that physical reactions were superficial. Distillation, being based upon separation by heat according to degrees of mobility, was an incomplete method for decomposition, a partial analytical tool. Solution analysis, in contrast, since based on essentially vital and fermentative processes, represented chymical deep-level resolution.

---

[39]On Dodart and the various research tensions within the Natural History of Plants project see Holmes 2004, esp. 277–288, where he claims that by early 1674 Dodart was "leader of the project of plant analysis" (277). See also Holmes 2004; Kim 2003, 53. For some general information on Dodart and the Academy see Sturdy 1995, 184–189.

[40]Stroup 1990, 83.

[41]Dodart 1676, *avertissement*: "Ce Livre est l'ouvrage de toute l'Academie… Il est donc de mon devoir d'avertir le public, qu'il doit à M. du Clos & à M. Borel, presque tout qu'il y a de Chymie… Que M. Bourdelin a executé & conduit presque tout les operations Chimiques, donné plusieurs avis, fait plusieurs remarques, & tenu la pluspart des Registres, d'où j'ay tiré les experiences Chymiques dont il est parlé dans ce Livre."

By way of analogy, whereas physico-mechanical decomposition processes, such as separation by heat, were the opposite of mixtion or physical aggregation of constituents, chymico-vital resolutive processes, such as the ones generated by radical solvents were the opposite of generation. Such notions can be traced back in time to Paracelsian and especially Helmontian doctrines, closely identified with alchemical, Platonic and Hermetic precepts, all of which were controversial, especially during the 1670s and 1680s. The young Academy, hosting members from a wide variety of backgrounds and closely dependent on royal funding for its existence preferred to distance itself from any apparently subversive views associated with natural magic, Platonism, or occultism.

In his second memoir on chymical analysis, treating resolutive menstrua, Duclos observed that such

> menstrua are either universal or particular. The universal [ones] must originate from the less specific salts, of a mixt and tempered nature. Such is the salt out of which Paracelsus produced his great solvent [Paracelsus' alkahest], which he named Sel Circulé.[42]

Two months later, in August 1668, Duclos discussed at length this Paracelsian salt, alongside Van Helmont's alkahest, the utopian universal solvent, presenting the assembly with a detailed survey and interpretation of alchemical literature concerning menstrua and seminal principles.[43] Duclos repeatedly drew the distinction between vulgar (distillatory, mechanical, physical) and philosophical (vegetal, chymical, transformative) practices.

In this sense Bourdelin was a vulgar chymist, a lifelong supporter of distillation analysis, a practice he doggedly refused to abandon, performing for the Academy thousands of distillations until his death 14 years after Duclos. The task of interpreting Bourdelin's results fell first on Duclos and later on Dodart. The wealth of records and information was copious: from the weight of distillants to temperature records, to accounts of colors, tastes and odors, to a range of distillation techniques such as subsequent replacements of recipients for each fraction, or the practice of heat control by employing a double boiler, to name a few. This abundance of data left academicians, and especially Dodart, puzzled. Nonetheless, the Academy remained largely committed to distillation throughout the rest of the seventeenth-century. Duclos did not live to see the application of his visionary views of solution analytical chymistry.[44]

---

[42]AdS, PV, 4: 65v: "Les menstrües sont ou universels, ou particuliers. Les universels doivent estre tirez des sels les moins spécifiez, mais de nature mixte et tempérée. Tel est le sel commun, duquel Paraclese a fait son grand dissolvant, qu'il nomme Sel Circulé."

[43]Duclos delivered two lengthy and detailed memoirs on the subject of solution chymistry and the alkahest (18/15 August 1668). See AdS, PV, 4: 134r–175r. Paracelsus and Van Helmont are mentioned more than anyone else, followed by a discussion of Deiconti's solvent; Glauber, Duchesne and Trismosin are mentioned too. On salts in the Paracelsian and Helmontian contexts see, for instance, Roos 2007, especially 10–107. For an introductory historical study of alkahest see Reti 1968; for an interesting exploration of the relation between alkahest and chymical theory in the second half of the seventeenth-century see Joly 1996; see also Newman and Principe 2002, *passim*.

[44]Shortly after Bourdelin's death Simon Bouldoc proposed the use of solution over distillation analysis. Kim 2003, 79. See also Stroup 1979.

Distillation, or fire analysis, was challenged as a legitimate means for the extraction of elementary constituents. But while research into the ultimate principles of mixts had been mostly Duclos' aim, Dodart was mainly interested in the medical virtues and uses of plants and in their nutritive values and mechanisms. This is not to say that Dodart had no qualms about distillation, which he attempted to scrupulously defend and justify in his *Mémoires*. Forced to support increasingly convoluted interpretations of experimental instances, he still remained committed to fire analysis. In his 1676 *Mémoires*, more than half of which had to be dedicated to justifying the shortcomings and inaccuracies of distillation, Dodart stated clearly why the alternative, long since promoted by Duclos, could not be tolerated:

> for knowing that which plants are, we do not have to make the great efforts of resolving them into that which the chymists call their primary principles; that is, to irreversibly resolve them into a simple solution, containing their virtues, by means of allegedly universal solvents, enigmatically described by Paracelsus, Van Helmont, Deiconti, etc.[45]

This reference is unmistakable since these are the very three authors whom Duclos discussed at length in 1668 in the context of the *sel circulé*, alkahest, and universal solvents.[46]

Dodart proceeded to dismiss any such ideas, claiming that, "these solvents are found only in books." He further noted that it would probably be more difficult to grasp the nature of these solvents than the nature of the plants themselves.[47] Judging these chymical views harshly, Dodart struggled to find ways to compensate for the deficiencies of fire analysis, undermining its empirically inaccurate and controversial nature while arguing for the inadequacy of chymistry – in the traditional Helmontian and Duclosian sense – to the pursuit of natural knowledge. Clearly, this was a far cry from Huygens' early vision of a natural history providing a "supreme way for attaining knowledge of [natural] causes," or of Duclos' similar allusions.

4 "We Must Stay Within These Limits": Empiricism and Natural History

Dodart's overall natural historical scheme consisted of four parts: "1. the description of plants, 2. their figures. 3. their culture [development and growth]. 4. their virtues, and the studies one can pursue, and those that we have pursued, for knowing the constitution of plants."[48] The study of plants' virtues comprised, by far, the most extensive section; it was divided into two segments: the knowledge (or study) of plants in themselves and the knowledge of plants through their effects. While the former spanned 34 pages (out of a total of fifty), the latter occupied

45 Dodart 1676, 13.
46 AdS, PV, 4: 128r–175r.
47 Dodart 1676, 13.
48 Dodart 1676, 2.

only four; distillation was part of the former. In his conclusion to a section entitled "Reflexions particulieres sur l'usage du feu dans les analyses des Plantes," Dodart surmised that,

> it is *not completely impossible* to arrive, by the [chymical] analysis, at a *certain* degree of knowledge, which may serve *at least* for forming *some* conjectures, reasonable *enough* to be examined, and *possibly* incorporated into physics, *almost* as ordinary descriptions... 2. that it is very difficult, not to say impossible, to attain by analysis an accurate and certain knowledge of the natural constitution of each plant; 3. that in employing chymistry, we do not actually engage in the pursuit of the principles of natural mixts, or as the chymists call them simple and inalterable... It is not that we do not seek greater certitude, but we believe that *we must stay within these limits*, hoping that more reasonable [capable] persons, well aware of the great difficulties in attaining the knowledge of the simpler things, and who know that even the possession of such knowledge will not remove all difficulties, will be content with what we have to offer... chymical inquiries which one day, we hope, will be employed as the basis for *reasonable*, even if *not certain*, conjectures.[49]

At first glance, Dodart's disillusionment with chymical analysis seems striking. With efficient solvents "found only in books" he accorded distillation a "certain" and "reasonable" degree of legitimacy. His unrelenting commitment to distillation is at odds with his explicit admission of its profoundly limited nature. Should this passage be read as expounding a methodologically convoluted and epistemologically impoverished discourse, bordering on simplistic skepticism? Should it be read as decisive evidence of the utter failure of the Natural History of Plants project? Or, should Dodart's stance be interpreted as a stubborn reaction against any and all applications of metaphysical and speculative aspects to natural philosophy, such as Duclos' alkahest or vital salts? Or, yet from a different perspective, should Dodart be understood as an avid proponent of natural historical empirical research, in the received traditional Baconian vein?[50]

[49]Dodart 1676, 14, 17: "qu'il ne pas évidement impossible de parvenir par l'analyse à un certain degré de connoissance, qui pourra servir au moins à former des conjectures assez raisonnables pour estre examinées, & peut-estre receuës en Physique, à peut prés comme les descriptions ordinaires... 2. qu'il est fort difficile, pour ne pas dire impossible, de tirer de l'analyse une connoissance sance precise & certaine de la constitution naturelle de chaque Plante ; 3. que nous servant de la Chymie, nous ne nous engageons ny à recevoir les principes des corps naturels, selon les Chymistes, comme principes, c'est à dire, comme généraux, ny comme simples, ny comme inaltérables... Ce n'est pas que nous ne desieassions une plus grande certitude, mais nous croyons devoir demeurer dans ces bornes, & nous espérons que les personés équitables, & qui sçavent combien les moindres choses sont difficiles à connoistre, & combien en a connu, nonobstant tout les difficultez, se contenteront de ce que nous pouvons leur promettre... la Chymie les recherches sur lesquelles on peut espérer de fonder un jour quelques conjectures raisonnable, encore qu'on ne s'y puisse promettre une entière certitude" (italics added).

[50]I here refer to the still widely-held view of Bacon as the establisher of the New Science's scientific method: inductive and careful empiricism, closely linked to a supposed unbiased collection of natural phenomena and their compilation into vast 'value-free' natural histories. For the best overall treatment of Bacon's natural philosophy see Gaukroger 2001; for a challenging and thought-provoking alternative to this trend see Rees' studies on Bacon's debt to Paracelsian matter theory: Rees 1975a, Rees 1975b, 1996.

A comparison between Dodart's and Duclos' approaches to plants' composition, *inner* nature, or "constitution," set against the background of their corresponding epistemological motivations within the natural historical project provides insights into these questions. In this respect, one of the crucial differences between the two natural philosophers derives from their different research goals. Simply put, Duclos was interested in the nature of matter, elements, and natural change; Dodart focused on the medicinal virtues of the plants and their corresponding uses. Duclos hoped that the analysis of plants will throw light on the ultimate (elementary) principles of vegetable substances and will subsequently bear upon the constitution of matter in general. Dodart sought to establish an accurate classification of plants, along various coordinates, one of which was their constitution, which further pointed to their potential medicinal uses.

For Duclos, the chymical constitution of a plant preceded its other qualities, both epistemologically and ontologically, since its properties and virtues derived from its constitution. The constitution would also reveal a mixt's place within a cosmological web of correspondences and connections, the framework of which drew on Paracelsian and Platonic ideas.[51] In this sense, a radical or complete chymical analysis promised not only to reveal the inner virtues of a plant (or mixts in general) but also to causally *explain* them. Once the ultimate principles or elementary constituents are made empirically known, the essence of the plant can be understood, since a plant could be theoretically resolved into one "single salt, which will contain all the virtues of that plant." This type of resolution provides the investigator with essential knowledge, not readily accessible by the senses, knowledge of organic and vital *principles* of growth and generation. Distillation, according to Duclos, was merely akin to physical separation between non-elementary constituents and hence was not considered a chymical process. Duclos employed this distinction in drawing the boundaries between the two domains, claiming the autonomy and specificity of chymical knowledge.

Unlike Duclos, Dodart was uninterested in the epistemological status of chymical knowledge in itself. Nor was he interested in the metaphysical role of chymical agents and entities. On distinct empirical grounds, seeking to establish an accurate and detailed natural history of plants, with their myriad properties, figures, cultures, and virtues, Dodart rejected the postulation of solvents. Although Duclos rejected the appeal to authorities in describing plants, in his examinations of the alkahest and various salts, he discussed at length Paracelsus, Van Helmont, and others. But Dodart rejected solution analysis for yet another significant reason, arguing that, "if we could possess these [utopian solvents] they would not further our knowledge of the nature of each plant, since it would be reduced into a certain [state of] universality." Dodart went on to deny any means that might "render general that which we would like to particularize." Subsequently, Dodart preferred the extraction from plants of the different substances of which they are composed and since we can know these substances "only through our senses," even if we

[51] Duclos 1680.

cannot perceive "that which is more intimate," we still gain a certain degree of knowledge. Distillation, Dodart added, enables us to see that which was hidden before, exposing it so that we can "separately examine its taste, smell, and other sensible properties," which would ordinarily be mixed with other substances and hence be unrecognizable.[52]

For Dodart, then, distillation was akin to a microscope, revealing hidden details and exposing them to the senses; it is, in a sense, analogous to dissection. The notion of fire as the chymist's scalpel is not new but it has been obscured by the forceful early modern critiques concerning the accuracy of this analytic tool (culminating with Boyle's *Skeptical Chymist*). Despite all the shortcomings of this method, Dodart defends it as the only legitimate means for pursuing the natural historical examination precisely because it is in epistemological and methodological full accord with his goals. Distillation is for him an essentially *descriptive* tool and not an explanatory one in itself. It is analytical in that it *shows* various features of the plant, rendering them empirically accessible to the human senses. Distillation and its products are meant as an internal description, a complementary addition to the external image of the plant. This is why Dodart had defended it persistently and this is the background against which it should be interpreted concerning its place within Dodart's natural historical pursuits. The result appears less limited and skeptical when considered in relation to the overall epistemological consistency of the project, its pragmatic goals, and Dodart's strict belief that "we must stay within these limits."

References

Boantza, Victor D. 2007. Reflections on Matter and Manner: Duclos Reads Boyle, 1668–1669. In *Chymists and Chymistry: Studies in the History of Alchemy and Early Modern Chemistry*, ed. Lawrence M Principe, 181–92. New York: Science History Publications.

Brown, Harcourt. 1934. *Scientific Organizations in Seventeenth Century France (1620–1680).* [History of Science Society Publications, new series, 5.] Baltimore: Williams & Wilkins Co.

Clericuzio, Antonio. 2000. *Elements, Principles and Corpuscles: A Study of Atomism and Chemistry in the Seventeenth Century*. Dordrecht: Kluwer Academic Publishers.

Debus, Allen G. 1967. Fire Analysis and the Elements in the Sixteenth and the Seventeenth Centuries. *Annals of Science* 23: 127–147.

Debus, Allen G. 1977. *The Chemical Philosophy*. New York: Dover Publications.

Debus, Allen G. 1991. *The French Paracelsians: The Chemical Challenge to Medical and Scientific Tradition in Early Modern France*. Cambridge: Cambridge Univesity Press.

[52] Dodart 1676, 13–14: "quand on les pourroit avoir, ils ne nous feroient pas mieux connoistre la nature de chaque Plante, qui se trouveroit par là reduite à une certaine universalité... render general ce que nous voudrions particulariser... nous ferions mieux de tirer des Plantes, autant qu'il nous fera possible, les matieres differentes dont elles sont composées... nous ne puissions connoistre ces matieres que par les sens, qui n'appercoivent jamais ce qu'il y a de plus intime, c'est tousjours un degré de connoissance, dans ce que les Plantes sont, que de voir ce qu'on ne voyoit pas, & d'en pouvoir examiner separement la saveur, l'odeur, & les autres proprietez sensibles qui estoient auparavant aussi meslées que les matieres ausquelles elles appartiennent."

Debus, Allen G. 2001. *Chemistry and Medical Debate: Van Helmont to Boerhaave*. Canton, MA: Science History Publications.

Dodart, Denis. 1676. *Mémoires pour Servir à l'Histoire des Plantes*. Paris: Imprimerie Royale.

Dorveaux, Paul. 1929. Les Grands Pharmaciens Apothicaires Membres de l'Académie Royale des Sciences. 1. Claude Bourdelin. *Bulletin de la Société de l'Histoire de la Pharmacie* 17: 289–298.

Duclos, Samuel C. 1675. *Observations sur les eaux minérales de plusieurs provinces de France, faites en l'Académie royale des sciences en l'année 1670 et 1671* . Paris: Imprimerie royale. Reproduced in Académie royale des sciences. (1729–33), 4, 31–90.

Duclos, Samuel C. 1680. *Dissertation sur les principes des mixtes naturels, faite en l'an 1677.* Amsterdam: Elsevier. Reprinted in Académie royale des sciences. (1729–33), 4: 1–30.

Franckowiak, Rémi. 2003. *Le développement des théories du Sel dans la chimie française de la fin du XVIe à celle du XVIIIe siècle*. Université Charles de Gaulle, Lille III: Unpublished PhD Thesis.

Franckowiak, Rémi. 2008. La Chimie du XVIIe Siècle: Une Question de Principes. *Methodos*, 8 [Chimie et mécanisme à l'âge classique]. Retrieved 31 August 2009, from http://methodos.revues.org/document1823.html.

Franckowiak, Rémi. 2009. Du Clos, un Chimiste Post-Sceptical Chemist. In *La Philosophie Naturelle de Robert Boyle*, ed. Ramond, Charles and Dennehy, Myriam, 361–378. Paris: Vrin.

Gaukroger, Stephen. 2001. *Francis Bacon and the Transformation of Early-Modern Philosophy.* Cambridge: Cambridge University Press.

Hahn, Roger. 1971. *The Anatomy of a Scientific Institution: the Paris Academy of Sciences, 1666–1803*. Berkeley: University of California Press.

Hirschfield, John M. 1981. *The Académie Royale des Sciences (1666–1683): Inauguration and Initial Problems of Method.* New York: Arno Press.

Holmes, Lawrence F. 1971. Analysis by Fire and Solvent Extractions: The Metamorphosis of a Tradition. *Isis* 62: 128–148.

Holmes, Lawrence F. 2003. Chemistry in the Académie Royale des Sciences. *Historical Studies in the Physical and Biological Sciences* 34: 41–68.

Holmes, Lawrence F. 2004. Investigative and Pedagogical Styles in French Chemistry at the End of the Seventeenth Century. *Historical Studies in the Physical and Biological Sciences* 34: 277–309.

Huygens, Christiaan. 1888–1950. *Œuvres complètes*, 22 vols. Amsterdam: Swets & Zeitlinger/ The Hague: Martinus Nijhoff. (Referred to in the footnotes as *Œuvres*).

Jacob, Margaret C. 2006. *Strangers Nowhere in the World: The Rise of Cosmopolitanism in Early Modern Europe*. Philadelphia: University of Pennsylvania Press.

Joly, Bernard. 1996. L'Alkahest, Dissolvant Universel ou Quand la Théorie Rend Pensable une Pratique Impossible. *Revue d'Histoire des Sciences* 49: 305–344.

Kim, Mi Gyung. 2001. The Analytical Ideal of Chemical Elements: Robert Boyle and French Didactic Tradition of Chemistry. *Science in Context* 14: 361–395.

Kim, Mi Gyung. 2003. *Affinity, That Elusive Dream: A Genealogy of the Chemical Revolution.* Cambridge, MA: MIT Press.

Laissus, Yves and Monseigny, Anne-Marie. 1969. Les Plantes du Roi: Note sur un Grand Ouvrage de Botanique Préparé au XVIIe Siècle par l'Académie Royale des Sciences. *Revue d'Histoire des Sciences* 22: 193–236.

Maindron, Ernest. 1888. *L'Académie des Sciences : Histoire de l'Académie, Fondation de l'Institut National, Bonaparte Membre de l'Institut National.* Paris: F. Alcan.

Mallon, Adrian C. 1983. *Science and Government in France, 1661–1699. Changing Patterns of Scientific Research and Development.* Queen's University, Belfast: Unpublished PhD Thesis.

Metzger, Hélène. 1969. *Les Doctrines Chimiques en France du Début du XVIIe à la Fin du XVIIIe Siècle*. Paris: A. Blanchard. (First published in 1923.)

Meynell, Guy G. 2002. *The French Academy of Sciences, 1666–1691: A Reassessment of the French Académie Royale des Sciences Under Colbert (1666–1683) and Louvois'*. Retrieved 31 August 2009, from http://www.haven.u-net.com/6text_7B2.htm.

Newman, William R. & Principe, Lawrence M. 1998. Alchemy vs. Chemistry: The Etymological Origins of a Historiographic Mistake. *Early Science and Medicine* 3: 32–65.

Newman, William R. & Principe, Lawrence M. 2002. *Alchemy Tried in the Fire: Starkey, Boyle, and the Fate of Helmontian Chymistry*. Chicago: University of Chicago Press.

Partington, James R. 1961–1970. *A History of Chemistry*, 4 vols. London: Macmillan.

Principe, Lawrence M. 1998. *The Aspiring Adept: Robert Boyle and his Alchemical Quest*. Princeton: Princeton University Press.

Rees, Graham. 1975a. Francis Bacon's Semi-Paracelsian cosmology and the Great Instauration. *Ambix* 22: 81–101.

Rees, Graham. 1975b. Francis Bacon's Semi-Paracelsian cosmology. *Ambix* 22:161–173.

Rees, Graham. 1996. Bacon's Speculative Philosophy. In *The Cambridge Companion to Francis Bacon*, ed. Markku, Peltonen, 121–145. Cambridge: Cambridge University Press.

Reti, Ladislao. 1968. Van Helmont, Boyle and the Alkahest. In *Some Aspects of Seventeenth-Century Medicine and Science: Papers Read at a Clark Library Seminar*, eds. Ladisalo Reti & William. C. Gibson, 3–19. Los Angeles: William Andrews Clark Memorial Library.

Roos, Anna M. 2007. *The Salt of the Earth: Natural Philosophy, Medicine, and Chymistry in England, 1650–1750*. Leiden/Boston: Brill.

Saunders, Elmo S. 1980. *The Decline and Reform of the Académie des Sciences à Paris, 1676–1699*. Ohio State University: Unpublished PhD Thesis.

Stroup, Alice. 1979. Wilhelm Homberg and the Search for the Constituents of Plants at the 17th-Century Academie Royale des Science. *Ambix* 26: 184–201.

Stroup, Alice. 1990. *A Company of Scientists: Botany, Patronage, and Community at the Seventeeth-Century Parisian Academy of Science*. Berkeley: University of California Press.

Stroup, Alice. 2002. Censure ou querelles scientifiques: l'affaire Duclos (1675–1685). In *Règlement, Usages et Science dans la France de l'Absolutisme*, eds. Christiane Demeulenaere-Douyère and Eric Brian, 435–452. Paris: Tec & Doc.

Sturdy, David. 1995. *Science and Social Status: The Members of the Académie des Sciences: 1666–1750*. Woodbridge: Boydell.

Taton, René. 1966. *Les Origines de l'Académie royale des sciences*. Paris: Palais de la découverte.

Todericiu, Doru. 1974. Sur la vraie biographie de Samuel Duclos (Du Clos) Cotreau. *Revue d'Histoire des Sciences* 27: 64–67.

John Locke and Helmontian Medicine

Peter R. Anstey*

Abstract This paper examines the sources and nature of Locke's medical thought. It is argued from a sampling of entries in Locke's medical notebooks and his correspondence, that Locke was a chymical physician. His medical thought contains two interlocking strands: he was an adherent of mercurialist transmutational alchemy and Helmontian iatrochemistry. The major, though not the only, influence on these aspects of Locke's thought was Robert Boyle.

1 Introduction

What sort of physician was John Locke? In asking this question I am not inquiring as to how effective he was in his medical practice, but rather how best can we characterise his approach to physic. An accurate answer to this question requires a thorough survey of Locke's medical milieu, the influences on Locke, his medical remains and other relevant writings. Surprisingly, such an appraisal of Locke as a physician has never been undertaken.[1] One of the reasons for this, I believe, lies in the general neglect of what today are known as the 'life sciences' in the interplay between the historiography of the scientific revolution and the neo-Kantian categories of Rationalism and Empiricism through which much early modern philosophy has been interpreted. A second related reason is the poverty of our understanding of early modern chymistry until the last two decades.

* I would like to thank Michael Hunter, J. R. Milton, William Newman and Lawrence Principe for advice on an earlier version of this paper.

[1] Osler 1900 provides a sketch of Locke's biography, his relations with Sydenham and transcriptions of some of the manuscript materials relating to Ashley's operation. Dewhurst 1963a provides a more detailed medical biography, but needs to be approached with caution because of its many errors. Its value lies in the transcription of the medical notes in Locke's Journal. Frank 1980 discusses Locke's involvement with the Oxford physiologists. See also Romanell 1984.

P.R. Anstey
Department of Philosophy, University of Otago

C.T. Wolfe and O. Gal (eds.), *The Body as Object and Instrument of Knowledge: Embodied Empiricism in Early Modern Science*, Studies in History and Philosophy of Science 25,
DOI 10.1007/978-90-481-3686-5_6,

There is no need to substantiate the claim about the effect of recent historiography on Lockean interpretation here. But it is important that a positive alternative set of terms of reference be substituted for Locke the empiricist and Locke the mechanical philosopher. I have argued elsewhere that the historical distinction, a distinction which informed Locke's own thought, between experimental and speculative natural philosophy, provides an illuminating and historically grounded way of ordering and highlighting the methodological, natural philosophical and even medical doctrines which flourished in mid-seventeenth-century England.[2] It can also help to illuminate the medical thought of John Locke.

In this paper I intend to use this distinction between experimental and speculative natural philosophy as the background terms of reference for a survey of the salient aspects of Locke's medical thought. The survey will not be uniform in its coverage because detailed treatment of some of the issues has appeared in other studies which can simply be summarised here. But the cumulative picture will, I hope, provide an insight into the salient features of Locke's medical thought. To this end, after a summary of the medical context, the survey covers Locke's medical methodology, his chymistry, his nosology (or theory of disease) and therapeutics, and his approach to physiology.

2 Medicine in England in the 1660s

From around 1659 through to early 1667, during the time when he was at Christ Church, Oxford, Locke invested considerable effort to equip himself in the cognate fields of medicine and chymistry. This period of self-directed study and practice coincided with an extraordinarily vexed phase in the history of English medicine. The institutional status of the College of Physicians and the theoretical status of Galenic medicine as a whole, were subject to serious challenges, and much of the debate was centred on physicians, chymists and natural philosophers within Locke's ambit. It is, therefore, impossible to understand the contours of Locke's medical remains from this seminal period of his development without first coming to grips with the storm that erupted around him.

In mid-seventeenth-century England qualified physicians received their training in the universities where they were taught the principles of medical theory and practice. This included the physiological, pathological, semiotical, hygenical and therapeutical parts of medicine. The dominant authority in the teaching of each of these subjects was Galen whose medical methodology was founded upon a largely Aristotelian natural philosophy. To be sure, medical students read widely amongst the recent medical authorities, but the underlying conceptual framework remained that of Galen, while the *methodus medendi* studied included both ancient authorities and the methods of treatment that had built up since the rediscovery of Galen's works in the sixteenth century.

---

[2] See Anstey 2005 and Anstey forthcoming.

Thus, for example, disease was conceived in terms of imbalances of the humours of the body and diagnosis and therapy were carried out by the application of the Aristotelian theory of qualities. From the time of Paracelsus there had been calls for the reform of medicine and these gathered in intensity in the writings of Francis Bacon and Joan Baptiste van Helmont. Pressure for medical reform intensified as new discoveries in anatomy and physiology undermined the authority of Galen, as alternative natural philosophies emerged as rivals to Aristotelianism, and as new chymical remedies were developed by Paracelsian and Helmontian chymists. A group of outspoken chymists and chymical physicians emerged in London and Oxford in the 1650s who began to challenge the Galenists or Methodists (who applied the *methodus medendi*). One of the severest critics of the traditional practice of physic was the American émigré George Starkey who accused Galenic medicine of being "erroneous and defective, dangerous and impotent, partly lame and ridiculous, partly lamentable and desperate."[3]

The situation came to a head in 1664 when a bill to approve a new charter for the bastion of the Galenists, the College of Physicians, was defeated in Parliament. Soon a rival Society of Chymical Physicians was proposed and the 'chymical physician' Thomas O'Dowde published an 'engagement' giving notice of the intention to incorporate the new society.[4] Reinforcing moves by the chymical physicians on the political front, Marchamont Nedham published a scathing attack on the College in 1665 called *Medela medicinae*. A bitter and highly charged pamphlet war immediately broke out between the chymical physicians and the Galenists which was to continue into the early 1670s.

The parties to the ongoing dispute can be divided into at least five groups. First there were the traditional Galenists, often members of the College (Henry Stubbe, Robert Sprackling, John Twysden); second there were the chymical physicians, most of whom had not received university training in physic (George Starkey, Thomas O'Dowde); third there were the promoters of the Royal Society, the new rival institution of the natural philosophers (Thomas Sprat, Joseph Glanvill); fourth there were those who are styled the 'virtuoso-physicians' who were sympathetic to the new natural philosophy and often members of both the Royal Society and the College (Timothy Clarke, Daniel Coxe); and fifth there was a small group of natural philosophers and physicians (Robert Boyle, Sir George Ent, Thomas Willis) who somehow transcended the debates and whose views and authority were appealed to by both sides.[5]

This was the highly charged and complex medical milieu in which Locke undertook to equip himself as a physician. Locke was not merely fully apprised of these disputes,[6] but was actually acquainted, in some cases intimately, with many of the

[3] Starkey 1657, Sig. A6v–7r.

[4] O'Dowde 1665, 93–5. The secondary literature on the Society of Chymical Physicians is now extensive. For seminal studies see Thomas 1953, Cook 1986, 1987.

[5] For the relation of the chymical physicians to the unlettered, entrepreneurial chymists, such as Lionel Lockyer and Richard Matthew, see Newman 2003, 201–3. Ent and Willis were both Fellows of the Royal Society.

[6] Locke's library contains most of the major publications in the pamphlet war. See Anstey and Burrows 2009, 12 and Harrison and Laslett 1971.

main actors, including some of the signatories to O'Dowde's 'engagement'. The key evidence for this connection has only recently come to light in the correspondence of the itinerant chymist John Read. Read had an altercation with Locke around April 1665 over Read's and Thomas Williams' refusal to reveal the recipe of a secret substance to Locke, a substance by which Locke thought, according to Read, 'all nature might be discovered'.[7] Locke was angered by Read's refusal, especially in view of the fact that Locke had introduced him to Williams, one of the agitators for the establishment of a Society of Chymical Physicians and a signatory to O'Dowde's 'engagement'. Read later wrote to Locke via Boyle, apparently in early 1666, in an attempt to appease him by revealing to him something of the 'holy water' which he had previously refused to divulge. The whole incident is of great interest, because of what it reveals about Locke's and Boyle's knowledge of the Society of Chymical Physicians in this critical stage of plans for its establishment. Apart from the fact that Read should have chosen to communicate to Locke about the holy water and the Society through Boyle, thus reinforcing the evidence for their close association in chymical matters during this period, we also learn that Read included for Locke a letter to Marchamont Nedham which contained not only information about the holy water, but also many details about the Society and Read's hopes to be affiliated with it. Before this correspondence came to light there was no known connection between Locke and the Society, nor of Locke's early connection with Williams who went on to become chymical physician to Charles II.[8] Nor was there such clear evidence of Locke' interest in alchemical secrets or his knowledge of Nedham and the latter's central role in the Society.

3 Methodology

The chymical physicians were a diverse bunch in both their theoretical commitments and the practical applications of their chymistry. They were, however, united by their opposition to Galenic medicine and by a loose cluster of methodological doctrines. The chymical physicians, to a man, called for an instauration in physic, but what each one thought should replace the hegemonic Galenism varied. Few, for example, would have subscribed to the somewhat bizarre 'wormatick' theory of disease espoused by Nedham.[9] As for their methodological views, many denied the utility of anatomy; some even denied the usefulness of botany. In keeping with the new experimental philosophy they all privileged observation and experiment over learning based upon authority, and they decried the use of speculative theories and hypotheses in physic. Many were highly critical of some of the mainstays of the

[7] John Read to Robert Boyle, 1666, Boyle 2001, 3, 3. The various letters are printed in ibid., 2–14.

[8] It has been long acknowledged that Locke recorded chymical notes from a Dr Williams, but he does not seem to have been identified. For more on Locke and Williams see below.

[9] See Nedham 1665, chap. 5.

Galenic *methodus* such as phlebotomy and purging, and most of them based their criticisms of the *methodus medendi* on their critique of the humoral theory of disease and its concomitant theory of qualities.

Locke's medical writings from the 1660s subscribe to most of these methodological doctrines. His essay "De arte medica" is an attempt at the reform of physic and contains a strong denunciation of speculative hypotheses and champions observation and experiment. His essay "Anatomia" is a critique of the efficacy of gross anatomy for physic. His essay on disease, "Morbus," contains a seminal theory of disease that is typical of the writings of the chymical physicians such as George Thomson.[10] Thus, it is clear that Locke aligned himself with the chymical physicians and against the Galenists.

However, once we descend into the details of Locke's views, we find that his position has its own subtleties and emphases which are best set in sharp relief by contrasting them with the views of other chymical physicians with whom he was broadly aligned. For example, in "Morbus" he does not merely develop a pathogenic theory of disease in opposition to the humoral theory, but announces that he is seeking a *via media* between Galenists and Paracelsians – in fact he espouses a Helmontian theory (see below). Furthermore, in contrast to some chymical physicians Locke maintained an active interest in botany, assembling his own herbarium in the summer months of 1664 and 1665,[11] and he never opposed bloodletting.

Now by the 1670s much of the heat had dissipated from the debate between the chymical physicians and the Galenists, and while Locke's other intellectual and vocational pursuits competed with and often obstructed his interest in physic, he continued to practise physic intermittently for the rest of his life. It would be wrong, therefore, simply to characterise Locke as a chymical physician if this appellation only connotes his polemical stance and affiliations in the formative 1660s. Locke remained a chymical physician throughout his life. Furthermore, it would be wrong to characterise Locke as an open critic of the College of Physicians. As it happens, arguably the most important experience of clinical medicine that he was to have in his life involved the advice of a handful of the most eminent members of the College.[12] This was his close involvement in Anthony Ashley Cooper's operation in June 1668 to drain a large hydatid cyst above his liver. Furthermore, it appears that Locke remained on good terms with John Micklethwaite whose advice he sought on a later occasion and who remained faithful to Shaftesbury during his time in the Tower.[13] Thus, we must resist defining Locke the physician solely in terms of the polemical context in which his views were forged and examine more closely the actual contents of his medical and chymical remains.

---

[10] See Thomson 1666.

[11] See Anstey and Harris 2006.

[12] They were Francis Glisson, Sir George Ent, John (later Sir John) Micklethwaite and Timothy Clarke.

[13] See John Mapletoft to Locke, 3 December 1677, Locke 1976–, 1, 530–531 and the enclosure from Micklethwaite, ibid., 532. For Micklethwaite's role in aiding Shaftesbury in the Tower see Haley 1968, 662.

4 Chymistry

Mention of Locke's medical and chymical remains brings us to the important question of the relation between physic and chymistry in Locke's day. By the early seventeenth century the preparation of medical remedies was one of the primary applications of chymistry. In fact, the Wittenberg chymist Daniel Sennert claimed,

> Chymistry is not a peculiar Art, but belongs to Physick, and is the perfection of it, for it is the part only of the Physitian to use and apply Chymical medicines for cure, and [he] may be called then a Chymical Physitian, and the Medicines Chymical, which are the perfection of Physick.[14]

The theory of disease, its diagnosis and treatment, as well as many facets of animal physiology, such as respiration and digestion, were founded upon chymical theories of one sort or another. Almost every leading chymist in England in the 1660s practised physic.[15] The diversity amongst them can be accounted for, in part, by the differences in the respective chymical theories and techniques that they deployed.

Locke's involvement in chymistry, and therefore in chymical medicine, probably began around 1659 and continued well into the 1690s. There were two important strands to Locke's chymistry. First, it is clear that he was deeply influenced by and practised chymistry in conformity to the mercurialist school who believed that the Philosophical Mercury, an essential ingredient in preparing the Stone, could be derived from common mercury. But there is a second, Helmontian strain in Locke's chymistry which is evident in his chymical notebooks, his correspondence and his medical receipts, and which is easily accounted for in terms of important sources of influence such as Boyle and his good friend, the Helmontian physician David Thomas. To be sure it is slightly artificial to separate out these two strands of Locke's chymical thought and practice, and yet it is true to say that the mercurialist strand is the most fascinating in terms of Locke's chymical practice, while the Helmontian elements are most important for understanding Locke's approach to physic. I will deal with each of these strands of Locke's chymistry in turn. The evidence presented is illustrative rather than exhaustive.

4.1 Locke and Mercurialist Chymistry

The derivation of the Philosophical or 'Sophic' Mercury involved two processes: first the removal of 'external' impurities using well-established purification techniques such as washing, grinding and distillation; second the removal of internal impurities.

[14] Debus 1990, 174. See ibid. for further discussion on the relation between chymistry and medicine.

[15] The one notable exception is Boyle. For a discussion of the relation between his chymistry and medicine see Principe 1998, 186–188, 305, 307. For Boyle on the Galenists see Hunter 2000. For Starkey see Newman 2003, 176–177 and Starkey's 'A Book on Paper of George Starkey' in Newman and Principe 2004, 319–27.

It was this process that was believed to be essential for the animation of the Sophic Mercury. Once purified the Sophic Mercury would then be able to liberate and nourish the seeds of gold and so enable transmutation.[16] The actual process that Boyle used to develop the Sophic Mercury involved combining mercury with an alloy of pure metallic antimony and silver. William Newman has shown that Boyle learnt the technique from George Starkey who in turn had derived it from Alexander von Suchten. Where does Locke fit into all of this?

On 20 May 1660 Dr. Ayliffe Ivye wrote to Locke at Oxford hoping for Locke's assistance:

> I hope Sir, you will lett slippe noe occasion whereby you may better your selfe, and soe me, by your aquaintance with Mr. Boyle, I longe to have an accounte of my Quaeries; I made Panacaea[17] Last weeke and have sent you two dragmes, tis the First preparation calcined via *humida*, liquore alkahestico then, washed and dryed you may go higher and with spirite of wine acuated etc. drawe off his perfect tincture; but truelye this worked admirable well and noethinge standeth in his way, and tis most safe to administer it secundum Glauberi modum ...[18]

Ivye's request suggests that he had sent some chymical queries to Boyle and that he hoped that Locke would follow them up for him. It is clear from this letter that Ivye believes that Locke is interested in his chymical preparations, not least because he had sent Locke two drachms of his Panacea and some details of his method of preparation of the expectorant: he refers to the *via humida* (the use of liquid solvents), alludes to van Helmont's alkahest and refers to the method of the German chymist Glauber. Moreover, this is the earliest known connection between Locke and Boyle and it suggests that they were already discussing chymistry together by May 1660.

Three months later we find Locke writing to one J. O. that he had failed to find in the study of Mr. B, that is Robert Boyle, the second part of a work by Alexander von Suchten (which he had promised him) and that instead he was sending a newly arrived work by Glauber.[19] In fact, Locke seems to have promised J. O. a manuscript translation of the second treatise of von Suchten's work on the secrets of antimony, the *Tractatus secundus de antimonio vulgari* (1604).[20] Two copies of a Latin translation of von Suchten's *Concordantia chymica* (1606) survive amongst the Boyle Papers.[21] It is therefore of great interest to note that Locke seems to have

[16] Summarising Principe 1998, 153–5.

[17] Glauber's panacea is probably Antimony pentasulphide, an expectorant. See Bodl. MS Locke f. 25, 279 and Bodl. MS Locke d. 9, 288 for entries on panacea. John Ward notes in his diary, 'There is one Mr Ivy a friend of Mr Locks of our house that lies some 2 miles from Bristow that canne make Glaubers panacea: hee practices not much', Folger MS V.a.290, fol. 58v (Michael Hunter provided the transcription).

[18] Ivye to Locke, 20 May 1660, Locke 1976–, 1, 146–7. For notes from Ivye see Bodl. MS Locke d. 9, 6; Bodl. MS Locke f. 27, 24, 38; Bodl. MS Locke f. 25, 117, 185, 262, 271, 305, 317a.

[19] Locke to J. O., August 1660, Locke 1976–, 1, 151. Esmond de Beer suggests that J. O. might be John Oliver who was mentioned in a previous letter, ibid., 150.

[20] In von Suchten 1604.

[21] Royal Society Boyle Papers (BP) 34, fols 1–152 and BP 14, fols 47–74.

believed that Boyle also had a copy of a translation of von Suchten's tract on antimony. It is important not to be misled by Locke's comment in this letter that "Mr. B prefers Glauber to Suchten" because, as mentioned above, von Suchten's method of preparing his Sophic Mercury as spelt out in the second treatise to which Locke refers, was absolutely crucial to George Starkey's preparation of the Philosophical Mercury; a preparation that Starkey transmitted to Boyle.[22] And, of course, we now know that Boyle was preoccupied with the preparation of the Sophic Mercury for over four decades because the preparation of Philosophical Mercury was thought to be a necessary preliminary to the preparation of the Philosophers' Stone.[23]

Was the young Locke cognizant of all of this? His chymical remains reveal that indeed he was. For in Bodleian Library (Bodl.) MS Locke f. 18 which he used from 1659–1660, that is at the time of his correspondence with Ivye and J. O., Locke records his opinion of a substance called Hews Powder as described by his chymical friend and physician Dr. William Currer (d. 1668). Currer, it should be noted, who was later to be a signatory to O'Dowde's 'engagement', had had an acrimonious dispute with Starkey in 1657–1658.[24] Locke records his view as follows,

> My opinion is Hews his powder is noething but mercury of antimony fixed with gold which workth as this doth witnesse Suchtenius[25]

We learn from this entry that Locke knew of Suchten's mercury of antimony either from Boyle's Latin translation of Suchten's work and/or from conversation with Boyle himself. But was Locke, under the influence of Boyle, to become a covert chrysopoeian seeking the recipe for the Sophic Mercury? Was he a mercurialist when it came to developing techniques for generating the Stone? Did he also seek the alkahest, the universal solvent, through the purification of salts after the manner of van Helmont, Starkey and Boyle? Was Locke a philosopher by fire? And most importantly, what light does this shed on Locke's views on the nature and practice of physic?

Around 1660 there was clearly a need in Oxford for some sort of instruction in chymistry for the clutch of young talented physicians that were associated with those practising the new philosophy there. To this end Boyle arranged for the German chymist Peter Stahl to teach a course in chymistry and Locke attended a course from 23 April to late May 1663.[26] Detailed notes from this course survive in Bodl. MS Locke f. 25 and, together with similar notes recorded by others, we can glean a fairly clear picture of what Locke was taught. Stahl was a physician and he

---

[22] Newman 2003, 135–41; Newman and Principe 2002, 50–6.

[23] See Principe 1998. Interestingly, in a diary entry of 1681, during a period or renewed chymical activity between Locke and Boyle, Locke records the following comment by Boyle, 'Suckten a very good chymist', Dewhurst 1963a, 207.

[24] For Currer's dispute with Starkey see Newman 2003, 190.

[25] Bodl. MS Locke f. 18, 52. The entry begins on 43 and continues on 52. It was later copied to Bodl. MS Locke f. 25, 317a–b. Interestingly, an entry in John Ward's diary deriving from Currer is almost identical to that of Locke. See the transcription in Wellcome MS 6175, 11.

[26] See Walmsley and Milton 1999.

taught basic laboratory techniques for preparing chymical remedies many of which, according to John Ward, derived from Oswald Croll a student of the German chymist Johann Hartmann.[27] If there is any truth to Anthony Wood's charge that during the course Locke was "prating and troublesome" it might be because Locke had already been inducted into the chymical arts and had well developed chymical views of his own.[28]

But Bodl. MS Locke f. 25 contains far more than simply notes from the course with Stahl. Interestingly, it includes extensive entries relating to antimony and mercury and in particular the mercury of antimony, the alkahest and even the Philosophical Mercury.[29] Their presence is strong evidence that an important seam in Locke's own chymical outlook was mercurialist and that the various preparations relating to antimony were conceived within the theoretical framework deriving from the lineage of von Suchten→Starkey→Boyle.[30] On the whole Locke's preoccupation was with chymical preparations that had some application in physic and the entries in Bodl. MS Locke f. 25 coincide with the period in which Locke was equipping himself as a physician. So, for example, the fact that the chymist John Read could write to Locke asking him "What it is in mettells & Minnerall that is Medicinal,"[31] indicates that Locke was associated with the medicinal applications of chymistry. Moreover, it is, as we have seen, but a small step from the medical applications of the mercurialist theoretical framework to a full engagement in the teleological structure of this approach to chymistry, that is, the quest for the Stone. Yet soon after this Read and Williams were to run into trouble over a chymical substance through which Locke feared "all nature might be discovered." Is it surprising then, that Locke supplied Boyle with his own recipe for an antimony of mercury *sub sigillo*, in confidence, which still survives in the Boyle Papers?[32]

Locke's chymical notebook (Bodl. MS Locke f. 25) also contains records of recipes derived from various chymists and chymical physicians within his

[27] 'Stall hath most of his praeparations out of Crollius and others notwithstanding what hee pretends of their being left him by his father', Diary of John Ward, Folger MS V.a.291, fol. 90r. (Michael Hunter provided me with this reference.)

[28] Wood 1891–1900, 1, 472.

[29] The same is true of the chymical entries in Bodl. MS Locke f. 27. See for example the entry on *Aurum* on 43–4: 'to purify gold, melt the gold & put to it 2 or 3 times its weight of antimony that takes away the impuritys of the gold, & to carry away the Antimony put to it a quarter of its weight of sublimate to know whether the sublimate be good put it upon the tongue & it will stick to it'.

[30] See for example Locke's entry on Mercury of antimony from Boyle on 76: 'Mercury of antimony: Take as much vitriol as you like and distill it in the usual manner. Pour the chaos over very finely powdered antimony and distill in a retort to dryness. Take what remains in the retory and sublime. Grind the sublimate upon a marble with oil of tartard and let the mercurial globules combine together into larger ones. Mr. Boyle' (translated by L. M. Principe).

[31] John Read to Locke, Boyle 2001, 3, 14.

[32] BP 26, fol. 102.

ambit, including William Currer and Thomas Williams (with whom Locke had had the altercation).[33] The most important of these is Boyle himself, and indeed, Boyle remains the single most important source of chymical advice and opinions in all of Locke's notebooks and many of Locke's chymical connections involve Boyle as well. One important example is Johann Schard whom Locke met on his visit to Cleves during the winter of 1665–1666. Locke had complained to Boyle of Cleves that "their physicians go the old road, I am told, and also easily guess by their apothecary's shops, which are unacquainted with chymical remedies. This, I suppose, makes this town so ill furnished with books of that kind, there being few here curious enough to enquire after chymistry or experimental learning."[34] But his meeting with Schard was productive. The many chymical recipes from Schard in Bodl. MS Locke f. 25 derive from the memorandum book Bodl. MS Locke f. 27, which Locke took with him to Cleves and this latter notebook reveals that Locke almost certainly spent time in Schard' chymical laboratory and that in addition to the preparation of various chymical remedies, they worked on mercury of antimony.[35]

After his return from Cleves in February 1666 Locke engaged in another period of chymical experimentation for which notes also survive in Bodl. MS Locke f. 25. 1666 was also an important year in Locke's development as a physician and we will have cause to examine his medical writings from this period below. Locke seems to have been practising chymistry with his friend Dr. David Thomas during the period from his return from Cleves on 18 February 1666 to the time of his departure for the household of Anthony Ashley Cooper in early April 1667. On 18 November Thomas wrote to Locke, who was visiting Lord Ashley in London, saying "If you bring with you mercury … we will make mercury sublimate our selves which wilbe much cheaper then to buy it."[36] Once established in Ashley's household he wrote to Boyle that his fingers still itched to practise chymistry again.[37] In fact, there was a chymical laboratory at Exeter House, but there is little evidence of his engaging in chymical experimentation. He did, however, practise physic and the period from September 1667 to September 1670 proved to be the most intense years of medical practice that Locke was ever to experience. During these years Locke was

33 For William Currer see Bodl. MS Locke f. 25, 23, 63, 277, 317b, Bodl. MS Locke f. 27, 47, 51 bis. For Thomas Williams see Bodl. MS Locke f. 25, fols 39, 60, 249, 316, 359, 361, Bodl. MS Locke f. 19, 207, Bodl. MS Locke c. 44, 60–1. Locke even received some receipts from Prince Rupert such as a recipe for *Auri tinctura* 'communicated by his own mouth', Bodl. MS Locke d. 9, 16. For Locke and Prince Rupert see Dewhurst 1963b.

34 Locke to Boyle, 12/22 December 1665, Locke 1976–, 1, 228.

35 Bodl. MS Locke f. 27, 75, copied into Bodl. MS Locke f. 25, 201/309. Locke's drawings of glassware and descriptions of apparatus (72, 80–1, 90) render it very likely that his notes were made in Schard's laboratory.

36 David Thomas to Locke, 18 November 1666, Locke 1976–, 1, 296. For Thomas' process for mercury of antimony see Bodl. MS Locke f. 25, 31.

37 Locke to Boyle, 12 November 1667, Locke 1976–, 1, 315.

able to start to apply some of the vast knowledge of chymical remedies which he had begun to accumulate in the late 1650s.

Locke's chymical interests never ceased. There are chymical preparations, notes and observations in his journal that testify to his ongoing interest in chymistry during his travels in France. Of particular interest is his discussion of Samuel Cottereau Duclos' potable gold, one of the most prized medicines in the armoury of chymical medicine. Locke's entry for 22 June 1678 provides the recipe for Duclos' potable gold which had apparently "cured quartans and dropsy," though Locke records that "Mr. Briot told me Duclos was a great liar."[38] He goes on to enter a query intended for Thomas Williams with whom he had clashed over the *sal circulatum* in 1665.

> Q. of Sr T. Williams concerning the processe of this aurum potabile of Dr. F. Anthony found amongst the papers of the Bishop of Winchester.[39]

A decade later, near the end of Locke's exile in the Netherlands Thomas wrote to him again, this time about the potable gold.

> I receaved the chymicall processes and have read Philalethes[40] more then once and doubt whether water in the receaver in the purification of the mercury may prejudice it That being in other processes prescribed. I desire you to read Jodocus Greverus in Theatr Chym vol 3: p. 699 to the same purpose. I entend and am now prepareing materialls for the potible gold which m Boyle assures though formerly of a contrary opinion is of very greate use and efficacy in physicke.[41]

Boyle had indeed been sceptical of the potable gold in his *Usefulness of Natural Philosophy*, II, i, published in 1663.[42] However, his view changed, perhaps due to his involvement with the mysterious ambassador of the asterism Georges Pierre who sent him a recipe for the potent medicine. In fact, Boyle finally published his own recipe for the potable gold in his *Observationes physicae* in 1691.[43]

After Locke's return from France, David Thomas wrote to him on 16 January 1682 hoping to see Locke because "this wilbe a convenient time for chemestry" and expressing to Locke that he believed that he could make a "principall remedy" approaching the alkahest.[44] The next notable episode in Locke's chymical

---

[38] Dewhurst 1963a, 128.

[39] Quoted from Dewhurst 1963a, 128. A copy of this entry is in Bodl. MS Locke f. 28, 184. Francis Anthony (1550–1623) had a run in with the College of Physicians over his potable gold. See Debus 2002, 184–5 for a summary discussion and references.

[40] Thomas refers to Eirenaeus Philalethes' *Introitus*, 1st edition, Amsterdam, 1667. Unbeknownst to Thomas and Locke, this was written by George Starkey (LL 1554).

[41] David Thomas to Locke, 20 October 1688, Locke 1976–, 3, 511.

[42] *Usefulness of Natural Philosophy*, II, i, Boyle 1999–2000, 3, 381–2. For Starkey's attempts to make von Suchten's potable gold see Newman and Principe 2002, 107–14.

[43] *Observationes physicae*, Boyle 1999–2000, 11, 419. For Georges Pierre and the potable gold see Pierre to Boyle, 3/13 May 1678, Boyle 2001, 5, 77–8. For further discussion see Principe 1998, 145 and Malcolm 2004 and Principe 2004a.

[44] David Thomas to Locke, Locke 1976–, 2, 474.

engagements involved Francis Mercurius van Helmont, the son of the great Flemish chymist, whom Locke met in the Netherlands in late 1686 while in exile. Francis Mercurius van Helmont had edited his father's literary remains which were published as *Ortus medicinae* in 1648. Locke was reading them carefully as early as 1657/8.[45] Francis Mercurius van Helmont was a peripatetic chymist and theologian whose most distinctive views concerned the transmigration of souls. In December 1690 after Locke had returned to England, Francis Mercurius, Locke's "Chymicall friend,"[46] provided him with a new furnace.[47] This signals a return to chymical trials on Locke's part that is almost certainly tied to his involvement with Boyle.

We now know that some months before Boyle's death, Locke received from Boyle part of a recipe for the Sophic Mercury. It is recorded in shorthand and code in Locke's Journal entry for 25 September 1691. A month later he mentioned in a letter to Boyle that "I have water, and I have vessels, I only want soap to be at work."[48] Lawrence Principe has shown that the 'soap' is the cleansing alloy for the internal process of purification of common mercury. After Boyle's death Locke corresponded with Newton concerning the preparation of the Sophic Mercury.[49] He made sure too that while he had access to Boyle's chymical papers he had copies made of some of them totalling about two-hundred pages, including the crucial letter from Starkey to Boyle of early 1651 outlining the method of von Suchten for preparing the Sophic Mercury.[50] There is no doubt then, that in the early 1690s Locke was still keenly interested in Boyle's chrysopoeian ambitions and that he himself took measures to record them and to try them out experimentally.

Finally, in October 1694 in a letter to Locke, James Tyrrell says, "I have no more but to assure you that as for the Manuscript you mention of the Course of Chymistry I doe not remember I ever so much as saw it."[51] This suggests that even as late as 1694 Locke remained interested in chymical matters. It also seems likely that the manuscript containing the "Course of Chymistry" is Bodl. MS Locke f. 25 which we have discussed above. All of this evidence shows that Locke maintained a keen interest in chymistry and specifically in mercurialist transmutational alchemy over four decades.

---

[45] See Bodl. MS Locke e. 4, 43–4 and 123–5.

[46] David Thomas to Locke, 7 July 1688, Locke 1976–, 3, 481. For chymical notes from F. M. van Helmont see Bodl. MS Locke d. 9, 156 (Rugæ), 205 (Ferrum).

[47] Locke 1976–, 4, 177.

[48] Locke to Boyle, 21 October 1691, Locke 1976–, 4, 321. Locke carried out a chymical experiment from 28 April to 6 July 1691, see Bodl. MS Locke d. 9, 223.

[49] Locke to Newton, 26 July 1692, Locke 1976–, 4, 4856. See Principe 1998, 175–77. See also Principe 2004b.

[50] Newman and Principe 2004, 12–31; Bodl. MS Locke c. 44. For early laboratory notes by Starkey that Locke almost certainly acquired from Boyle see Bodl. MS Locke c. 29, fols 115r–18v, transcribed and translated in Newman and Principe 2004, 3–11.

[51] James Tyrrell to Locke, 16 October 1694, Locke 1976–, 5, 163.

4.2 Locke and Helmontian Chymistry

Van Helmont died in 1644 and his substantial manuscript remains were published by his son Francis Mercurius van Helmont in 1648. His writings exerted a significant impact on the Hartlib circle including George Starkey and the young Robert Boyle. By 1660 when Locke was beginning in chymistry, Helmontian iatrochemisty was embedded in and had become one of the mainstays of reformist English medicine.[52]

The story of the transmission and assimilation of Helmontian ideas in England from the 1650s is multifaceted. However, as is the case with the mercurialist approach to chrysopoeia, Locke's main instructor in the application of Helmontian ideas was Robert Boyle. But Locke also read widely amongst the works of van Helmont and counted a number of Helmontian physicians amongst his friends, including, as we have seen, David Thomas whom Locke seems to have befriended in the mid-1660s and who remained one of his closest friends.

Van Helmont's chymistry is predicated upon the view that the fundamental elements are water and air, and that water is the primal principle to which all substances can be reduced. Van Helmont claimed that there is a universal solvent, the alkahest, which can reduce vegetables and minerals to their constituents and then to primal water. He identified Paracelsus' *sal circulatum* with his own alkahest. The alkahest operates by stripping substances of their forms, which forms are produced by the seminal principles that reside in the substance. Once the reduction has taken place the alkahest can then be separated off and reused because it is not affected by that which it works upon. Of particular importance for medicine was the fact that the alkahest could work on the essence of a substance and isolate the active ingredient within a substance from its inert and noxious matrix. This in turn enabled the development of more powerful specific medicines which, stripped of their noxious matrix, were able to work in harmony with the *archeus* of each person and be absorbed by the intestines and thence ameliorate the diseased condition of the patient. By contrast, Galenic remedies were rejected by the *archeus* which purged the body of their gummous poisons.

According to van Helmont, other substances over and above the alkahest had medicinal value. Of particular importance was the volatile salt of tartar which worked, not on the essence or crasis of the substance to be ingested, but on its noxious impurities. This process converted natural substances into perfected sulphurs which were by this process prepared for ingestion by the patient. The volatile salt of tartar was, therefore, an important succedaneum to the aklahest in Helmontian medicine. Other Paracelsian medicines promoted by van Helmont include the tincture of Lili and *Mercurius diaphoreticus*.[53]

---

[52] See Clericuzio 1993. Newman and Principe claim that 'Van Helmont's writings constituted what was probably the most wide-ranging and influential chymical theory of the second half of the seventeenth century', 2002, 296.

[53] See Porto 2002, 16. For Mercurius diaphoreticus dulcis in Locke see Bodl. MS Locke f. 25, 124, 126. For Helmont's Mercury diaphoreticus see Bodl. MS Locke f. 27, 46.

It was mentioned above that van Helmont identified his alkahest with Paracelsus' *sal circulatum* and thus regarded it as a special species of salt. In fact, van Helmont had a well-developed classification of salts and a theory concerning their manner of interacting with other chymical substances. Van Helmont's tripartite distinction between acid, alkaline and urinous salts was taken up and adapted by Boyle who also derived a further tripartite division of spirits from Starkey. Boyle also adopted the Helmontian theory of exantlation by which it was supposed that acids lost their corrosive power when acting upon other substances, which Boyle explained in terms of the mechanical affections of the subtle bodies involved rather than in the more vitalistic explanatory categories of van Helmont.[54]

Now van Helmont deployed two features derived from Renaissance chymistry in the service of his iatrochemistry. First, he developed and applied quantitative techniques of gravimetrics for his chymical analyses of substances. Second, he used corpuscular explanations of the sub-microscopic material changes that gave rise to the chymical phenomena that he observed. He also developed the Paracelsian ontological conception of disease as pathogenic *semina* which have their own *archeus* which comes into conflict with the *archeus* of the patient. This conception of disease was naturally tied to the chymical theory of the therapeutic applications of the alkahest and volatile salt of tartar, but also had radical implications for the traditional Galenic *methodus medendi*. No longer was disease to be considered in terms of humoral imbalance and treatment determined in terms of addressing excesses and privations of the Aristotelian primary qualities, hot, cold, wet and dry. Instead, van Helmont decried the use of venesection and other traditional therapeutic techniques and advocated the development of chymical remedies based upon his conception of the operation of the *archei* and the transformative power of his solvent and salts.[55]

Van Helmont's was not the only conception of the alkahest, nor is it clear that the term 'alkahest' referred to one determinate substance in his *oeuvre*.[56] Others such as the German chymist Glauber developed and applied their own alkahests.[57] There was also a plethora of seminal theories of disease deriving from Paracelsus. In this regard, the views of the Dane Severinus provide a nice counterpoint to the Helmontian seminal theory of disease.[58] However, Helmont's ideas and laboratory techniques were undoubtedly the most influential in mid-seventeenth-century English medicine, and the most important locus of their development and deployment was in the work of Starkey and Boyle.

As it happens, all of these Helmontian substances and notions (the efficacy in physic of the alkahest and the volatile salt of tartar; the theory of salts; the seminal

[54] Newman and Principe 2002, 289–96.

[55] For a thorough treatment of van Helmont's theory of disease see Pagel 1982, 141–61.

[56] Porto 2002.

[57] For Glauber's alkahest see Roos 2007, 33–46.

[58] For Severinus see Shackleford 2004 and for a survey of theories of *semina* from the Renaissance to the period up to Boyle see Hirai 2005.

theory of disease; a concern with quantitative chymical experimentation and even the *archeus*) are to be found in Locke's chymical and medical notebooks and correspondence.[59] Locke also sought out specific receipts deriving from van Helmont. Let us first examine the trail of Helmontian ideas Locke's chymical notebooks and his correspondence before turning to his more focused treatments of the nature of disease and animal physiology. We turn first to the alkahest. In late 1666 Locke wrote,

> Sal Circulatus Paracelsi est Alkahest Cellarius p. 26 61[60]

Clearly he is aware of the relation between van Helmont's solvent and Paracelsus' circulatory salt. In the same year he recorded a long note on Schard's recipe for the alkahest in Bodl. MS Locke f. 25, pp. 194/301. Another entry from around the same time records Boyle's view of the medicinal value of the alkahest or similar substance,

> Alkahest   Or a menstruum like it dissolvd crud antimony, & when drawn of <f> left christall of very great efficacy in physick, pourd upon salt of tartar & drawn of<f> & the remainder dissolvd in water afforded strange chrystalls Mr. Boyle

Just before leaving for London to join Ashley's household Locke wrote to Boyle concerning one of van Helmont's recipes for the use of warts cut from horses. Locke wonders

> [w]hether they are to be taken from live horses, since (if I forget not) Helmont some where says, that if in histerical fits, (for in that disease he commends them) you use those that are taken from an horse, *æstuante venere*, they have different effects from others.[61]

Locke also seeks from Boyle advice on the correct dosage of *sal ammoniac*.[62] Years later Thomas told him

> I thinke a principall remedy may be made by Armoniacke salts satiated with acid salts and volatilized which I beleeve may be by a short way effected and farther advanced to allmost the Alkahest.[63]

Of course the alkahest was not the only Helmontian substance in which Locke took an ongoing interest. He also seems to have adopted, probably via Boyle, Helmont's theory of salts. For example, in Bodl. MS Locke d. 9 we find a signed entry implying a belief in the Helmontian tripartite division.

---

[59] For Locke's extensive notes on Salt of Tartar see Bodl. MS Locke f. 25, 106–17/320/339 (N.B. the solidus is used to indicate the page numbers of continuous entries).

[60] Bodl. MS Locke f. 25, 33. The work by Andreas Cellarius is *Harmonia macrocosmica*, Amsterdam, 1661 (not listed in Locke's library). In Bodl. MS Locke d. 9, 132 Locke has notes on Paracelsus' 'Hilech', 'Paracelsus vocat Hilech magnum et sal circulatis minor' and 'Alkahest Paracelsi index ce mercurius philosophorum est non circulatum minus quod est Alkahest Helmontii'.

[61] Locke to Boyle, 24 March 1667, Locke 1976–, 1, 309.

[62] Ibid., 310.

[63] David Thomas to Locke, 11 January 1682, Locke 1976–, 2, 474.

> Whether volatil or urinous salts, acid & alkali may by any art of chymistry be changed one into another & what difference is to be found amongst the particulars of each of these 3 species JL[64]

Locke is clearly aware of the Helmontian origins of this theory. The very next entry concerns the derivation of volatile salts from herbs with a reference to van Helmont's *Ortus*:

> How the oyls of hearbs may be turned into volatil salts v. Helmont de Feb. c. 15 §7 52.[65]

Moreover, as will become apparent below, the Helmontian view of salts and spirits plays an important role in Locke's views on the use of respiration in animals and humans.

It may be objected that much of what Locke appropriated from van Helmont was undergirded by a speculative theory that included abstruse ontological categories such as *gas*, *blas* and ferments and that this is inconsistent with the experimental philosophy with its opposition to speculation and hypotheses. However, this is to miss three crucial features of the Helmontian legacy in Locke's thought. First, it must be said that almost all of Locke's chymical notes concern practical chymistry and there is no sustained discussion of underlying ontological categories. Locke's Helmontianism was practically and therapeutically oriented. Where Helmontian notions do appear, as in his theory of seminal disease (to be discussed below), there is no detailed explanation of what these categories actually are.

Second, on the rare occasions when Locke actually does report explanations of what is happening at the sub-microscopic level in chymical reactions, he is either reporting Boyle's corpuscular explanations or providing corpuscular musings of his own. For example, a marginal comment in Bodl. MS Locke f. 25, p. 309 for an entry on "Mercury of Antimony made by ascending at an intense heat by distilling in a retort" says "This calx fixes the oyly parts, & fastens imbibes them to itself Mr. Boyle."[66] The important point here is that, as Newman and Principe have shown,[67] Boyle tended to give mechanical explanations of Helmontian processes and it is this 'de-vitalised' Helmontianism that he transmitted to Locke.

5 Nosology and Therapeutics

Turning now to Locke's views on the nature of disease, we find that they are typical of those held by the chymical physicians of the 1660s and that the salient doctrines were widely held amongst Locke's peers. The *locus classicus* of Locke's view of

---

64 Bodl. MS Locke d. 9, 30. See ibid., 222 for a reference to Daniel Coxe's 1674 article on alkali and fixed salts in the *Philosophical Transactions*.

65 Bodl. MS Locke d. 9, 30. For the process of converting oils into salts in the work of Starkey, Starkey's 'grand design for medicine', see Newman and Principe 2002, 136–46.

66 Entries on Mercury of antimony from Schard and Petrus de Nicol begin at Bodl. MS Locke f. 25, 201 and continue on 309/310/313.

67 Newman and Principe 2002, 289–96.

disease is his "Morbus" entry in British Library Add. MS 32554 about which much has been written. This espouses a seminal theory of some diseases as a *via media* between the views of the Paracelsians and the Galenists. Locke's treatment of disease in this long entry from c. 1666 is strongly Helmontian and illustrates the manner in which Helmontian medical ideas were seen as being in opposition to Paracelsian notions.

Of particular interest in "Morbus" is the strong corpuscularism that underlies the theory. Locke delimits generation to two kinds: generation by seed and generation by the mixtion of parts,

> some things are producd by seminall principles, & some other by bare mistion of the parts, to which might be added the circumstantiall assistances of heat & cold &c, by seminall principles or ferments I meane some small & subtile parcelles of matter which are apt to transmute far greater portions of matter into a new nature & new qualitys, which change could not be brought about by any other knowne means, soe that this change seems wholy to depend upon the operation or activity of this seminall principle, & not on the difference of the matter its self that is changd.[68]

Note too the reference to the Helmontian notion of ferments which for van Helmont are equivalent to the formative power of seeds. Locke also speaks of the *archeus* of the disease,

> How these small & insensible ferments, this potent Archeus works I confesse I cannot satisfactorily comprehend, though the effects are evident but yet I believe 'twould be worth considering, to finde what deseases spring from these ferments, such as I beleive are contagions.[69]

This is strongly Helmontian in tone and is typical of the theories of disease which Locke was reading in the mid-1660s. The seminal ferments are regarded, following van Helmont, as invasive pathogenic agents which have their own *archeus*. Interestingly, Locke's discussion contains a residue of Galenism in his references to temperaments:

> it may be observd that in many deseases of this nature, the particular constitution of the body doth not make the deseases though some tempers be better fitted to be wrought on by this & some by that ferment, though if the seminall virtue be strong enough it will lay hold on any soe most seeds will grow almost in any soyle, though in some they thrive much better & others starve & dwindle. soe sanguine complexions are observd most easily to admit the seminall principles of the plague easily melancholy tempers more difficultly.[70]

The seminal theory of disease combined naturally with the miasmic view of Fracastorius and others that was promoted by Boyle, Hooke and later by Sydenham.[71] Locke's own later foray into environmental medicine with Charles Goodall was predicated on this miasma theory and is evidence for Locke's continued

---

[68] British Library Add. MS 32554, 232. The transcription is my own and will appear in Locke forthcoming. 'Morbus' is transcribed in Romanell 1984, 207–9 and Walsmley 2000, 390–3.

[69] British Library Add. MS 32554, 237.

[70] British Library Add. MS 32554, 237.

[71] See Nutton 1990 and Keele 1974.

belief in the seminal theory itself. And in fact, we find in his Journal entry for 22 July of 1678 that he claims "certain body types may carry seeds of certain diseases, or are more predisposed to contract them."[72]

As many scholars have pointed out, this ontological conception of disease lends itself to the view that there are different species of pathogenic agents. Locke seems to have been particularly attracted to Thomas Sydenham's classification of fevers in the 1670s and its susceptibility to the method of natural history which, probably under Locke's own influence, became the hallmark of Sydenham's medical methodology and to which Locke had, under the influence of Boyle, long subscribed.[73] It is important to stress, however, that Locke concedes that not all diseases have this cause,

> Other deseases I suppose may probably be conceivd to be producd by a bare mistion of two unfitt ingredient<s>, as when acid & volatile salts are mixd, there presently is producd an ebullition, & then the two differing salts coagulate into a 3^d substance far enough different from either of the ingredients. which I suppose not to be donne by any seminall principle.[74]

One cannot help but note the implicit commitment to a Helmontian theory of salts which seems to have played an important role in Locke's understanding of his own medical receipts as he went on and practised physic in later years. The depth of this commitment to this common theory of salts at this early stage in Locke's development as a physician is perhaps best illustrated in his short disputation on the use of respiration and it is to Locke's physiology that we now turn.

6 Physiology

In 1666 Locke drafted a medical disputation on the use of respiration. It now survives in the Shaftesbury Papers in the National Archives in Kew. It is entitled "Respirationis usus" and is written in small difficult Latin.[75] Before we examine some of its contents, a note of caution is in order. Locke despised the disputations of the Schools which he regarded as vacuous performances designed to titillate rather than to instruct. It seems most likely that he composed this disputation in order to cover himself, or at least his conscience, in his bid to have the degree of Doctor of Medicine conferred on him without fulfilling the requirements of the degree. With this in mind, we can surmise that there was in Locke's heart a degree of reluctance as he composed this draft and that its contents were designed more for the occasion than to record his own precise theoretical reflections on a very vexed problem in animal physiology. In other words, we should not take all of what

72 Romanell 1984, 139 and Dewhurst 1963a, 136.

73 For further discussion see Anstey 2002a and Anstey and Burrows 2009.

74 British Library Add. MS 32554, 248.

75 National Archives PRO 30/24/47/2, fols 71r–4v.

Locke says in the "Respirationis usus" too seriously, at least not as the definitive statement of his views. It was not composed as a record of research findings, but as an academic exercise, a necessary evil in order to secure a Medical Studentship at Christ Church.

With these preliminaries in hand, let us turn to the text. The first thing to notice is the dramatic disputational form that it takes. While this is most evident to the reader of Latin, there is enough in translation to capture the theatricality of the prose. It opens in Heraclitean tones,

> Nature never hides and flees from us more than when she seems to come forth openly and to show herself to anyone as obvious and easy. The vital breath of air that we draw in and expel with continuous labor from the first moment of life to its final extent seems merely to jeer at us. It pours itself into our inner breast only to slip away, and it cheats the embraces into which it rushed at first, and with the same subtlety it escapes the sharpness of both mind and eyes.[76]

And it is not long before the Helmontian themes start to leap off the page. After speaking of the "vestal fire of life" that nourishes us we are told that it is not the function of respiration to cool this fire but rather

> there are so many kitchens [*culinae*] of digestion and coction in the body, hence there are such various ferments of the internal parts all of which appear to work together so that there is finally something that can be inflamed and so that the vital flame may have tinder; to this purpose above all else, respiration seems to be devoted.[77]

The kitchen metaphor is classic van Helmont and the reference to the "various ferments of the internal parts" speaks of the Helmontian theory of digestion. Locke then moves on to the theory of the role of fermentation in the generation of animal spirits in terms redolent of Van Helmont,

> It is now generally acknowledged by everyone that the life of animals consists in the continuous generation and flow of subtle spirits. It is evident that these spirits of the heart are generated either by heat or fermentation by means of a previous digestion of ingested substances in the stomach, the intestines, the mesentery, and other workshops [*officinis*].[78]

It is the next stage of the process that brings us to the Helmontian theory of salts:

> animal life turns upon this hinge, that a continuous and constant supply of animal spirit be produced, that is, that the parts of the blood be exalted into a subtle and volatile material.[79]

---

[76] NA PRO 30/24/47/2, fol. 73r. The transciption and translation of this text are by Lawrence Principe and will appear in Locke forthcoming.

[77] NA PRO 30/24/47/2, fol. 73v.

[78] 'Vita animalium in continua spirituum subtilium generatione et fluxu consistere apud omnes jam in confesso est. Istos spiritus cordis sive calore sive fermentatione generari praevia ingestorum in stomacho intestinis mesenterio aliisque officinis digestione constat', ibid. For *culinae* in van Helmont see, for example, *Ortus medicinae*, 'Sextuplex digestio alimenti humani', numbers 67–8, Van Helmont 1648, 222–3.

[79] 'cum enim eo in cardine versatur vita animalis, ut continuus constansque fiat proventus spirituum animalium, hoc est ut sanguinis partes in materiam subtilem et volatilem exaltentur ', NA PRO 30/24/47/2, fol. 73v.

This "subtle and volatile material" is diffused throughout the body and when it has "played its role"

> these mature effluvia of the blood finally transpire and fly off into the breeze, thus furnishing a place for the spirits following behind them. In this way, by repeated circuits through the heart and lungs the mass of blood furnishes material to the vital flame, and finally the whole mass of blood, having been made volatile (leaving no residue behind) and transmuted into the nature of spirits, is breathed out and vanishes through sweat; this could never be done without the air's fellowship.[80]

What is the solvent in the air that brings enables it to "agitate, subtilize, volatilize, and finally kindle" the body? Following Robert Hooke, Locke speculates that

> It would seem to be a certain highly volatile nitrous spirit, <for> some have not unaptly observed that saltpetre is the proper menstruum for sulphureous and inflammable bodies. Especially since it is well known that the volatile salts of animals (for example, of blood and of urine) produce niter when cofermented with earth exposed to solar rays.[81]

This is, in fact, the widely held Helmontian account of the role of the volatilisation of blood venous blood in respiration. Locke even deploys the corpuscular terminology so characteristic of Van Helmont:

> it is probable that it is air which consumes bodies and makes them burn, not fire, which seems to be nothing other than the greatest agitation of the minute parts, while the air loosens their texture and shatters them.[82]

The point here is not that Locke was a thorough-going Helmontian, but that these were the tropes, the turns of phrase, the theoretical framework that Locke had at hand to work with. To be sure, in his own notebook entries concerning respiration and air he is more circumspect and less flowery, but even there the stamp of Helmontian chymistry and physiological theory are everywhere apparent.[83]

---

[80] 'haec sanguinis effluvia exoleta tandem transpirent et in auras evolent, succedentibus spiritibus locum praebentia. et ita sanguinis massa repetitis per cor et pulmones circuitionibus flammae vitali materiam praebeat et tota tandem volatilis facta nulla relicta faece in spirituum naturam transmutata per διαπνοην difflatur et evanescat, quod nunquam fieri possit sine aeris commercio', NA PRO 30/24/47/2, fols. 73v–72r. For van Helmont's view that the purpose of respiration is the volatalisation of the blood see van Helmont 1648, *Ortus medicinae*, 'Blas humanum', 178–92, especially numbers 22, 31, 34–37 and 46.

[81] 'esse spiritum quendam nitrosum summe volatilem, qui non inepte nonnulli observarunt salem petrae corporum sulphureorum et inflammibilium esse menstruum appropriatum. praesertim cum constet sales animalium volatiles sanguinis puta et urinae terrae radiis solaribus expositae confermentata nitrum progignere', NA PRO 30/24/47/2, fol. 72r. For the connection with Hooke see Bodl. MS Locke f. 19, 158. The notion of an aerial nitre was widely discussed amongst the Oxford physiologists; see Frank 1980. For the origins of the notion see Debus 1964.

[82] Ibid. For Locke's notes on nitre see Bodl. MS Locke f. 25, 147–51.

[83] See, for example, British Library Add. MS 32554, 91 and 93; Bodl. MS Locke f. 19, 158 (adapted from Bodl. MS Locke f. 27, 3–4 (rear)).

7 Conclusion

Let us turn again to the question with which this paper began: What kind of physician was Locke? It seems undeniable that he was a chymical physician, a mercurialist and a Helmontian. Locke was almost certainly inducted into the chymical arcana by Robert Boyle at least by early 1660 when he had access to Boyle's chymical papers and laboratory. It is little wonder then that Locke was one of three physicians to whom Boyle entrusted his chymical papers on his death and that it is through Locke's copies of some of them that one of the keys to unlocking Boyle's own quest for the Sophic Mercury was discovered. It is somehow so appropriate that Locke should be a key to the chymical Boyle because so much of Locke's own development as a chymical physician was inspired by Boyle.

Five things are worth noting in conclusion. First, while Locke was a mercurialist and was tantalised by the prospect of securing the Sophic Mercury, there is, to my knowledge, no evidence of Locke as a crypto-chrysopoeian. Unlike Boyle, who secretly pursued transmutation and was even duped by unscrupulous pretenders, such as Georges Pierre, Locke remained first and foremost one who sought to apply mercurialist and Helmontian chymistry in physic. Locke was not attempting to be an adept of chrysopoeia.

Second, note the absence of Thomas Sydenham from this study. Many have and do hold that the seminal influence on Locke's formation as a physician, both methodologically and practically was Sydenham. Yet Sydenham was not, in spite of the efforts of Daniel Coxe,[84] attracted to chymistry, and before he met Locke, he seems to have had little if anything in common with the physicians within Locke's ambit. It was, I believe, Locke who influenced Sydenham in the matter of methodology, while Locke learnt from him much about the treatment and classification of fevers.[85] More importantly, however, what I hope that this study shows is that the distinctive character of Locke as a physician was set before Locke met Sydenham.

Third, Locke was not uncritical of his sources, teachers and those whom he read. He would have had no truck with the pretensions to revelation of the likes of Starkey and van Helmont himself. This, for Locke, would smack of enthusiasm, though, to my knowledge, there is no record of his response to the revelatory claims of the chymists.

Fourth, it should be noted that there is a natural fit between Locke and van Helmont. Van Helmont was a serious experimenter who contributed significantly to experimental method in chymistry. In this he was a significant source for the experimental philosophy: his gravimetric techniques were deployed by Boyle who also performed some of his experiments, most famously the willow tree experiment.[86] Now from the early drafts of *An Essay concerning Human Understanding* (*Essay*) composed in 1671 we know that Locke was deeply concerned with standards of

[84] Daniel Coxe to Boyle, 14 October 1666, Boyle 2001, 3, 249.

[85] See Anstey and Burrows 2009.

[86] See Webster 1965.

measure and that his ideal natural philosophy was to be based upon a corpusuclar metric.[87] The corpuscular matter theory of Boyle was conducive both to Locke's ideal for a fully quantified natural philosophy and at the same time was part and parcel of the Boylean application of Helmontian ideas. If we add to this the iatrochemical focus of van Helmont's writings, it is clear that the whole package fits nicely with Locke's natural philosophical and medical outlook. Observation, experiment, measurement and cure were the hallmarks of this approach to medicine and these were central to the experimental philosophy.

Note too that Locke's appropriation of Helmontian medicine dovetails nicely with his other natural philosophical interests and methodological views. A life-long interest in botany fed into Locke's training and practice as a physician, as did his involvement in Boyle's researches into the air. All of this was subsumed under the Baconian rubric of the need to assemble natural histories, and the classification of plants, the classification of diseases, as well as the assembling of meteorological readings, and even his foray into environmental medicine, form parts of an integrated whole that preoccupied Locke for four decades.

And this brings us fifth and finally to the historiography underlying so much Lockean interpretation. How are we to square Locke's commitment to seminal principles and the teleological substructure of mercurialist matter theory with a commitment to the sparse ontology of a corpuscular matter theory and the primary and secondary quality distinction? There are three considerations that need to be brought to bear on this question. First, as is typical of Locke in many areas of natural philosophy, he declares his nescience concerning the manner of operation of the seeds: "[h]ow these small & insensible ferments, this potent Archeus works I confesse I cannot satisfactorily comprehend."[88] This reflects Boyle's own claims of ignorance about the mode of operation of seminal principles and is consistent with Locke's reflections in "Anatomia" of 1668 and the corpuscular scepticism of the drafts and the *Essay* itself. A second important consideration is that Locke's commitment to transmutational mercurialist chymistry is predicated upon a homogeneous matter that underlies the corpuscular concretions that provide the stable materials with which the chymists worked. It is this homogeneous, solid matter which Locke describes in the *Essay* as "every where the same, every where uniform"[89] that when modified with determinate shape, size and motion, gives rise to the conceptual problems for the corpuscular philosophy that Locke discusses in Book Two of the *Essay*.

However, and this is the third consideration, not only did Locke bring corpuscular analyses to bear wherever he could upon matters of chymical analysis and interpretation, but he clearly thought that such analyses would eventually be constituents of a demonstrative natural philosophy. Here is how he puts it in the chapter on universal propositions in Book Four:

[87] See Anstey forthcoming and Locke 1990.

[88] See Clericuzio 1990 and Anstey 2002b.

[89] Locke 1975 (hereafter *Essay*) III. x. 15, 498.

> [c]ould we begin at the other end, and discover what it was, wherein that Colour consisted, what made a Body lighter or heavier, what texture of Parts made it malleable, fusible, and fixed, and fit to be dissolved in this sort of Liquor, and not in another; if (I say) we had such an *Idea* as this of Bodies, and could perceive wherein all sensible Qualities originally consist, and how they are produced; we might frame such abstract *Ideas* of them, as would furnish us with matter of more general Knowledge, and enable us to make universal Propositions, that should carry *general Truth* and *Certainty* with them.[90]

Such universal propositions would not be merely general and certain truths, but like the truths of mathematics, they would be *instructive*.

Locke left no legacy in chymistry or in physic, but his philosophical legacy is unchallenged. And it is possible when reading the *Essay concerning Human Understanding* to be in blissful ignorance of Locke's involvement in chymistry and medicine. However, apprised as we now are of his mercurialist commitments is it any wonder that the most common illustrations in Book Three, in which Locke discusses the nature of species, are chymical ones; gold, antimony and vitriol? Perhaps there is an autobiographical element in Locke's claim

> [t]hat we find many of the Individuals that are ranked into one Sort, called by one common Name, and so received as being of one *Species*, have yet Qualities, depending on their real Constitutions, as far different one from another as from others, from which they are accounted to differ *specifically*. This, as it is easy to be observed by all, who have to do with natural Bodies; so Chymists especially are often, by sad Experience, convinced of it, when they, sometimes in vain, seek for the same Qualities in one parcel of Sulphur, Antimony, or Vitriol, which they have found in others.[91]

References

Anstey, Peter R. 2002a. Locke, Bacon and natural history. *Early Science and Medicine* 7: 65–92. Reprinted in *John Locke: Critical Assessments*, 4 vols., ed. Peter Anstey, 4, 299–324. Abingdon Oxon.: Routledge, 2006.

Anstey, Peter R. 2002b. Boyle on seminal principles. *Studies in History and Philosophy of Biological and Biomedical Sciences* 33: 597–630.

Anstey, Peter R. 2005. Experimental versus speculative natural philosophy. In *The Science of Nature in the Seventeenth Century: Patterns of Change in Early Modern Natural Philosophy*, eds. Peter R. Anstey and John A. Schuster, 215–242. Dordrecht: Springer.

Anstey, Peter R. forthcoming. *John Locke and Natural Philosophy*. Oxford: Oxford University Press.

Anstey, Peter R. and Burrows, John. 2009. John Locke, Thomas Sydenham, and the authorship of two medical essays. *electronic British Library Journal*.

Anstey, Peter R. and Harris, Stephen A. 2006. Locke and botany. *Studies in History and Philosophy of Biological and Biomedical Sciences* 37: 151–171.

Boyle, Robert. 1999–2000. *The Works of Robert Boyle*, 14 vols., eds. Michael Hunter and E. B. Davis. London: Pickering and Chatto.

Boyle, Robert. 2001. *The Correspondence of Robert Boyle*, 6 vols., eds. Michael Hunter, Antonio Clericuzio and Lawrence M. Principe. London: Pickering and Chatto.

---

[90] *Essay* IV. vi. 10, 584.

[91] *Essay* III. vi. 8, 443.

Cellarius Andreas. 1661. *Harmonia Macrocosmica.* Amsterdam.
Clericuzio, Antonio. 1990. A redefinition of Boyle's chemistry: and Corpuscular Philosophy. *Annals of Science* 47: 561–589.
Clericuzio, Antonio. 1993. From van Helmont to Boyle: a study of the transmission of Helmontian chymical and medical theories in seventeenth-century England. *British Journal for the History of Science* 26: 303–334.
Cook, Harold J. 1986. *The Decline of the Old Medical Regime in Stuart England.* Ithaca: Cornell.
Cook, Harold J. 1987. The Society of Chymical Physicians, the new philosophy, and the Restoration court. *Bulletin of the History of Medicine* 61: 61–77.
Coxe, Daniel. 1674. A Discourse Denying the Prae-Existence of Alcalizate or Fixed Salt in Any Subject, before It Were Exposed to the Action of the Fire. *Philosophical Transactions* 9:107: 150–158.
Debus, Allen G. 1964. The Paracelsian aerial nitre. *Isis* 55: 43–61.
Debus, Allen G. 1990. Chymistry and the universities in the seventeenth century. *Estudos Avançados* 4: 173–196.
Debus, Allen G. 2002. *The Chymical Philosophy, Paracelsian Science and Medicine in the Sixteenth and Seventeenth Centuries.* 1st edition 1977. Mineola, NY: Dover.
Dewhurst, Kenneth. 1963a. *John Locke (1632–1704). Physician and Philosopher.* London: Wellcome.
Dewhurst, Kenneth. 1963b. Prince Rupert as a scientist. *British Journal for the History of Science* 1: 365–373.
Frank, Robert G. Jr. 1980. *Harvey and the Oxford Physiologists: A Study of Scientific Ideas.* Berkeley: University of California Press.
Haley, K. D. H. 1968. *The First Earl of Shaftesbury.* Oxford: Clarendon.
Harrison, John and Laslett, Peter. 1971. *The Library of John Locke.* 2nd edition. Oxford: Clarendon.
Helmont, J. B. van. 1648. *Ortus medicinae.* Amsterdam.
Hirai, Hiro. 2005. *Le Concept de Semence dans les Théories de la Matière à la Renaissance: de Marsile Ficin à Pierre Gassendi.* Turnhout: Brepols.
Hunter, Michael. 2000. Boyle versus the Galenists: a suppressed critique of seventeenth-century medical practice and its significance. In *Robert Boyle (1627–1691): Scrupulosity and Science*, 157–201. Woodbridge: Boydell Press.
Keele, Kenneth D. 1974. The Sydenham–Boyle theory of morbific particles. *Medical History* 18: 240–248.
Locke, John. 1975. *An Essay concerning Human Understanding,* ed. Peter H. Nidditch. 4th edition 1700. Oxford: Clarendon.
Locke, John. 1976–. *The Correspondence of John Locke*, 9 vols., ed. E. de Beer. Oxford: Clarendon.
Locke, John. 1990. *Drafts for the* Essay concerning Human Understanding *and other Philosophical Writings, Vol. 1 Drafts A and B,* eds. Peter H. Nidditch and G. A. J. Rogers. Oxford: Clarendon.
Locke, John. forthcoming. *John Locke: Writings on Natural Philosophy and Medicine*, eds Peter R. Anstey and Lawrence M. Principe. Oxford: Clarendon.
Malcolm, Noel. 2004. Robert Boyle, Georges Pierre des Clozets, and the Asterism: a new source. *Early Science and Medicine* 9: 293–306.
Nedham, Marchamont. 1665. *Medela medicinae.* London.
Newman, William R. 2003. *Gehennical Fire: The Lives of George Starkey An American in the Scientific Revolution.* Chicago: University of Chicago Press (first published in 1994, Harvard: Harvard University Press).
Newman, William R. and Principe, Lawrence M. 2002. *Alchemy Tried in the Fire: Starkey, Boyle, and the Fate of Helmontian Chymistry.* Chicago: University of Chicago Press.
Newman, William R. and Principe, Lawrence M. eds. 2004. *George Starkey: Alchemical Laboratory Notebooks and Correspondence.* Chicago: University of Chicago Press.

Nutton, Vivian. 1990. The reception of Fracastoro's theory of contagion: the seed that fell among thorns. *Osiris*, 2nd series 6: 196–234.
O'Dowde, Thomas. 1665. *The Poor Man's Physician*, 3rd edition. London.
Osler, Sir William. 1900. An address on John Locke as a physician. *Lancet* 156:1115–1123.
Pagel, Walter. 1982. *Joan Baptista Van Helmont: Reformer of Science and Medicine*. Cambridge: Cambridge University Press.
Porto, Paolo A. 2002. Summus atque felicissimus salium: the medical relevance of the Liquor Alkahest. *Bulletin of the History of Medicine* 76: 1–29.
Principe, Lawrence M. 1998. *The Aspiring Adept: Robert Boyle and His Alchemical Quest*. Princeton: Princeton University Press.
Principe, Lawrence M. 2004a. Pierre des Clozets, Robert Boyle, the Alchemical Patriarch of Antioch and the Reunion of Christendom: further new sources. *Early Science and Medicine* 9: 307–320.
Principe, Lawrence M. 2004b. Lost Newton manuscript recovered at CHF. *Chymical Heritage* 22: 6–8.
Romanell, Patrick. 1984. *John Locke and Medicine: A New Key to Locke*. Buffalo: Prometheus Books.
Roos, Anna Marie. 2007. *The Salt of the Earth: Natural Philosophy, Medicine, and Chymistry in England 1650–1750*. Leiden: Brill.
Shackleford, Jole. 2004. *A Philosophical Path for Paracelsian Medicine: The Ideas, Intellectual Context, and Influence of Petrus Severinus*. Copenhagen: Museum Tusculanum Press.
Starkey George. 1657. *Nature's Explication*. London.
Starkey George. 1667. [Eirenaeus Philalethes, pseud.] *Introitus apertus as occlusum regis palatium*. Amsterdam.
Suchten, Alexander von. 1604. *Mysteria gemina antimonii*. Leipzig.
Suchten, Alexander von. 1606. *Concordantia chymica*. In *Cabala chymica*, ed. F. Kieser. Mülhausen.
Thomas, Sir Henry. 1953. The Society of Chymical Physicians: an echo of the great plague of London, 1665. In *Science, Medicine and History*, ed. E Ashworth Underwood, 2 vols., 2, 56–71. London: Oxford University Press.
Thomson, George. 1666. *Anatomizing the Pest*. London.
Walsmley, J. C. 2000. 'Morbus' – Locke's early essay on disease. *Early Science and Medicine* 5: 366–393.
Walmsley, J. C. and Milton, J. R. 1999. Locke's Notebook "Adversaria 4". *The Locke Newsletter* 30: 85–101.
Webster, Charles. 1965. Water as the ultimate principle of nature: The background to Boyle's *Sceptical Chymist*. *Ambix* 13, 96–107.
Wood, Anthony. 1891–1900. *The Life and Times of Anthony Wood*, 5 vols., ed. Andrew Clark. Oxford Historical Society/Clarendon Press.

Part II
The Body as Instrument

Empiricism Without the Senses: How the Instrument Replaced the Eye

Ofer Gal and Raz Chen-Morris

Abstract The optical instruments developed through the seventeenth century allowed peering into the very far and the very small; a spectacle never before experienced. The telescope, and later the microscope, was now expected to answer fundamental questions and resolve cosmological riddles by direct observation into the foundations of nature. But this ability came at an unexpected price and with unexpected results. For Kepler and Galileo, the new instruments did not offer extension and improvement to the senses; they replaced them altogether. To rely on their authority was to admit that the human eye is nothing but an instrument, and a flawed one at that. Rather than the intellect's window to the world, the human senses became a part of this world, a source of obscure and unreliable data, demanding uncertain deciphering. Accurate scientific observation meant that we are always wrong.

1 Introduction

On receiving news of Galileo's observations of the four satellites of Jupiter and the rugged face of the moon through his newly invented *perspicillum*, Kepler in great excitement exclaimed:

> Therefore let Galileo take his stand by Kepler's side. Let the former observe the moon with his face turned skyward, while the latter studies the sun by looking down at a screen (lest the lens injure his eyes). Let each employ his own device, and from this partnership may there some day arise an absolutely perfect theory of the distances.[1]

This Hollywood-like scene of the two astronomers marching hand in hand toward the dawn of a new scientific era was no attempt by Kepler to appropriate Galileo's

[1]Kepler 1965 [1610], 22.

O. Gal
Unit for History and Philosophy of Science
The University of Sydney, Bar Ilan University

R. Chen-Morris
Science, Technology and Society Program
The University of Sydney, Bar Ilan University

C.T. Wolfe and O. Gal (eds.), *The Body as Object and Instrument of Knowledge: Embodied Empiricism in Early Modern Science*, Studies in History and Philosophy of Science 25,
DOI 10.1007/978-90-481-3686-5_7, © Springer Science+Business Media B.V. 2010

success or to diminish the novelty of the telescope.[2] On the contrary, Kepler repeatedly asserted how short sighted he was in misjudging the potential for astronomical observations inherent in lenses, and how radically Galileo's instrument transformed the science of astronomy. It was a deep sense of recognition that beyond their different scientific temperaments and projects, they shared a common agenda of a new mode of empirical engagement with the phenomenal world: the instrument. For Kepler and Galileo, empirical investigation was no longer a direct engagement with nature, but an *essentially* mediated endeavor. The new instruments were not to assist the human senses, but to replace them.

2 Galileo: An Instrument for an Eye

The eye, in this new scheme, was to become a part of the instrument. When Kepler describes the nipple shaped lens, which he suggests as an improvement to Galileo's telescope, he explains that the position of the observer's eye depends on the "spot where the rays from all points of the object under observation converge at a common focus (this is the function of the hyperbolic nipple)."[3] The rays coming parallel from a distant point are converged by the nipple shaped lens at the crystalline humor and are continually refracted within the eye until "they strike the retina." The resulting image is "quite confused" and in order to set it right, Kepler suggests another lens to manipulate these inner refractions by adjusting the coming of the rays as if "they come from some nearby point." The result will be that the rays will "find their points of convergence on the retina itself. This is the definition of clear vision."[4] Kepler does not differentiate between the "natural" ocular membranes – the crystalline humor and retina – and the artificial lenses and screens; they are all integrated into a continuous process that creates from the rays of light a clear picture on a flat opaque surface.

Kepler's account of his suggested improvement to Galileo's invention accords well with Galileo's own instructions for its use. Such special instructions are required, and provided in the first pages of his *Sidereus Nuncius*, because unlike

[2] Kepler's ulterior motives constitute the crux of Biagioli's recent reconstruction of this exchange (Biagioli 2006, esp. 33–39). Biagioli queries Kepler's hasty and enthusiastic reply to Galileo's request to assess his telescopic observations, when in fact Kepler had not yet seen a telescope or observed personally the celestial phenomena Galileo claimed to have seen. Biagioli claims that the distance between Galileo and Kepler and the partial information Kepler had regarding Galileo's status in the Florentine court contributed to this exchange. Albeit, one should notice that Kepler's inability to build a high power telescope was due to specific practical problems concerning lens production and not to lack of theory about how telescopes work and how lenses magnify. Kepler's embraced Galileo's telescope as a confirmation of his optical theories and of his proposed observational practice using instruments in preference of a naked eye.

[3] Kepler 1965 [1610], 20.

[4] Kepler 1965 [1610], 21.

optical predecessors such as the spectacles, the scientific instrument does not simply strengthen the weak eye. It thoroughly replaces the 'natural' function of the sense organ with reasoned procedure of observation, into which measurement and comparison are built and into which the eyes should be assimilated. Galileo first cautions the readers that they have to "prepare a most accurate glass that shows objects brightly, distinctly, and not veiled by any obscurity." He then asks them to verify the power of magnification through an exercise that demands both the ability to produce accurate geometrical drawings of two circles, one 400 times larger than the other- together with the ability to force the eyes to gaze each at different objects, yet still be able to compare their sizes. After adjusting the eyes to the instrument, Galileo proceeds to his sketchy instructions as to how to measure distances viewed through the telescope's lenses.[5]

These instructions did not reveal to Galileo's readers how fundamentally novel was the status he assigned to his instrument. The telescope was accepted with great fanfare and immediately embraced by the scholarly community.[6] In fact, it seemed like an excellent way to conduct innovative astronomy without being entangled with the cosmological ramifications of the new astronomical theories, and was enthusiastically endorsed for this reason by the Jesuits.[7] "Now no part of the sky escapes our glance," they rhapsodized about "the lynx eyes:"

> nor is the beauty of the moon so great as it was for us formerly. We have been able to distinguish the circular motions of Venus and Mercury, and who does not blush to see the sun occasionally disfigured? We have laid bare the stratagems of Mars in approaching the earth and we have exposed the attendance of Jupiter and Saturn, hitherto hidden away to no purpose.[8]

But these very words, written a decade after the introduction of the telescope, did nothing but set up the inevitable clash between the radical epistemology that Galileo attached to the new instrument and the traditionalist project to which the Jesuits were attempting to harness it. The occasion for drawing clear lines between the modes of instrument-aided observation was introduced by the appearance of a sequence of three comets in the European sky during 1618 (Fig. 1).

The three comets – especially the third – were most impressive and had been carefully observed in late November by astronomers all over Europe, initiating a masquerade of astronomical treatises in which anonymous writers, pseudonyms and false authorship paraded through the European astronomical community, examining the demarcation lines between appearances and substances, optical illusions and authentic events.

---

[5]Galilei 1989 [1610], 38–39.

[6]See Malet 2005, Shea 1972, and van Helden 1974.

[7]van Helden 1974, 53.

[8]Grassi 1619, 5–6.

COMETA ORIEN-
TALIS,

Kurtze vnd eygentliche Beschreibung deß newen Cometen/so im November deß abgelauffenen 1618. Jahrs in Orient/oder gegen Auffgang der Sonnen allhie erschienen/vnd von menniglich gesehen worden.

Auß warhafften Astrologischen vnd Historischen Gründen menniglich zur Nachrichtung gestellt/

Durch

M. Gothardum Arthusium Dantiscanum, P.C.Histor. & Philo-Mathem. zu Franckfurt am Mayn.

2. Timoth. 3.

Instabunt tempora periculosa.

Franckfurt am Mayn/bey Sigismundo Latomo.

M. DC. XIX.

Fig. 1 The Comets of 1618

Galileo's entrée into this parade and the positions he expresses in it are very surprising. Never having observed the comets himself, he waited until an anonymous treatise by the Jesuit mathematician Horatio Grassi, *De Tribus Cometis Anni 1618 Disputatio Astronomica* (1619) introduced the issues of the scholarly discussion about these spectacular objects. Grassi's views are ones that Galileo should have been delighted with. Not only does he begin with the tribute to Galileo and his instrument cited above, his main conclusion is one that should have particularly appealed to Galileo: "our comet was not sublunar" he declares, "but clearly celestial."[9] Tycho Brahe had already arrived at this conclusion concerning the comet of 1577, as well as the *Nova Stella* of 1572, but the idea that comets and other transient objects are superlunary did not lose its revolutionary cosmological significance. The sharp dichotomy between the realm under and above the moon, which the heavenly position of comets undermined, was not an arcane scholarly conviction; it was well entrenched in religion and commonsense.[10] If Galileo was engaged in "a polemic against the Aristotelian and scholastic physics,"[11] as is commonly assumed, Grassi and Tycho should have been his close allies and the superlunary position of the comets his cherished weapon.

Yet Galileo chooses to join the discussion with a fierce assault on Grassi's *Disputatio*. He does it stealthily, dictating a *Discourse on the Comets* in Tuscan through his disciple and fellow academician Mario Guiducci (1619).[12] Grassi, however, is not fooled. He mocks "Galileo [for] order[ing] the matter to be discussed through intermediaries and interpreters" and joins the play of "secrets of mind" by replying with *Astronomical and Philosophical Balance*, apparently composed with other members of the *Collegio Romano*, under the anagram Lothario Sarsi of Siguenza (Sigensano 1619). For the celebrated climax of the debate Galileo unmasked and answered 'Sarsi's' Latin *Balance* (*Libra*) with his own – again Tuscan – *The Assayer* (*Il Saggiatore*, 1623), simultaneously denying that he was behind Guiducci *and* complaining about Sarsi's lack of manners in exposing him.[13] This was more than literary playfulness. In *The Assayer* Galileo not only discloses his own true identity and points to that of his rivals, but urges the reader to unmask Nature herself to see beyond "the bounty [of] her effects."[14]

Modern historians and philosophers of science, ensnared by these playful polemics, usually read *The Assayer* as a defense of Copernican astronomy against a complex of theological and conservative views of nature. But Galileo's arguments and those of Grassi's he chooses to refute demonstrate that the polemics are about the power and significance of instrumental observation. Yet Galileo does not defend the value of the telescope, which Grassi, we saw, has never doubted. Galileo takes

---

[9]Grassi 1619, 14.

[10]Concerning the significance of the superlunary position of comets see Van Nouhuys 1998. On the difficulties the Jesuits had with Tycho's cosmology see: Baldini 1992, 217–250; Blackwell 1992, esp. 148–153; Lattis 1994, 94–102, 211–216.

[11]Cassirer 1942, 316.

[12]Favaro asserts that "the pages of the first part … have corrections and additions in Galileo's handwriting. A second part … is entirely in Galileo's writing. The third part in Guiducci's hand … but there are correction by [Galileo]" and concludes that "the entire discourse may be said to be essentially his work." Quoted in Drake and O'Malley 1960, xvi–xvii.

[13]Galileo 1623, 169.

[14]Galilei 1623, 236. "La ricchezza della natura." In Galilei 1890–1909, 6: 281.

the opportunity of the discussion about comets to pursue his (and Kepler's) new empiricism, in which instruments are to replace the human senses.

Galileo reveals these intentions by targeting Grassi's two main arguments for the superlunary position of comets. The first one is based on the same method Tycho had perfected a generation earlier to make his claims about the celestial position of comets:

> If the comet was observed from different places and compared with the stars of the firmament, and if it preserved the same distance from them, it must be regarded as either in the firmament or certainly not far removed from it. But if it underwent parallax, it must be placed below the firmament in proportion to the amount of the difference of aspect.[15]

Parallax calculations – the comparison of the angles of sight from different positions – was a traditional and most venerable method of observational astronomy. For Grassi and the Jesuits, it gained further support through their intricate network of scholars and institutions deployed all over Europe. When observers in "Milan and in Parma … from Innsbruck in Germany and from France and Belgium" report the same position for the comet, and when finally two particularly accurate observations from Rome and Cologne completely coalesce in time and place, Grassi feels validated. The claim for the superlunary position of the comets does not "exceed the boundaries of our knowledge" and remains within the realm of empirical modesty.[16]

Yet what for Grassi is a carefully measured factual conclusion is completely unacceptable for Galileo. "Parallax operates reliably," he lets Guiducci claim on his behalf, "in real and permanent things whose essence is not affected by anyone's vision; these do not change place when the eye is moved. But parallax does not function in mere appearances," and comets are among those

> reflections of light, images, and wandering simulacra which are so dependent for their existence upon the vision of the observer that not only do they change position when he does, but … would vanish entirely if his vision were taken away.[17]

Galileo's real target should not be mistaken. It is neither the case that he harbored some deep aversion to parallax considerations, nor that he was particularly committed to Aristotelian meteorology or convinced in the errors of Tycho's way, or even that he was indeed so careful in his own parallax observations as to limit them only to unquestionably solid celestial bodies. In a series of public lectures following the supernova of 1604, for example, he took a clear Tychonic position and used the lack of parallax to argue that the new star was super lunar.[18] What Galileo is really assaulting is a particular type of empiricism, encapsulated in Grassi's knock-down parallax argument:

---

[15]Grassi 1619, 11.

[16]Grassi 1619, 6.

[17]Guiducci 1619, 36–37.

[18]Galilei 1890–1909, 2: 277–284. Cf. Dupré 2003, 373.

> If at the same time from other regions the same star was also observed very near to the comet, no stronger and clearer argument could be hoped for by which it might be demonstrated that the comet had very little or no parallax, since this could be observed without any instrument and by *observation with the unaided eye*.[19]

This last clause captures for Galileo the Jesuit-Tychonic position on parallax, an approach to scientific observation which he is obliged to confront and against which he defines his own empiricism. For Grassi and the Jesuits, the final arbitrator and the measure of all observations is the "unaided eye." Their parallax observations are particularly trustworthy, they assume, because they can be conducted "without any instrument." Galileo's arguments aim at the conditions of visibility necessary for parallax measurements because he wishes to undermine the fundamental assumption of the Jesuits' epistemology – the reliability of observation by the naked eye:

> in order to have the comet appear as without parallax to all observers, and still originate in the elemental sphere, it would suffice for vapours ... to be diffused on high and to be capable of reflecting the sun's light through distances and spaces equal to ... those from which the comet is perceived.[20]

Parallax is liable to the baffling effects of appearances and optical illusions because it is dependent on the naked eye. This is the crux of Galileo's surprising assault: observation "without any instrument" is not preferable, it is, on the contrary, fundamentally suspect.

Galileo further focuses the attack on the naked eye in answering Grassi's other main argument for the superlunary position of comets. "It has been discovered by long experience," claims Grassi,

> and proved by optical reason that all things observed with this instrument [the telescope] seem larger than they appear to the naked eye; yet according to the law that the enlargement appears less and less the farther away [the observed things] are removed from the eye, it results that fixed stars, the most remote of all from us, receive no perceptible magnification from the telescope. Therefore, since the comet appeared to be enlarged very little it ... is more remote from us than the moon, since when [the moon] has been observed through the telescope it appears much larger.[21]

This is the one argument that Galileo would not forgive. He would not allow that one can conclude about the distance of the comets from the failure of the telescope

---

[19]Grassi 1619, 14. Italics added. "Si enim in aliis etiam regionibus eodem tempore eadem stella cometae proxima observaretur, nullum maius atque evidentius optari poterat argumentum, quodemonstraretur nullum aut perexiguam parallaxim cometae fuisse, cum hocabsque ullo instrumento, unico oculorum intuitu, observari posset." Grassi in Galilei 1890–1909, 6: 31.

[20]Guiducci 1619, 40. One should note how much Galileo's argument is as removed from "opposition to the closed system of the schools," as Drake presented "The Assayer" in the preface to Drake and O' Malley, 1960, xxiii. Galileo's rejection of the parallax is based on proto-Aristotelian concept of the comets as resulting from exuding vapors, and furthermore, preserves "the Aristotelian duality" between heavens and the "elemental sphere."

[21]Grassi 1619, 17.

to magnify them, because he will not admit to this failure; the telescope magnifies regardless of distance:

> If a surface of a ball seen through the telescope at a distance of half a mile increases a thousand times, then so will the moon's disc increase a thousand times and no less; so will that of Jupiter and finally that of a fixed star.[22]

It is important to stress again that Grassi is an avowed supporter of the telescope and "the lynx eyes." The telescope is a legitimate, marvelous extension of the eye, strengthening its weaknesses and repairing its errors. Indeed, interpreting Galileo as arguing for this legitimacy, Grassi bitterly protests (under the name of Sarsi and on behalf of his Jesuit colleagues) what he perceives to be his portrayal as a scientific reactionary. Nothing is farther from the truth, the author of *Libra* complains; he is a champion of progress and a staunch defender of Galileo and his instruments:

> There were not lacking those who … asserted that … the telescope carries spectres to the eyes and deludes the mind with various images; therefore it does not display genuinely and without deception even those things which we observe close at hand, much less those which are far removed from us, except it will show them bewitched and deformed. We … publicly confuted the ignorance of those for whom this instrument was of no significance … we hoped that by protecting from invidious calumnies this telescope … we might therefore deserve well of [Galileo] rather than ill.[23]

There is no particular reason for Grassi to reject the telescope. Traditional mathematical optics provides him with a clear and trustworthy account of the principles of the instrument's operation:

> Objects are enlarged by the telescope because these objects are carried from it to the eye under a greater angle then they are observed without this instrument … according to optics, whatever things are observed under a larger angle seem larger.[24]

This analysis also provides him with an explanation, again in terms taken directly from traditional eye-centered, Euclidean-based optics, why the fixed stars, and presumably comets, should elude magnification:

> Be the visible objective whatever it may, the more it is removed from the eye the smaller and smaller the angle at which it is seen … thus, the angle of incidence of the images at the telescope scarcely vary after the objects have reached a very great distance, for then it is just as if all the rays fell perpendicularly on the lens.[25]

Note how 'modern' Grassi is, how well entrenched in the most contemporary cosmology: Copernican or Tychonian, Grassi's world offers "very great distance" for the fixed stars. Yet Kepler and Galileo's move towards the abolition of the dichotomy between eye and instrument, if Grassi is aware of either, has

---

[22] Galileo 1623, 220.

[23] Sarsi 1619, 80–81.

[24] Grassi 1619, 79.

[25] Grassi 1619, 82.

left him completely unaffected. The "visible object," for him, is seen *by the eye*. The telescope is of a different status altogether; it is a part of the medium *through* which vision occurs, and subject to the same mathematical analysis. It helps like any instrument might – hence the title *Libra* – but it does not change the principal onus of evidence and argument. This lies, always, with what "could be observed without any instrument and by observation with the unaided eye."[26]

With this loyalty to the naked eye, as far as Galileo is concerned, Grassi completely misses the import of the telescope. Grassi, we saw, was taken aback by the vehemence of Galileo's replies, and understood them as a defense of telescopic observation. But his baffled defense of his credentials in this respect was misplaced. Galileo had even less patience for his hearty support than for his mild criticism: "Sig. Sarsi, give up trying to exalt this instrument with this admirable new properties of yours unless you wish to throw it into utter disrepute."[27] His sarcasm aside, what is clear is that Galileo was not disturbed by Grassi's empirical claims but by his analysis and arguments, and disturbed enough that he felt compelled to reject both Grassi's support and his conclusions.

Galileo has no qualms about Grassi's geometrical analysis of magnification "for objects seen naturally." In that case, "the diminution of the angle is made in a continually greater ratio the more the object is removed."[28] But Galileo has little respect for the way objects are "seen naturally." From Galileo's point of view, the instrument does not extend the sense organ, but replaces it altogether, and in the process it is the naked eye that loses its legitimacy as a source of knowledge:

> The naked eye distinguishes none of these shapes [of the heavenly bodies] without the telescope. [29]

Galileo, then, does not reject Grassi's cosmological conclusions because he nurtures some deep-held belief in the sublunary nature of comets. Rather, he finds himself placing the comets in the "elemental sphere" because he is adamant to reject the implications of Grassi's main arguments to the contrary. Namely: that fixed stars, like comets, are not magnified by the telescope, which implies that the telescope does not magnify *all* objects; and that absence of parallax is *the* unassailable testimony for the great distance of comets, which implies the supremacy of what "could be observed without any instrument and by observation with the unaided eye."

It will be a mistake to think of Galileo as defending the telescope or apologizing for its failure to magnify very distant objects. There is no such failure, he insists:

[26]Grassi 1619, 14.

[27]Galilei 1623, 209.

[28]Galilei 1623, 221.

[29]Galilei 1623, 321. "Senza il telescipio, l'occhio libero niuna di cotali figure distingue." Galilei 1890–1909, 6: 359.

comets and fixed stars do not appear magnified *to the eye* because of a distortion produced *by the eye* and repaired by the telescope:

> what we meant by saying the telescope 'robs the stars of irradiation' … is that it operates upon the stars in such a way [as to circumvent] the irradiation which disturbs the naked eye and impedes precise perception.[30]

The eye introduces a spurious splendor around stars and comets that makes them appear larger. This is not real magnification, of course: the body of the celestial objects remains invisible to the naked eye. Because the telescope removes this "irradiation," the eye fails to notice that it has also magnified and made visible the celestial bodies themselves. This is not an *apologia* for the instrument. Rather, it is a charge against the eye, which errs twice: once in introducing the false "wig" and once in failing to notice the correction and magnification.[31] Grassi's claim that fixed stars and comets "suffered scarcely any enlargement" meant that they did not appear larger *to the eye*. But the eye, in Galileo's new radical instrumentalism, is no longer the main point of reference for visual phenomena, and definitely not the final adjudicator of their trustworthiness.

3 Looking at the Sun

Kepler's excitement over Galileo's instrument was genuine. The coming of the telescope facilitated a new level of accuracy that far exceeds the abilities of the human naked eye. The resulting, artificially-produced image made Tycho Brahe's accurate and meticulous observations obsolete. Kepler, who regarded Brahe's observations as "the pinnacle" of human scientific enterprise, had to admit that "your telescope, Galileo, surpasses these attainments."[32] It is true that Tycho measured celestial degrees most accurately, but with the new instrument the astronomer "subdivides [Tycho's celestial degrees] with the utmost nicety into minutes and fractions of minutes." Consequently, the intellect, which until the invention of the telescope could have only abstractly fathom certain generalities about the heaven, could now be assisted with exact and concrete observations of the distant realms. The intellectual imaginations of ancient philosophers as well as of Copernicus,

---

[30]Galilei 1623, 324.

[31]This is a recapitulation of his arguments (including the hairy metaphor) in the *Sidereus Nuncius:* "The reason for this is that when the stars are observed with the naked eye, they do not show themselves according to their simple and, so to speak, naked size, but rather surrounded by a certain brightness and crowned by twinkling rays … Stars are therefore seen unshorn in the midst of darkness, but daylight can shear them of their hair … The spyglass likewise does the same thing: for first it takes away the borrowed and accidental brightness from the stars and thereupon it enlarges their simple globes." Galilei 1989 [1610], 57–58.

[32]Kepler 1965 [1610], 21–22.

Bruno and Kepler himself were disassociated from the limitations of sense experience. The telescope allowed the new philosopher to compare these fantasies on the shape and character of the planets to real observations.

The excitement was genuine, because by the time he was presented with Galileo's instrument, Kepler himself was already pursuing the astronomical use of optical (that is – lens based) instruments for some time. Even more important, Kepler shared with Galileo an understanding of the significance of the telescope: optical instruments were to take primacy in observation, and the eye would need to be integrated into them. This epistemological novelty, that Galileo would hint at in the *Sidereus Nuncius* of 1610 and attempt to explain in the controversy over the comets a decade later, was already explicated in Kepler's 1604 optical *opus magnum*, the *Optical Part of Astronomy* (*Ad Vitellionem Paralipomena*).[33] Kepler provides a vivid example for this new instrumental empiricism in describing his observation of a lunar eclipse through a *Camera Obscura*. Introducing an artificially-produced image, Kepler turns his instrument of observation into the locus of astronomical knowledge, and lets the human observer slip out of his optics:

> On 1602 21/31 December at 6h in the morning, through a device described in Ch. 2 [camera obscura] and an instrument made for this purpose, a description of which is furnished below, the moon made an image of itself brightly upon the paper lying below, inverted in situation, just as it was in the heavens, gibbous … You should not think that what I would consider to be in the moon's ray was in the paper, for both the gibbous face and the spot in its middle were carried over to all parts of the paper whatever that was placed beneath it; rather, indeed, it was from moving the paper that the spot was first discovered.[34]

The observation, Kepler stresses, is not *his*. It is no-one's. The image of the moon is not the culmination of a cognitive process. It does not require an observer; a piece of paper is enough. In fact, even the paper is not necessary: it can be moved around without affecting the production of the image.[35] A decade later and a few years before entangling himself in the debate over the comets, Galileo observed the sun

---

[33] Kepler 1937– [1604]; Kepler 2000 [1604]. We will use *Ad Vitellionem* (1937– [1604]) to refer to the original Latin and *Optics* (2000 [1604]) to refer to Donahue's translation.

[34] Kepler 2000 [1604], 259.

[35] The relation between the camera obscura and the eye is at the heart of the historiographic debate concerning Kepler's optics. For Straker (1971, and cf. Crombie 1953) the instrument represents Kepler's novel commitment to the mechanization of the eye and his indebtedness to the artisanal tradition. Kepler's claim that the locus of images is the retina rather than the crystalline humor, Straker argues, is an immediate consequence of comparing the eye to a camera obscura. Lindberg, in contrast, argues for Kepler's reliance on the perspectivist tradition, stresses that "only on one occasion did [Kepler] explicitly compared the eye to a camera obscura" (Lindberg 1976, 206). As we claimed above and will argue below, this debate is somewhat misdirected: Kepler's main motivation in equating the eye and the camera obscura is legitimating the instrument no less than understanding the eye.

spots in a similar way; only instead of the simple *Camera Obscura* he posited a telescope between the observed object (the sun) and the paper:

> Direct the telescope upon the Sun as if you were going to observe that body. Having focused and steadied it, expose a flat white sheet of paper about a foot from the concave lens; upon this will fall a circular image of the Sun's disk, with all the spots that are on it arranged and disposed with exactly the same symmetry as in the Sun. The more the paper is moved away from the tube, the larger this image will become, and the better the spots will be depicted … In order to picture them accurately, I first describe on the paper a circle of the size that best suits me, and then by moving the paper towards or away from the tube I find the exact place where the image of the Sun is enlarged to the measure of the circle I have drawn. This also serves me as a norm and rule for getting the plane of the paper right, so that it will not be tilted to the luminous cone of sunlight that emerges from the telescope. … By tilting the paper the proper position is easily found, and then with a pen one may mark out the spots in their right sizes, shapes, and position.[36]

As Galileo does in his lunar observations, Kepler relegates the eye to a secondary role, as the sunlight imprints (*stampata*) the image on the paper. The observer is turned into a draughtsman whose role is to stabilize and trace the outlines of the image.

Galileo's telescope thrust into public attention the assumptions implicit in Kepler's highly professional new optics with its direct challenge to the traditional eye-oriented modes of astronomical observation. It compelled astronomers to reconsider their epistemological manual as to what is a valid observational practice, but not many were willing to adopt Kepler and Galileo's epistemology together with their instruments. The Jesuits, loyal to their mandate to employ the novelties of the New Science in the service of the mores of counter-reformation church,[37] were particularly conspicuous in their efforts to implement the new means and techniques of observation while completely ignoring the marginalization of the human eye that for Kepler and Galileo was a necessary implication. Their debate with Galileo over the sunspots was, in this respect, analogous to the comets debate to follow. Their representative, Christoph Scheiner, fiercely competed with Galileo for credit on the empirical discovery, and between them, Scheiner's interpretation of it was, metaphysically, the more radical: the spots, he suggested, were shadows of hitherto unknown planets. Epistemologically, however, he remained unwavering in his commitment to the superiority of the naked eye. Ignoring the obvious hazards, he insistently used his eyes to look at the sun. This in spite of the different filtering devices he had to apply in order to protect his eyes that distorted and obscured the observations. Like Grassi, Scheiner was proficient with the new instruments and techniques, and experimented with projecting images from the telescope on a white surface (Fig. 2). These projections, however, were not intended to produce pictures of the sunspots but merely to examine any flaws in the lenses as well as other optical effects produced by them. Scheiner invested much energy

[36]Galilei 1890–1909, 5: 136–137. Quoted and translated in Biagioli 2006, 190.

[37]*Cf.* Dear 1995, esp. Chapter 2, and Feldhay 2000.

Fig. 2 Chistophoro Scheiner, Rosa Ursina (1626–1630)

to minimize the role of the instrumental mediation so the sunspots will be "seen without a tube, by the eye of any man."[38]

Like the ensuing one over the comets, the controversy over the sunspots was not primarily over observations and their interpretation. Rather, it revealed a deeper level of contention concerning what the role of vision is in the production of knowledge. The Jesuits, following Scheiner, devoted most of their efforts to preserving the status of the eye as the guarantor of any knowledge of nature. Disregarding Kepler's rejection of the traditional distinction between direct and mediated vision, Jesuit mathematicians held on to the Pauline belief in the supremacy of "face to face" acquaintance. In his 1613 large volume containing *Opticorum libri sex: Philosophis iuxta ac Mathematicis utiles* the Jesuit Fransciscus Agvilonius most emphatically reaffirmed traditional hierarchy of sight:

> All the things that are contained in Optics are considered under a triple reason, [compared] to the triple mode through which creatures come to know God. First direct [vision], that is our eye, as it turns towards the thing in front of it, so it is compared to the cognition of the minds of the blessed contemplating the presence of God, as St. Paul said: face to face. The second [part] is reflection [repercussion] that is the perception of those things, whose

[38] Galilei 1890–1909, 5: 59–61. Quoted and discussed in Biagioli 2006, 200–201.

images come back to us from mirrors, this is not unlike that cognition, that through faith we see God in the created things as in a kind of mirror or in enigmas. Thence the third, that we call infraction; this is how the species of things are transmitted through dissimilar diaphanous [media], and from them [the species] enter the eye as if deformed and fractured. Thus some of the divinatory notions of the heathens, corrupted by many errors, are affected by the light of nature only.[39]

Whether he is ignorant of Kepler and Galileo's new concept of the eye and its place in visual perception or consciously rejecting it, Agvilonius' commitment to the traditional cognitive hierarchy and especially the distinction between mediated and direct vision remains unwavering: any mediation is a source of enigmas, distortions or heathen delusions and errors.

4 Kepler: the Eye as an Instrument

For Kepler, the introduction of the telescope provided another demonstration of the poverty of this hierarchy and the need to undo the supremacy of "face to face" vision that Grassi, Scheiner and Agvilonius assumed. Galileo's observations made it ever clearer that true knowledge of heavenly bodies was produced through the mediation of lenses and their complex and multiple refractions. Kepler was very explicit that his own treatise on optics was to provide optical foundations for a new, instrument-based astronomical observation: *artificiosa observationes*, as he calls them in the subtitle.

Published 5 years before the advent of the telescope, the main artificial instrument of observation to occupy Kepler's attention in the *Ad Vitellionem* is, as we saw, the *camera obscura*. Kepler first establishes its legitimacy and efficiency by demonstrating that the image obtained through it is indeed that of the observed object.[40] He goes on to elucidate by way of physical simulation its underlying principle, namely – the formation of an image on a screen behind a small aperture:

[39] Agvilonius 1613, 3. "Continetur omnis Optice triplici fere videndi ratione … triplici etiam modo quo deum creaturae cognoscunt, … comparauit. Prima directa, quae est oculi nostri, sic, ut in rem propositam intendit, cum illa cognitione componitur qua beatorum mentes praesentem Deum, facie ad faciem, ut D. Loquitur Paulus, contemplatur. Altera repercussa, sive earum rerum perceptio, quarum a speculis ad nos imagines revertuntur, cui non absimlis est illa cognitio, qua Deum per fidem in rebus creatis, veluti quodam speculo aut aenigmate videmus. Tertia denique, quam infractam vocant, ea est, qua rerum species per dissimilia diaphana transmissae, et ab iisdem quasi deformatae ac fractae in oculos immittuntur. Sic Ethnici divinitatis notionem aliquam, sed multis erroribus vitiatam, naturae solius lumine affecuti sunt."

[40] For the import of the camera obscura in the study of Kepler's optics see f.n. 35.

> I set a book in a high place, which was to stand for a luminous body. Between this and the pavement a tablet with a polygonal hole was set up. Next, a thread was sent down from one corner of the book through the hole to the pavement, falling upon the pavement in such a way as to graze the edges of the hole, the image of which I traced with chalk. In this way a figure was created upon the pavement similar to the hole. The same thing occurred when an additional thread was added from the second, third and fourth corner of the book, as well as from the infinite points of the edges. In this way, a narrow row of infinite figures of the whole outlined the large quadrangular figure of the book on the pavement.[41]

The threads from the book's corners pass through the polygonal hole, grazing its edges and projecting images in the shape of the hole – a hole-shaped image for each corner of the book. The four images of the book's corners are arranged on the floor in reverse order, and when this process is repeated from (ideally) every point of the book, a multitude of hole-shaped images will be projected on the floor, arranged in the (reversed) pattern of the book.

This is a neat solution to an age old mystery, but the solution is not where the novelty of Kepler's optics rests. Neither the phenomenon of pinhole images, on which the *camera obscura* is based, nor its account in terms of intersecting rays is new to the optical tradition.[42] Explanations of the phenomena based on geometrical analysis of rays were available to the optical tradition at least since Levi ben Gershon (Gersonides) in the beginning of the fourteenth century.

Kepler's novelty is in setting the stage to the radical instrumentalization of observation he would share with Galileo by eradicating from his explanation any references to the eye and human vision. For the perspectivist account from Gersonides to Maurolyco in the first half of the sixteenth century, the pinhole image was not just a reliable projection of its source. It was a unique *re-presentation* of the sun.[43] The circular image was not *caused* by the sun and by light; it was the true form of the sun or the perfect dissemination proper of light, as John Pecham in the late thirteenth century explains:

> The spherical shape is associated with light and is in harmony with all the bodies of the world as being to the highest degree conservative of nature, all parts of which join together most perfectly within itself. This is why a raindrop assumes roundness. Therefore, light is naturally moved toward this shape and gradually assumes it when propagated some distance.[44]

Understood this way, the circularity of the image does not simply testify to a property of its source; it is a sign of the image's indubitable authenticity. This essential relation between source and image completely disappears from Kepler's account, together with the exactness of representation it insures. There is nothing unique to the

[41] Kepler 2000 [1604], 56.

[42] See Aristotle 1984, Problems Bk 15, Ch. 6, 911b1, 2:1417; Pecham 1970, 67. See also Lindberg 1968, 1969; Thro 1996, 20–54.

[43] *Cf.* Lindberg 1969, 303 ff. For Maurolyco on pinhole images see Zik and Hon 2007.

[44] Pecham 1970, 70–71.

circularity of the pinhole image: a rectangular body will produce a rectangular image, as the experiment with the book shows. Neither does the pinhole image *represent* light: it is light that is simulated by the threads pulled through the hole, but the image projected on the pavement can be of any object, not necessarily luminous – a book. The trustworthiness of the projection, for Kepler, does not rest on its perfect loyalty to the object projected but on understanding the physical process of projection. Indeed, Kepler discovers, one cannot hope for such loyalty: The book-pattern on the floor is created by a "narrow row" of partially overlapping "figures," so not only is the image reversed, its boundaries are fuzzy. Moreover, these stains are reflections of the *aperture*. Even for those perspectivists, like Maurolyco, who appear to suggest a similar account, the image cast through the aperture is composed of many images *of the luminous body*. These are merged together as the distance of the screen from the aperture grows and the images of the source grow respectively.[45] Kepler's "figures" bear no resemblance to the light source. The complete, smooth, upright perception of the book on the pavement is a construct.

What Galileo would try to affect with his fierce rhetoric – the takeover of astronomy by *artificiosa observationes* – Kepler attempts to legitimize by turning optics into a mathematical-physical study of the production of images by light:

> from the Sun, and from the colors illuminated by the Sun, species exactly alike are flowing, diminished by the flow itself, until for whatever reason, they fall on an opaque medium, where they paint their source: and vision is produced, when the opaque screen of the eye is painted this way[46]

In any act of visual perception, light is a necessary mediator in communicating visual data. It is light that carries images, bouncing of "an opaque medium" and falling on an "opaque screen." If the screen happens to be the eye, "vision is produced," but there is nothing unique to the eye: any screen will do.[47] Even though there can be no doubt that Kepler is deeply indebted to the perspectivist tradition,[48] indebtedness he generously acknowledges by titling his book after Witelo, Kepler's transformation of optics is a fundamental.

The subject matter of traditional optics was human vision. Vision, so was its basic assumption, is a direct acquaintance of the visual faculty with visible objects, and optics is the study of the agents whose function is to communicate these objects to the eye.[49] This communication – the optical process – has always been self-evidently teleological. It was aimed at providing adequate images of visible objects for the intellect: "a species produced by a visible object has the essential property

[45]See Lindberg 1985, esp. 37–40; and Lindberg 1984, esp. 134–135, also Zik and Hon 2007, esp. 561.

[46]Kepler 1937– [1604], 41–42.

[47]Kepler 1937– [1604], 41–42.

[48]This is the central argument of Lindberg 1976.

[49]For the role of these visual impressions in medieval spirituality see: Park 1998, 254–271; Hamburger 2000, 47–69.

of manifesting the object of which it is the likeness" says Pecham.[50] Kepler was well aware of this: "Aristotle defines light," he writes, "not ... in its nature, but to the extent that it is characteristic of the process of vision."[51] The teleology survived throughout the Renaissance, for practical as well as theoretical optics: "Alberti's picture," Alpers points out, "begins not with the world seen, but with a viewer who is actively looking out at objects."[52] And indeed, summarizing scholastic optics for his audience of painters, humanists and art patrons, it is this teleology that Alberti chooses to stress:

> Philosophers ... Say that surfaces are measured by certain rays, ministers of vision as it is (*quasi visendi ministris*), which they therefore call visual rays, since by their agency the images of things are impressed upon the senses.[53]

The physical nature of the "ministers of vision" was debated since antiquity: simulacra or forms, visual rays or species, but their teleology and authenticity were never in doubt. Grosseteste, for instance, founds them on the premise that it is an essential property of the visible object itself, its agency or 'virtue', which 'multiplies' itself until it made itself present to the eye:

> A natural agent continuously multiplies its power from itself to the recipient, whether it acts on sense or on matter. This power is sometimes called species, sometimes a likeness, and it is the same thing whatever it may be called.[54]

Following Grosseteste's teachings, Roger Bacon underscores the essential relation which assures the fidelity of the visual agents, the multiplied species, to the visible object: "species is similar in essence and definition to the agent and the things generating it." [55] The authenticity of species was a fundamental assumption not only of optics but of medieval Aristotelianism as a whole; optics legitimated natural philosophy by accounting for the fundamental knowability of His Creation.[56] Visual rays guarantied the veracity of vision and the geometrical analysis of their propagation was always subordinate to the assumption of their intentionality and their consequent indubitability.[57] So was the analysis of the eye, as Pecham stresses: "vision takes place by the arrangement of the species on [the surface of] the glacial humor *exactly* as [the parts] of the object [are arranged] outside."[58] This is so, precisely because "*unless* this

---

[50]Pecham 1970, 161.

[51]Kepler 2000 [1604], 45. Cf. Smith 1981.

[52]Alpers 1983, 41.

[53]Alberti 1972, 41.

[54]Grosseteste 1912, 60.

[55]Bacon 1983, 7. "species sit similes agenti et generanti eam in essential et diffinitione." Bacon 1983, 7. For an extensive treatment of species in medieval optical theory see especially: Smith 1981; Spruit 1994. Tachau 1982 provides an authoritative treatment of the issues involved in medieval theory of species, and also Tachau 1988. See also Denery II 2005, esp 82–96.

[56]Smith 1981, 569.

[57]For the teleological nature of the Aristotelian theory of perception cf. Descartes 1998, 159–161.

[58]Pecham 1970, 121. Italics added.

were so, the eye would not see the object distinctly."[59] Optics, Pecham assumes, is a theory of visual perception, and any such theory that failed to account for the adequacy of the seen image is *ipso facto* false.

Kepler does away with this line of reasoning. The optical process, he declares, is strictly the effect of light: "genuine vision occurs when the folding door or pupil of the eye is exposed most closely to the arriving ray of light."[60] Gone are all intentional agents, and with them – the privileged import of the eye. Passively receiving "illumination" like any instrument, the eye is not merely comparable to "a closed chamber": the cornea is truly nothing but a lens; the retina nothing but a screen, essentially the same as the paper or the pavement; the pupil is just another aperture, "for the pupil takes the place of the window."[61]

This is Kepler's way of justifying the new observational astronomy. With light as the sole agent of all optical phenomena, there is no fundamental epistemological difficulty with observing the distant celestial phenomena: the mathematical nature of light and the assertion that the rays do not decay, only disperse (propositions 6 and 7 of *Ad Vitellionem*) turns distance into nothing but an element in the geometrical analysis of observation.[62] Even more important, with light there is no epistemological difficulty with *artificiosa observationes*. The image on the pavement is reversed and fuzzy, but so is the one on the retina. The instrument is trustworthy not because it does not interfere with the visual flow, but because it is no worse than the eye.[63]

5 Epistemological Considerations

Kepler was well aware of the primary rival to radical instrumentalism: the traditional, Aristotelian empiricism assumed by Galileo's Jesuit competitors, with its strict preference to the human sense,and he did not hesitate to challenge it directly and in detail.

According to Aristotle the initial state of cognition is perception; all that the intellect knows was previously in the senses.[64] This assumption, Kepler insists, does not agree well with astronomical practice; it does not seem that "the motions of the heaven come immediately into the perception of the eyes." The eye's physiology

[59]Pecham 1970, 121.

[60]Kepler 2000 [1604], 78.

[61]Kepler 2000 [1604], 184.

[62]For Kepler's mathematization of light cf. Gal and Chen-Morris 2005.

[63]We discuss in detail Kepler's transformation of optics from a teleological theory of human vision into a causal theory of the production of the images by light, as well as its far reaching epistemological ramifications, in Gal and Chen-Morris 2010 (in press).

[64]Aristotle 1984, *Posterior analytics*, II, 19, 100^{a}4–15, 165–166.

does not allow for such immediate perception: it is "attached to the posterior part of the head;" it has a spherical shape and it depends for its proper functioning on "the use of multiple refractions." There is no lack of evidence that the eye judges motion erroneously, especially motion of distant objects, and Kepler provides a handful.[65]

These shortcomings are unavoidable, Kepler claims, and in fact the eyes will fault even the most acute observer, such as Tycho Brahe. His own instrumental empiricism was much more radical in intellectual ambitions and commitments than that of his mentor in matters of astronomical observations, and Kepler did not shy away from the difference. Tycho utilized the *Camera Obscura* to observe solar eclipses, Kepler points out, under the assumption that the astronomer's eye remains the chief arbitrator and the instrument is but a prosthesis; a secondary aid to the sense of sight. Tycho thus "noticed that eclipses of the sun, whether the ray be allowed in through a notch or received by the eyes, always show the moon's diameter to be much less than it appears at oppositions." Trusting the accuracy of his visual perception, Tycho concluded that the real lunar diameter is smaller than was assumed. He was, however, wrong, because he did not take into account "the actual structure of the sense of sight" and its tendency to enlarge "the edges of luminous bodies, particularly in darkness." Kepler admonishes that the "astronomer should carefully take note of this … that unless he be endowed with the sharpest and most powerful sense of sight, he is not equal to measuring that moon's diameter at the full with the eyes without error." Therefore

> one has to distinguish carefully … between those things that happen to the sense of sight and those that happen when the consideration of the sense of sight is removed. For those things that happen to the sense of sight vary by individual cases, but those things that really happen are uniform within a single horizon. … Astronomers will now take note of this: that one must not trust the sense of sight… it cannot therefore be argued from this accident of the sense of sight to what happens outside the consideration of the sense of sight … For astronomers should not present anything other than those things that in actual fact occur. The sense of vision, however, we leave to the physicians to remedy.[66]

The remedy of astronomical observation and measurements is relegating the eyes in favor of an instrument. The careful application of the *Camera Obscura* allows "a most certain procedure for measuring the quantities of eclipses … If this device be correctly applied, the diameter of the moon appears decidedly greater than the amount that Tycho's table shows."[67]

In spite of his insistence that the application of instruments to astronomical investigation was the only way to avoid "the inadequacies of the eyes,"[68] Kepler was not promoting skepticism about the human senses. The purpose of his optics is to "subdue the hostile fortress of doubt,"[69] not to reinforce it, and much of

[65] Kepler 2000 [1604], 336.
[66] Kepler 2000 [1604], 298.
[67] Kepler 2000 [1604], 298.
[68] Kepler 2000 [1604], 57.
[69] Kepler 1937- [1604], 2:6.

Ad Vitellionem is dedicated to accounting for the reliability of the retinal image. Galileo, unperturbed by Kepler's careful optical and epistemological deliberations, is significantly more extreme in his stand for the instrument: *the eye* mediates and distorts; the instrument provides the standard of trustworthy perception against which the eye is to be judged. In a sense, Galileo re-introduces the distinction between direct and mediated vision that Kepler labored to abolish, reversing the epistemological order between eye and instrument. The two ways of observation, he argues during the controversy over the comets, provide data of entirely different value, not to be conflated or compared, and his adversary Grassi is fundamentally wrong to submit the telescope to the same analysis as ocular vision:

> Your error lies in comparing the star taken together with its irradiation when seen with the naked eye to the body of the star alone when seen with the telescope and distinguished from the irradiated regions.[70]

If Kepler was keen to hold the skeptical ramifications of his optics at bay, Galileo is unhesitant: his endorsement of the instrument comes at the expense and with the explicit distrust in the eye. The human organ is not merely weak but a positive source of various deceptions, which he makes a point to enumerate:

> There is another illumination here, made by refraction in the moist surface of the eye, and by this, the real object appears to us to be surrounded by a luminous circle ... there is a third vivid splendor here, almost as bright as that of the original light itself; this is produced by reflection of the primary rays in the moisture at the edges of the eyelids, and it extends over the convexity of the pupil ... this radiant crown [is] a sensation of the eye ... it does not depend upon the illumination of the surrounding area.[71]

Like Kepler, Galileo thrusts the eye into the outside world. From a veridical conduit of knowledge it becomes part of a causal process of material nature, producing phenomena to be studied and explained physically. And while the eye mediates, adds spurious and distorting brilliance, the telescope is not only a reliable source, but the standard against which to judge the observation made through the eye and the means by which to remove the errors it introduces:

> Fancy to yourself some definite size for [a] wig, and in the center of this imagine a very tiny luminous body. The shape of this will be lost, being crowned by excessively long hair ... the telescope, by enlarging the star but not the wig, makes the tiny disc which originally was imperceptible ... so that its shape may be well distinguished. [72]

[70] Galilei 1623, 326.

[71] Galilei 1623, 319–320.

[72] Galilei 1623, 322–323. "Figuratevi una determinate grandezza d'una capellatura; nel mezo della quale se voi intenderete essere un piccolissimo corpo luminoso, perderà la sua figura, coronato di troppo lunghi crini ... il telescopio, accrescendo la stella ma non la chioma, fa che, dove prima il piccolissimo disco tra sì ampio fulgore era impercettibile ... si può distinguere ed assai ben figurare." Galilei 1890–1909; 6 360–361.

6 The Eye of the Mind

The metaphor of the "wig" – this paradigm of artificiality – to denote the eye-added splendor stresses Galileo's deliberate inversion of natural and artificial, direct and mediated. Galileo encapsulates his argument for this inversion of epistemological roles and standing in his most famous phrase of the "Assayer:"

> Philosophy is written in this grand book – I mean the universe … in the language of mathematics.[73]

The eye had always been, and for Grassi, Scheiner, and Agvilonius still was, the divinely assigned instrument of visual knowledge. This is nicely and simply put by Scheiner in his *Oculus* published in 1619, with no immediate reference to any of the controversies: "in order to see, the eye of the animal fulfils the duty it was ordained by God, grasping the presence of visible things."[74] For Kepler and Galileo, the eye loses this independent "duty" and becomes part of the "things." It introduces error because it is immersed in the confusing nature to be observed and its passions and affects are causally-bound physical phenomena.

The telescope, on the other hand, is not bound to the physical world. It is mathematical *in essence*, argues Galileo; it is fully captured by the mathematical laws governing the shape and relative placing of its lenses; "the convex lens unites the rays, the concave glass expands them and forms an inverted cone." [75] The asymptotic diminution of the angle of vision to the eye, with which Grassi accounted for the apparent lack of magnification, is thus of no significance. Magnification is strictly a mathematical relation, and the telescope always magnifies, whether the eye is capable of perceiving it or not. In a pedantic mood, Galileo even insists that changing the mutual position of the lenses results in having a completely different instrument:

> [Sarsi] says that a telescope which is now long and now short may be called the same instrument though differently applied … Our case is just the reverse, for the use of the telescope is always the same … while the instrument itself is diversified by its alteration in one essential respect, which is the interval between its lenses.[76]

---

[73]Galilei 1623, 183–184.

[74]Scheiner 1619, 2.

[75]Galilei 1623, 209. Antoni Malet writes: "In our understanding of them, telescopes always work by producing geometrical optical images, real or virtual, regardless of whether or not any observer is peering through them. From our theoretical point of view, it does not matter whether an eye, or a screen, or just empty space gets the light rays coming out of the ocular lens, because the telescope *always* produces one geometrical image. However, in Kepler's time, and up to the last decades of the seventeenth century, when somebody looked through a telescope, it was not understood to work by producing images similar to the pictures projected upon screens" (Malet 2005, 239). We differ from Malet in arguing that in spite of the absence of theoretical grounding, the independence of the geometrical image from the observer is exactly the position Galileo formulates and defends.

[76]Galileo 1623, 225.

This pedantry is not merely a rhetorical maneuver. As Sven Dupré has recently shown against what has been long assumed, Galileo did develop a mathematical understanding of the telescope.[77] This was based on contemporary optics, which owed as much to new lens and mirror grinding techniques as to traditional perspective theory, and it was not informed by Kepler's innovations, but it provided Galileo with the confidence to insist on the mathematical nature of his instrument. Being thoroughly mathematical, the telescope is not an extension of the eye but of reason. As was beautifully put by his fellow Lyncean Johann Faber, Galileo,

> with marvellous skill so fit spectacles to an aging world that with mind still sound but eyes dimmed and body weakened it might see through two glasses. [78]

In an important sense, we saw, Faber was missing Galileo's point: the telescope was *not* a pair of spectacles. But he did capture the dream of an instrument of mind, superior to the eye and answering directly to the laws of reason, which a few years earlier Kepler put into more careful, less exhilarating prose:

> Certainly, the mind itself, if it never had the use of an eye at all, would demand an eye for itself for the comprehension of things outside it, and would lay down laws of its structure which were drawn from itself (if in fact it were pure and sound and without hindrance, that is, if it were only what it is).[79]

In Kepler's terms, Galileo could have said that the telescope is the sense organ that reason *would have* had. For this reason he finds it is very important to stress that in contrast to its Dutch predecessor, his telescope was "discovered by the way of reason." "The original contraption was accidentally discovered by a "simple maker of ordinary spectacles" (he does not honor Hans Lipperhey by name) who "in casually handling pieces of glass of various sorts happened to look through two at once, one convex and the other concave, and placed at different distances from the eye." His instrument, on the other hand, followed a "reasoning" which he cursorily recounts, allegedly to "render less incredulous those people who, like Sarsi, may wish to diminish whatever praise there is in it that belongs to me."[80] One can easily identify with Grassi's astonishment at the ingratitude of the one he dubbed "the Lynx," but it was less important for Galileo to gather supporters than to clarify that his instrument was no pair of spectacles (and he, of course, no "simple maker"). It does not assist the eye; it is an extension of reason, an embodied mathematical entity, and it can allow reason an unmediated approach to reality because *reality* is mathematical, "written in this grand book … the universe … in the language of mathematics."

[77]Dupré 2005. See also Zik and Van Helden 2003.

[78]Faber in Galileo 1623, 154. "[An], velut in vetulo languentes corpore ocelli, Mente tamen valida, per duo vitra vident, Forte senescenti tu sic OCULARIA mundo Aptasti, mirae dexteritatis opus?" Faber in Galilei 1890–1909, 6: 205.

[79]Kepler 1937– [1619], 6: 304.

[80]Galilei 1623, 212–213.

This is the import of the celebrated idiom on the mathematical language of nature: to re-construct observed reality so it can be approached by reason, through the instrument. It is not an ontological justification for mathematical theorizing but for radical instrumental empiricism; nature is written in a language legible *only* through the instrument. The telescope does not mediate – it reveals the real makeup of nature; shapes, figures, quantities – directly to Reason. *The senses* mediate, creating appearances which are not proper representations of the "external bodies:"

> I do not believe that for exciting in us tastes, odors and sounds there are required in external bodies anything but sizes, shapes, numbers, and slow or fast movements; and I think that if ears, tongues, and noses were taken away, shapes and numbers and motions would remain, but not odors or tastes or sounds. These, I believe, are nothing but names, apart from the living animal just as tickling and titillation are nothing but names when armpits and the skin around the nose are upset.[81]

"Sizes, shapes, numbers, and slow or fast movements;" nature is comprised of elements the instrument makes apparent, but the senses mask by with "tastes, odors and sound … tickling and titillation." And what is true for noses and armpits is just as true for the eye:

> I believe that vision, the sense which is eminent above all others, is related to light, but in that ratio of excellence which exists between the finite and the infinite, the temporal and the instantaneous, the quantity and the indivisible; between darkness and light.[82]

7 Conclusion: The Price

Some 50 years later, Robert Hooke, the seventeenth century's most definitive beneficiary from and prominent follower of the legacy of radical instrumental empiricism, would give the conquest of the artificial over the natural a religious aura. In terms of "observations" and the capacity to "behold the works of nature" through the senses, Mankind is essentially inferior to "Beasts:"

> As for the actions of our *Senses*, we cannot but observe them to be in many particulars much outdone by those of other Creatures, and when at best, to be far short of the perfection they seem capable of.[83]

This 'shortness of perfection" is our own doing, but deliverance is also within our grasp:

> By the addition of such *artificial Instruments* and *methods*, there may be, in some manner, a reparation made for the mischiefs, and imperfection, mankind has drawn upon it self, by negligence, and intemperance, and a wilful and superstitious deserting the Prescripts and Rules of Nature, whereby every man, both from a deriv'd corruption, innate and born with

[81]Galilei 1623, 311.

[82]Galilei 1623, 311–312.

[83]Hooke 1665, xvii–xviii.

> him, and from his breeding and converse with men, is very subject to slip into all sorts of errors.[84]

Hooke may be overstated, but he is completely sincere. It is appropriate to discuss instruments in terms of fall and redemption, "the only way which now remains for us to recover some degree of those former perfections" (Hooke 1665, xvii), because they represent the one divine advantage that humans enjoy over beasts:

> It is the great prerogative of Mankind above other Creatures, that we are not only able to *behold* the works of Nature, or barely to *sustein* our lives by them, but we have also the power of *considering, comparing, altering, assisting, and improving* them to various uses. And as this is the peculiar priviledge of humane Nature in general, so is it capable of being so far advanced by the helps of Art, and Experience, as to make some Men excel others in their Observations, and Deductions, almost as much as they do Beasts.[85]

For Hooke, the arguments that have turned proper observation to the work of the mind rather than the senses have become almost a commonplace. He ignores, or suppresses, the great tensions and anxieties with which these arguments came. Such indifference was not an option for Galileo and Kepler. Hooke can marvel at "Telescopes or Microscopes producing new Worlds and Terra-Incognita's to our view" (Hooke 1665, xxxii), but Galileo, anxious to defend the mathematical rapport by which he legitimized the inversion of epistemological standings, finds himself defending the simplicity of its revelation. Gone are the marvels and wonders of the *Sidereus Nuncius*; radical instrumentalism requires the bare bones representation of the heavens as the "triangles, circles and other geometrical figures" of Fig. 3. And Hooke can simply complain that "the eye cannot distinguish a smaller object then [*sic.*] appears within the angle of half a minute,"[86] but Kepler has to admit that vision, as a whole, has become a complete mystery:

> How this image or picture is joined together with the visual spirits that reside in the retina and in the nerve, and whether it is arraigned within by the spirits into the caverns of the cerebrum to the tribunal of the soul or of the visual faculty; whether the visual faculty, like a magistrate, given by the soul, descending from the headquarters of the cerebrum outside to the visual nerve itself and the retina, as to lower courts, might go forth to meet this image – this, I say, I leave to the natural philosophers to argue about.[87]

The mediation of light justified radical instrumentalism. It supported and explained the marvellous achievements of the *camera obscura*, the telescope and later the microscope, and provided a most convincing account of the function of the eye itself. But it came at a most difficult price, a bewilderment that would haunt the New Science: how is it that we see at all?

---

[84] Hooke 1665, xvii.

[85] Hooke 1665, xvii.

[86] Hooke 1674, 8.

[87] Kepler 2000 [1604], 180.

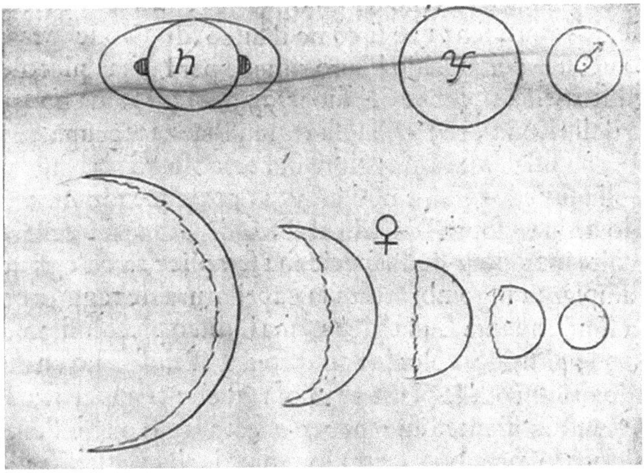

Fig. 3 The planets according to Galileo's "Assayer"

References

Alberti, Leon Battista. 1972. *On Painting and on Sculpture: The Latin Texts of De Pictura and De Statua*, trans and eds, introduction and notes by C. Grayson. London: Phaidon.

Alpers, Svetlana. 1983. *The Art of Describing: Dutch Art in the Seventeenth Century*. Chicago: University of Chicago Press.

Aristotle. 1984. Problems; posterior analytics. In *The Complete Works of Aristotle*, ed. Jonathan Barnes. Princeton: Princeton University Press.

Agvilonius, Franciscus. 1613. *Opticorum libri sex philosophis iuxta ac mathematicis utiles*. Antwerp: Plantin Press, widow and sons of J. Moretus.

Bacon, Roger. 1983. Roger Bacon's natural philosophy: a critical edition. Incl. *De multiplicatione specierum* and *De speculis comburentibus*, trans. and ed. David C. Lindberg. Oxford: Oxford University Press.

Baldini, Ugo. 1992. *Legem impone subactis: Studi su filosofia e scienza dei gesuiti in Italia, 1540–1632*. Rome: Bulzon.

Biagioli, Mario. 2006. *Galileo's Instruments of Credit: Telescopes, Images, Secrecy*. Chicago: The University of Chicago Press.

Blackwell, Richard. 1992. *Galileo, Bellarmine, and the Bible*. Notre Dame: University of Notre Dame Press.

Cassirer, Ernst. 1942. The Influence of language upon the development of scientific thought. *The Journal of Philosophy* 39,12: 309–327.

Crombie, Alistair C. 1953. *Robert Grosseteste and the Origins of Experimental Science 1100–1700*. Oxford: Clarendon.

Dear, Peter. 1995 *Discipline & Experience: the Mathematical Way in the Scientific Revolution*. Chicago: University of Chicago Press.

Denery II, Dallas G. 2005. *Seeing and Being Seen in the Later Medieval World: Optics, Theology and Religious Life*. Cambridge: Cambridge University Press.

Descartes, René. 1998. *The World and Other Writing*, ed. and trans. Stephen Gaukroger. Cambridge: Cambridge University Press.

Drake, Stillman and O'Malley, C.D. (eds. and trans.). 1960. *The Controversy on the Comets of 1618: Galileo Galilei, Horatio Grassi, Mario Guiducci, Johann Kepler.* Philadelphia: University of Pennsylvania Press.

Dupré, Sven. 2003. Galileo's telescope and celestial light. *Journal for the History of Astronomy* 34, 4: 369–399.
Dupré, Sven. 2005. Ausonio's mirrors and Galileo's lenses: the telescope and sixteenth-century practical optical knowledge. *Galilaeana: Journal of Galilean Studies* 2: 145–180.
Feldhay, Rivka. 2000. Mathematical entities in scientific discourse: Paulus Gulding and his *Dissertatio de Motu Terrae*. In *Biographies of Scientific Objects*, ed. Lorraine Daston. Chicago: University of Chicago Press.
Gal, Ofer and Chen-Morris, Raz. 2005. The archaeology of the inverse square law part I: Metaphysical images and mathematical practices. *History of Science* 43.4: 391–414.
Gal, Ofer and Chen-Morris, Raz. 2010 (in press). Baroque optics and the disappearance of the observer: From Kepler's optics to Descartes' doubt. *Journal of the History of Ideas*.
Galilei, Galileo. 1623. *Il Saggiatore (The Assayer)*. Rome: Giacomo Mascardi. In *The Controversy on the comets of 1618: Galileo Galilei, Horatio Grassi, Mario Guiducci, Johann Kepler*, Drake and O'Malley, eds. 1960. And trans. Stillman Drake and C.D. O'Malley. Philadelphia: University of Pennsylvania Press.
Galilei, Galileo. 1890–1909. *Le Opere di Galileo Galilei* (20 vols) 1890–1909, ed. Favaro, A. Florence: Barbera.
Galilei, Galileo. 1989 (1610). *Sidereus Nuncius, or, The Sidereal Messenger*, ed. and trans. Albert Van Helden. Chicago: University of Chicago Press.
Grassi, Horatio. 1619. *Tribus Cometis Anni M. DC. XVIII. Disputatio Astronomica*. Rome: Iacobi Mascardi. In *The Controversy on the comets of 1618: Galileo Galilei, Horatio Grassi, Mario Guiducci, Johann Kepler*, Drake and O'Malley, eds. 1960. And trans. Stillman Drake and C.D. O'Malley. Philadelphia: University of Pennsylvania Press.
Grosseteste, Robert. 1912. De lineis angulis et figuris. In *Die Philosophischen Werke des Robert Grosseteste, Bischofs von Lincoln*, ed. Ludwig Baur. Beiträge zur Geschichte der Philosophie des Mittelalters. Münster i. W.: Aschendorff.
Guiducci, Mario. 1619. *Discorso Delle Comete*. Firenze: Pietro Cecconcelli. In *The Controversy on the comets of 1618: Galileo Galilei, Horatio Grassi, Mario Guiducci, Johann Kepler*, 1960, eds. and trans. Stillman Drake and C.D. O'Malley. Philadelphia: University of Pennsylvania Press.
Hamburger, Jeffrey F. 2000. Seeing and believing: the suspicion of sight and the authentication of vision in late medieval art. In *Imagination und Wirklichkeit: Zum Verhältnis von mentalen und realen Bildern in der Kunst der Frühen Neuzeit*, eds. Alessandro Nova and Klaus Krüger, 47–69. Mainz: Von Zabern.
Hooke, Robert. 1665. *Micrographia*. London: John Martin.
Hooke, Robert. 1674. *Animadversions on the ...* Machina Coelestis *of ... Johannes Hevelius*. London: John Martin.
Kepler, Johannes. 1937. *Gesammelte Werke* 1571–1630, eds. Walther von Dyck and Max Caspar München: C. H. Beck.
Kepler, Johannes. 1965 [1610]. *Kepler's Conversation with Galileo's Sidereal Messenger*, trans. Edward Rosen. New York: Johnson Reprint Corp.
Kepler, Johannes. 2000 [1604]. *Optics: Paralipomena to Witelo and the Optical Part of Astronomy*, trans. William H. Donahue. Santa Fe, NM: Green Lion Press.
Lattis, James. 1994. *Between Copernicus and Galileo*. Chicago: University of Chicago Press.
Lindberg, David C. 1976. *Theories of Vision from Al-Kindi to Kepler*. Chicago: University of Chicago Press.
Lindberg, David C. 1968. The theory of pinhole images from antiquity to the thirteenth century. *Archive for the History of Exact Sciences* 5: 154–176.
Lindberg, David C. 1969. The theory of pinhole images in the fourteenth century. *Archive for History of Exact Sciences* 6: 299–328.
Lindberg, David C. 1984. Optics in 16th century Italy. In *Novità Celsti e Crisi del Sapere*, ed. P. Galuzzi, 131–148. Firenze: Giunti Barbera.

Lindberg, David C. 1985. Laying the foundations of geometrical optics: Maurolico, Kepler, and the Medieval Tradition. In *The Discourse of Light from the Middle Ages to the Enlightenment.* Los Angeles: William Andrews Clark Memorial Library, UCLA 1–65.

Malet, Antoni. 2005. Early Conceptualizations of the telescope as an optical instrument. *Early Science and Medicine* 10, 2: 237–262.

Park, Katherine. 1998. Impressed images: reproducing wonders. In *Picturing Science and Producing Art*, eds. Caroline A. Jones and Peter Galison with Amy Slaton, 254–271. New York and London: Routledge.

Pecham, John. 1970. *John Pecham and the Science of Optics: Perspectiva Communis.* trans. and ed. David Lindberg. Madison: University of Wisconsin Press.

Sarsi Sigensano, Lothario. 1619. *Libra Astronomica ac Philosophica.* Perugia: Marci Naccarini. In *The Controversy on the comets of 1618: Galileo Galilei, Horatio Grassi, Mario Guiducci, Johann Kepler*, 1960, eds. and trans. Stillman Drake and C.D. O'Malley. Philadelphia: University of Pennsylvania Press.

Scheiner, Christoph. 1619. Reprinted 1652. *Oculus, hoc est, fundamentum opticum.* Innsbruck, London.

Shea, William R. 1972. *Galileo's Intellectual Revolution.* London: Macmillan. Sigensano, Lothario Sarsio. 1619. *Libra astronomica ac philosophica qva Galilaei Galilaei opiniones de cometis ... examinantur.* Perusiae: Marci Naccarini.

Smith, A. Mark. 1981. Getting the big picture in perspectivist optics. *Isis* 72, 568–589.

Spruit, Leen. 1994. *Species Intelligibilis: From Perception to Knowledge* (2 vols.) Leiden, New York, Köln: E.J. Brill.

Straker, Stephen. 1971. *Kepler's Optics.* Unpublished Dissertation. Indiana University.

Tachau, Katherine. 1982. The problem of the species in Medio at Oxford in the generation after Ockham. *Medieval Studies* 44, 394–443.

Tachau, Katherine. 1988. *Vision and Certitude in the Age of Ockham: Optics, Epistemology and the Foundations of Semantics, 1250–1345.* Leiden, New York, Köln: E. J. Brill.

Thro, E. Broydrick. 1996. Leonardo's early work on the pinhole camera: the astronomical heritage of Levi ben Gerson. *Achademia Leonardi Vinci* 9: 20–54.

Van Helden, Albert. 1974. The telescope in the seventeenth century. *Isis* 65.1.

Van Nouhuys, Tabitta. 1998. *The Age of Two-Faced Janus.* Leiden: Brill.

Zik, Yaakov and Giora Hon. 2007. Geometry of light and shadow: Francesco Maurolyco (1494–1575) and the pinhole Camera. In *Annals of Science* 64: 4, 549–578.

Zik, Yaakov and Albert van Helden. 2003. "Between Discovery and Disclosure." In Beretta, Marco et al., *Musa Musaei: Studies on Scientific Instruments and Collections in Honour of Mara Miniati.* Firenze: L. S. Olschki, 173–190.

Mastering the Appetites of Matter. Francis Bacon's *Sylva Sylvarum*

Guido Giglioni*

Quindi si va lustrando la selva de le cose naturali dove son tanti oggetti sotto l'ombra e manto, e come in spessa, densa e deserta solitudine la verità suol aver gli antri e cavernosi ricetti; fatti intessuti de spine, conchiusi de boscose, ruvide e frondose piante: dove con le raggioni più degne et eccellenti maggiormente s'asconde, s'avvela e si profonda con diligenza maggiore, come noi sogliamo gli tesori più grandi celare con maggior diligenza e cura, acciocché dalla moltitudine e varietà de cacciatori (de quali altri son più exquisiti et exercitati, altri meno) non vegna senza gran fatica discuoperta.

Giordano Bruno, *Degli eroici furori*

Abstract Francis Bacon's *Sylva Sylvarum* (published posthumously in 1627) occupies a paradoxical place in the history of seventeenth-century medicine and natural philosophy. It is the work where Bacon expounded, at his clearest and best, in vernacular and not in Latin, his views on the material appetites of nature, and did so not by writing in the abstract, but by describing and performing experiments aimed at disclosing the appetitive nature of matter. However, such an original model of experimental investigations on the appetites of matter was abandoned by the great majority of Bacon's followers, especially those associated with the Royal Society, replaced with the more reassuring project to mechanise the natural forms and passions of matter. By doing so, man was restored as the proper subject of knowledge and appetite, whereas nature was left with its status of lifeless object of dispassionate study. In this paper I explore the theoretical and experimental strategies deployed by Bacon to investigate the appetites of matter. It will become apparent that a characteristic hermeneutical

* I am indebted to Julian Smith-Newman for his comments and stylistic improvements. Translations from Bacon's Latin works are mine.

G. Giglioni
Warburg Institute, University of London

C.T. Wolfe and O. Gal (eds.), *The Body as Object and Instrument of Knowledge: Embodied Empiricism in Early Modern Science*, Studies in History and Philosophy of Science 25, DOI 10.1007/978-90-481-3686-5_8, © Springer Science+Business Media B.V. 2010

circle underlies Bacon's natural philosophy, a circle that, depending on the chosen point of view, could be regarded at the time as either virtuous or vicious. On the one hand, Bacon was convinced that man's self-knowledge rested on the knowledge that nature has of itself, since nature is first and foremost appetite and man's essence is rooted in appetite. On the other hand, he was also convinced that knowledge of nature was based on knowledge of the self, since the best accounts concerning the nature of the appetites were to be found in the works of poets and historians (rather than in Renaissance systematisers of natural magic and natural philosophy). This is what Bacon meant by 'georgics of the mind': the understanding of the material appetites of nature cannot be separated from an ethical and political consideration of the mechanisms mediating knowledge and appetite in human societies.

1 Introduction

Francis Bacon's *Sylva Sylvarum* (published posthumously in 1627) occupies a paradoxical place in the history of seventeenth-century medicine and natural philosophy. Here we have a work in which Bacon expounded, unambiguously, in vernacular and not in Latin, his characteristic views on the material appetites of nature, and did so not by writing in the abstract, but by describing and performing experiments that aimed specifically to disclose the appetitive nature of matter. Despite his clear adherence to a theory of natural appetites, though, the great majority of natural philosophers who pledged their allegiance to Bacon's philosophy – especially those whose names are often associated with the birth of the Royal Society – decided very early on to ignore Bacon's original injunction to provide experimental descriptions of material appetites. Instead, they concentrated on the project of providing mechanical interpretations of material forms. By highlighting one part of Bacon's thought and disregarding another, post-Baconian experimental philosophers reinforced the idea of the human being as the sole subject capable of knowing and desiring while demoting nature to the status of a lifeless object of dispassionate study.

In this paper I will take *Sylva Sylvarum* as my principal focus of analysis and I will explore the theoretical and experimental strategies which Bacon deployed to investigate the appetites of matter. It is my contention that a characteristic hermeneutical circle underlies Bacon's natural philosophy, a circle that in his day could be seen as either virtuous or vicious depending on the chosen point of view. Bacon believed that human self-knowledge could not and should not sever its ties with nature's self-knowledge, that is, with the very knowledge that nature has of itself, for nature, in Bacon's view, was first and foremost appetite (or a form of unsentient perception driven by appetite)[1] and man's essence was rooted in appetite. At the same time, he

[1] Bacon 1857–1874, II, 602: 'It is certain that all bodies whatsoever, though they have no sense, yet they have perception: for when one body is applied to another, there is a kind of election to embrace that which is agreeable, and to exclude or expel that which is ingrate: and whether the body be alterant or altered, evermore a perception precedeth operation; for else all bodies would be alike one to another. And sometimes this perception, in some kind of bodies, is far more subtle than the sense; so that the sense is but a dull thing in comparison of it'. See also Bacon 1857–1874, I, 610.

also believed that the very knowledge which nature has of itself, despite its level of obscurity, implied a kind of primal and elementary self, a supple, quivering and yet stable subject, capable of accounting for all sorts of natural actions, from the formation of a bubble out of soap to the passions of a vainglorious man. Since the best accounts of the nature of the appetites were to be found in the works of poets and historians (rather than in Renaissance systematisers of natural magic and natural philosophy), a historical and political understanding of desires could find fruitful applications in the study of the appetitive life of nature. Bacon referred to this peculiar hermeneutic situation as the "georgics of the mind," meaning that the understanding of the material appetites of nature could not be separated from an ethical and political consideration of the mechanisms that mediate knowledge and appetite in human societies.

Three basic interpretative principles concerning Bacon's philosophy inspire my reading of *Sylva Sylvarum* and the project of experimental physics contained in it. First, Bacon's philosophy rests on a fully-fledged metaphysics of matter. This metaphysics contains a bold assertion of realism and materialism in a period when both philosophy and science were increasingly turning to forms of phenomenalism and idealism. Second, the notion of appetite is at the very centre of Bacon's metaphysics of matter. In contrast to the assumption that appetite is always driven by some degree of knowledge, obscure as it may be – an assumption that has prevailed in the history of philosophy – Bacon considers appetite as a blind and unrestrainable drive. Finally, Bacon is not the father of any of the following cultural movements: mechanical philosophy, empiricism and experimental science. Or, put in a less dogmatic way: Bacon is certainly not the father of mechanical philosophy. He might be seen as the putative father of empiricism (if by empiricism we mean a historiographical device useful for outlining tidy lineages in early modern thought) and might be considered the adoptive father of experimental science (if by experimental science we mean a programme of conducting experiments on the primordial appetites of matter – a statement with which many scientists and historians of science will certainly feel uncomfortable, and understandably so).

2 A Physics of Material Appetites

Sylva Sylvarum contains as it were Bacon's experimental physics. In it, to use Bacon's own words, he joined "the contemplative and active part together," the "materiate" and the "immateriate."[2] The book combines a view of method as a rigorous discipline of the mind with a positive consideration of manual and material experience. "I practise as I do advise," he tells his readers who are asked to follow his experiments; "which is, after long inquiry of things immerse in matter, to interpose some subject which is immateriate, or less materiate; such as this of sounds; to the end that the intellect may be rectified, and become not partial."[3] The very

[2]Bacon 1857–1874, II, 385.

[3]Bacon 1857–1874, II, 390.

word *sylva*, with all its rhetorical and metaphorical accretions, indicates a situation of analytical drifts and digressions, intractable farragoes and tangles of disparate topics – the "vast wood of experience," the "undigested heap of particulars."[4] In the *Sylva Sylvarum* Bacon also joined the high with the low of experience. In the preface to the reader, William Rawley, Bacon's chaplain and editor of many of his posthumous works, including the *Sylva*, reminded the reader that "his lordship" was fully aware of the "vulgar," "trivial," "mean," "sordid," "curious" and "fruitless" nature of the experiments he had collected and examined. In a programmatic way, Bacon declared his lack of interest in any form of conspicuous consumption of wonders. On the contrary, he undertook the project with the specific purpose of rendering the very notion of wonder banal. (Significantly, in the *Advancement of Learning*, Bacon had called wonder a kind of "broken knowledge," whereas in *Sylva Sylvarum*, he described the passion of wonder as a form of unilateral, unproductive focus, "the fixing of the mind upon one object of cogitation.")[5]

Bacon's main aim in his experimental study of nature was therefore to provide a description of the basic appetites of matter at work in the most common operations of nature. In his preface, Rawley reported that his lordship was particularly concerned that his readers would not understand "the difference between this Natural History and others":

> For those Natural Histories which are extant, being gathered for delight and use, are full of pleasant descriptions and pictures, and affect and seek after admiration, rarities, and secrets. But contrariwise, the scope which his lordship intendeth, is to write such a Natural History as may be fundamental to the erecting and building of a true philosophy; for the illumination of the understanding, the extracting of axioms, and the producing of nobles works and effects.[6]

Bacon – Rawley added in his account – saw his task in the realm of knowledge as that of an architect who, in planning his building, did not disdain "to dig the clay and burn the brick," to work as a "workman" and a "labourer."[7] He justified "the baseness of many of the experiments" as perfectly legitimate instances of "God's works," and vindicated their "vulgareness" by claiming that "true axioms must be drawn from plain experience and not from doubtful." Again: Bacon's "course" was "to make wonders plain, and not plain things wonders." Finally, "his lordship" intimated that experience needed to "be broken and grinded, and not whole, or as it groweth."[8] In other words, Bacon intended the experimental practice to be *prosaic*, *plain* and *parcelled*: its course should not be hampered by its material and undignified aspects, but it should concentrate on precisely such ordinary phenomena without its practitioner being mesmerised by isolated and rare occurrences. Finally, holistic

---

[4]Bacon 1857–1874, II, 385, 335, 337.

[5]Bacon 2000, 8; Bacon 1857–1874, II, 385, 570.

[6]Bacon 1857–1874, II, 335–336.

[7]Bacon 1857–1874, II, 336. And he continues: 'more than that, (according to the hard condition of the Israelites at the latter end) to gather the straw and stubble over all the fields to burn the brick withal'.

[8]Ibid.

and structural patterns of experience and dynamic processes should be reduced to their constituent, static elements.

For Bacon, then, the most 'sublime' works of both nature and art were in fact ordinary expressions of material cupidity. This is one of the fundamental reasons behind Bacon's decision in *Sylva Sylvarum* to focus on the cognitive significance of seemingly 'base' phenomena of nature. "The eye of the understanding," he claimed, "is like the eye of the sense; for as you may see great objects through small crannies or levels, so you may see great axioms of nature through small and contemptible instances."[9] And it is true that the experiments which Bacon performed or suggested in *Sylva Sylvarum* are extremely simple, if considered from the point of view of technical apparatuses or observational protocols. Their complexity belongs rather to the level of analytical scrutiny, the level that Bacon characterised as *interpretatio naturae*. For the prosaic, plain and parcelled character of Bacon's experimental inquiry depended on the very object of analysis, i.e., material cupidity. Appetites are by definition of a low and yet powerful nature. It is not by accident, therefore, that Bacon's observations and experiments often combined, in a strikingly vivid manner, heightened attention to the most prosaic details of material experience with a special focus on the intensity of material desires:

> houswives do find a difference in waters, for the bearing or not bearing of soap: and it is likely that the more fat water will bear soap best; for the hungry water doth kill the unctuous nature of the soap.[10]

Here there is no metaphorical transfer. Water's hunger is real, as it is the soap's desire to adhere to other bodies.

Bacon's metaphysics of matter – which is the rationale behind his "sylvan" model of experimental practice – is therefore a metaphysics of appetites. A cluster of interrelated concepts that are notoriously difficult to define (forms, simple natures, schematisms and natural motions) forms the backbone of his philosophical system. At the risk of simplifying a very complex subject, we might say that natural bodies are defined by forms, that forms are structural patterns determined by natural motions, and that natural motions result from the basic appetites of matter. Bodies, therefore, might be defined as crystallizations of natural appetites.[11] Conflicting tendencies deriving from conflicting appetites manage to coexist in each single body; indeed, such tendencies form bodies precisely as a result of the tension created by their antagonistic desires. Solid, fluid and aerial states of matter depend on the intensity of matter's appetites. The appetite of continuity, for instance, is weak in fluid bodies and stronger in solid ones.[12] The *Novum organum*

---

[9]Bacon 1857–1874, II, 377.

[10]Bacon 1857–1874, II, 471–472.

[11]Bacon 2004, 118–120, 382. See Bacon 2004, 278: "Cum enim omne corpus suscipiat multas naturarum formas copulatas et in concreto, fit ut alia aliam retundat, deprimat, frangat, et liget; unde obscurantur formae singulae."

[12]Bacon 1857–1874, III, 26: "constat appetitum continuitatis etiam liquidis inesse, sed debilem. At contra in rebus solidis viget, et motui naturali sive gravitati praedominatur."

(1620) records nineteen structural motions: motion of resistance, connection, liberty, hyle, continuity, of profit and want (*lucrum et indigentia*), of greater and lesser congregation, magnetic motion, motion of flight, of assimilation, of stimulation, of impression, of configuration, of passing through (*pertransitio*), royal or political, of rotation, of trepidation, of rest (*decubitus*).[13] Despite their large number, the first three and the last two seem to be the most important motions: matter is riddled with tendencies to resistance, liberty and union; as a result, matter finds itself in a constant state of agitation, always aiming at restoring the previous condition of balance. All the natural, basic motions enumerated in the *Novum organum* are manifestations of underlying primordial appetites. They are of a limited number and form the letters of the so-called alphabet of nature. These primordial appetites are what Bacon also calls "simple natures."

In *Sylva Sylvarum*, the "centuries" of experiments and observations aimed at disclosing the inner life of matter lay bare series of binary structures resulting from the ultimate contrarieties of nature. Motions of desire are primordial, caused by original tensions between opposing forces. Corruption and generation, for instance, are "nature's two terms or boundaries" and "the guides to life and death."[14] The basic oppositions of matter crystallise into what Bacon calls the "first congregations" of matter. They represent "two great families of things," identified by complementary names: "sulphureous" and "mercurial," "inflammable" and "not inflammable," "mature" and "crude," "oily" and "watery." These pairs, Bacon continues, "though they be unlike in the primitive differences of matter," nonetheless they "seem to have many consents." The study of these original oppositions and attractions in matter is "one of the profoundest inquiries of nature." Bacon compares "the first congregations of matter" to "a general assembly of estates" which "doth give law to all bodies."[15] From this point of view, the operations of nature are from the very beginning "political" and "economical," in that they result from a whole range of mediations and transactions among conflicting and competing spheres of interest and power.

This view of nature as a theatre of conflicts underpins the setting and the interpretation of Bacon's experiments. Here are some examples. The behaviour of bubbles reveals the presence of the "appetite of continuation," namely, "the appetite to resist separation or discontinuance," stronger in solid bodies, "fainter and weaker" in fluid bodies.[16] Experiments conducted on flames demonstrate that, in itself, "fire is not violent or furious." It becomes so only when "it is checked and pent."[17] The reason why kernels of grapes placed close to the root of a vine seem to help the growth of the vine itself might be that "the root being of greater strength,

[13]Bacon 2004, 382–416.

[14]Bacon 1857–1874, II, 451.

[15]Bacon 1857–1874, II, 459–460.

[16]Bacon 1857–1874, II, 347.

[17]Bacon 1857–1874, II, 353.

robbeth and devoureth the nourishment, when they have drawn it; as great fishes devour little."[18] Nowhere is the pervasive nature of material appetite more apparent than in the whole body of air surrounding the earth. Not only are the spirits (also called "pneumaticals" by Bacon) inside natural bodies "unquiet to get forth and congregate with the air" (a tendency that for Bacon is at the basis of numerous natural processes, such as colliquation, digestion, putrefaction, vivification, figuration, desiccation, induration and consumption)[19]; the very body of air, too, is made up of tendencies and desires. While experimenting with the nature of sounds, Bacon considers it very likely that "air doth willingly imbibe the sound as grateful, but cannot maintain it"; and the reason is that "the air hath (as it should seem) a secret and hidden appetite of receiving the sound at the first; but then other gross and more materiate qualities of the air straightways suffocate it; like unto flame, which is generated with alacrity, but straight quenched by the enmity of the air or other ambient bodies."[20]

Natural phenomena can therefore be seen as the result of elaborate motions caused by searching and unsettling appetites. Every time they are under pressure, material particles are eager to find the easiest way to release the tension accumulated in a particular arrangement of matter:

> whensoever a solid body (as wood, stone, metal, etc.) is pressed there is an inward tumult in the parts thereof, seeking to deliver themselves from the compression. And this is the cause of all violent motion. Wherein it is strange in the highest degree, that this motion hath never been observed nor inquired; it being of all motions the most common, and the chief root of all mechanical operations.[21] This motion worketh in round at first, by way of proof and search which way to deliver itself; and then worketh in progress, where it findeth the deliverance easiest.[22]

Bacon's description of material particles recalls that of a stampede, in which people who are caught in an unforeseeable disaster try every possible way to reach safety. He calls this structural tendency in matter the "motion of liberty."[23]

Another original appetite of matter – an appetite that is complementary to the motion of liberty – is the "appetite of union" and "evitation of solution of continuity." As already stated, Bacon claims that this appetite is present in three degrees: weak in fluids, strong in solids and intermediate in "bodies cleaving or tenacious." Bacon makes clear that "all solid bodies are cleaving, more or less," but "those

[18] Bacon 1857–1874, II, 354–355.

[19] Bacon 1857–1874, II, 451.

[20] Bacon 1857–1874, II, 436.

[21] Here "mechanical" means "violent."

[22] Bacon 1857–1874, II, 342. See also ibid., 382.

[23] Bacon 1857–1874, II, 342–343: "This motion upon pressure, and the reciprocal thereof, which is motion upon tensure, we use to call (by one common name) *motion of liberty*; which is, when any body, being forced to a preternatural extent or dimension, delivereth and restoreth itself to the natural."

bodies which are noted to be clammy and cleaving, are such as have a more indifferent appetite (at once) to follow another body, and to hold to themselves. And therefore they are commonly bodies ill mixed; and which take more pleasure in a foreign body, than in preserving their own consistence."[24]

The primordial tension caused by these two basic appetites of matter – the motion of liberty and motion of union – was the original spark which began the various natural processes of corruption and generation; as a result, these appetites engendered the "two great families of things," the two dynasties that have ruled nature since its creation. The original tension still affects the particles of matter by keeping them in a constant state of "trepidation."[25] Indeed, matter *qua* matter for Bacon "trepidates." Every time material corpuscles perceive external pressure as a condition that might imperil their safety, they react by fleeing or rebelling, depending on the particular situation. Motion of liberty is fundamental in Bacon's physics because matter is constantly reacting to situations of potential violence, constantly dealing with pressures to be assimilated. Processes of generation and corruption, therefore, seem to follow the same cycles of formation and dissolution that affect political structures. To accelerate motions of putrefaction in experimental situations, for instance, Bacon suggests to "dissolve" the "government" of the "principal spirits" so that every part abandons the current state of affairs and "returneth to his nature or homogeny." Bacon uses the same analogy of political volatility to explain such phenomena as the spread of infections and contagious diseases, "for that the malignity of the infecting vapour daunteth the principal spirits, and maketh them fly and leave their regiment; and then the humours, flesh, and secondary spirits, do dissolve and break, as in an anarchy."[26] In solid bodies, the general state of trepidation is less intense than in fluid or spirituous bodies, for these bodies are in a natural state of "torpor." This intrinsic torpor is caused by a "motion of gravity" and a "natural appetite not to move at all."[27] Bacon sees this motion of self-gravity as one of the most powerful and original in matter: "this same motion of weight or gravity (which is a mere motion of matter, and hath no affinity with the form or kind) doth kill the other motion, except itself be killed by a violent motion."[28]

As I said at the beginning of this paper, Bacon recommends that the experimental study of matter be "broken" and "grinded." Put differently, the inherently plural nature of material appetites demands an experimental approach that relies on a particulate view of matter. In such an atomistic view of material appetites, Bacon tends to regards all natural operations as combinations of primal appetites, which may cluster into larger units and then break down again into the original appetitive atoms. This atomism of the appetites has far-reaching implications for both the

[24] Bacon 1857–1874, II, 438.
[25] Bacon 1857–1874, II, 451.
[26] Bacon 1857–1874, II, 452.
[27] Bacon 1857–1874, II, 586.
[28] Bacon 1857–1874, II, 565.

organisation of the experimental strategies and the reshuffling of traditional divisions in the hierarchies of the senses. In Bacon's opinion, that is, the fact that sight has been the prevailing sense in leading natural investigations has caused natural philosophers to overlook such decisive factors as the "fineness of the body," the "smallness of the parts" and the "subtilty of the motion," although these are the "things that govern nature principally." This, warns Bacon, is not "a question of words." On the contrary, it is "infinitely material in nature," for philosophers, by overemphasising the sense of sight, have developed such dominant notions as vacuum, soul and incorporeal virtues.[29] Instead, a more material consideration of the senses, an emphasis on touch and tactile reactivity for instance, would have shifted the focus from visible shapes to invisible motions, from structures to dynamic processes. Bacon likens his view of atomic appetites – "the motions of the minute parts of bodies" – to Democritus' theory of material atoms.[30] Natural philosophers have ignored "the tumult in the parts of solid bodies when they are compressed, which is the cause of all flight of bodies through the air, and of other mechanical motions."[31] Not only is the motion of the minute parts invisible ("unobserved"); often it also "passeth without sound."[32] As already said, matter trepidates by its very nature. This constant state of trepidation is the result of innumerable appetitive drives, what Bacon calls the "secret" trepidation of the "minute parts" of matter.[33]

3 Experimental Practice and Discipline of the Appetites

Although it may seem like a trite statement, it is safe to say that Bacon was concerned with action more than contemplation. In metaphysical terms, the same statement can be rephrased by saying that Bacon's philosophy is an original attempt to find successful ways of intervening in and modifying the very fabric of matter. As he argued in aphorism 127 of the first part of the *Novum organum*, the new method of interpreting nature – induction – was not to be applied only to "the motions and discourses of the mind," but also to "the nature of things." Since for Bacon appetite constituted the innermost nature of things, acquisition of knowledge could not be separated from the mastery of appetites, both natural and civil. Bacon argued that there could not be a proper management of human affairs without a study of natural bodies. In the *Advancement of Learning*, he had already characterised civil knowledge as "a subject which of all others is most immersed in matter," meaning that the complexity and opaqueness of the appetitive life are particularly evident in that domain.[34] In the

[29] Bacon 1857–1874, II, 380–381.

[30] Bacon 1857–1874, II, 381.

[31] Bacon 1857–1874, II, 381–382.

[32] Bacon 1857–1874, II, 391.

[33] Bacon 1857–1874, II, 416.

[34] Bacon 2000, 156.

Sapientia veterum, in which Bacon chose Orpheus to represent philosophy, the Greek poet is characteristically portrayed as capable of attracting both infernal powers (i.e., of dealing with natural philosophy given the hidden and lower nature of the spirits) and wild beasts (i.e., of dealing with human passions, which are the basic constituents of moral and civil philosophy). In Bacon's sumptuously emblematic language, this means that in the domain of natural knowledge philosophy aims at the conservation of natural things ("restoration and renovation of perishable things"), while in the domain of civil knowledge it teaches men "to forget their unrestrained affects." In both cases, it is a matter of securing the control of the appetites.[35]

Orpheus had already made his appearance in the *Advancement of Learning* (1605). There he symbolised the power that learning exercises on both "relieving the necessities which arise from nature" and "repressing the inconveniences which grow from man to man":

> which merite was liuely set forth by the Ancients in that fayned relation of *Orpheus* Theater; where all beasts and birds assembled: and forgetting their seuerall appetites; some of pray, some of game, some of quarrell, stood all sociably together listening vnto the ayres and accords of the Harpe; the sound whereof no sooner ceased, or was drowned by some lowder noyse; but euerie beast returned to his owne nature.[36]

This mythological device, Bacon went on, throws light upon "the nature and condition of men," who "are full of savage and unreclaymed desires; of profite, of lust, of revenge." The moral behind Orpheus' myth is that the domestication of the appetites passes through the civilising effect of culture:

> as long as they [the appetites] giue eare to precepts, to lawes, to religion, sweetly touched with eloquence and perswasion of Bookes, of Sermons, of haranges; so long is societie and peace maintained; but if these instruments bee silent; or that sedition and tumult make them not audible; all things dissolue into Anarchie and Confusion.[37]

Given such a view of nature, Bacon's search for ways of both arousing and containing appetites should not therefore come as too much of a surprise. On many occasions, he investigated the power of laws, customs, habits and opinions in channelling the energy of human appetites. He no doubt assumed that laws exercised a certain influence over the appetites, but in general he considered laws to work like palliative remedies.[38] In the section of the *Advancement of Learning* devoted to the "culture and cure of the mynde," Bacon acknowledged the fact that fortune and nature are "without our commaund"; in cases like these, Bacon recommended proceeding

[35] Bacon 1857–1874, VI, 646–648.

[36] Bacon 2000, 39.

[37] Bacon 2000, 39.

[38] Bacon 1857–1874, VII (*Reading of the Statute of Uses*), 417–418: "new laws are like the apothecaries' drugs; though they remedy the disease, yet they trouble the body: and therefore they use to correct them with spices. So it is not possible to find a remedy for any mischief in the commonwealth, but it will beget some new mischief; and therefore they spice their laws with provisoes to correct and qualify them."

by "application" (or "accommodation"), that is, by gradually adjusting ourselves to each single situation determined by natural conditionings and contingent events. In "Morall and Ciuile matters," Bacon found particularly useful the application of strategies of mutual control based on sophisticated systems of checks and balances. In this way, he argued, we know how "to set affection against affection, and to master one by another" – "as we use to hunt beast with beaste and flye byrd with birde." He specified that, by definition, appetites were factional (and therefore atomic). For a shrewd politician, however, factions could also be seen as a resource, "for as in the gouernmente of states it is sometimes necessarye to bridle one faction with another, so it is in the gouernmente within."[39]

Bacon thought that custom – "the Principall Magistrate of Mans life"– was more powerful than the institution of laws and a cunning use of factional disputes.[40] In *A Letter and Discourse to Sir Henry Savile, Touching Helps for the Intellectual Powers*, Bacon expanded on this power of habit:

> as to the body of man, we find many and strange experiences how nature is overwrought by custom, even in actions that seem of most difficulty and least possible. As first in Voluntary Motion; which though it be termed voluntary, yet the highest degrees of it are not voluntary; for it is in my power and will to run; but to run faster than according to my lightness or disposition of body, is not in my power nor will. We see the industry and practice of tumblers and funambulos, what effects of great wonder it bringeth the body of man unto.[41]

Bacon added a list of other instances to demonstrate "how variously, and to how high points and degrees" the human body "may be (as it were) moulded and wrought." After the body, he turned his attention to the will. In this domain, he considered religion to be "the most sovereign of all," being able to change and transform the will "in the deepest and most inward inclinations and motions." Immediately after religion he placed "opinion and apprehension" ("whether it be infused by tradition and institution, or wraught in by disputation and persuasion"), then the use of examples and the technique of using one affect to heal and correct another. "And lastly" – Bacon concluded this section – "when all these means, or any of them, have new framed or formed human will, then both custom and habit corroborate and confirm all the rest."[42] In the *Essays* (1601), Bacon attributed such an exceptional power to habit that men, for all their protestations to the contrary, in fact act "[a]s if they were Dead Images, and Engines moved onely by the wheeles of *Custome*."[43] Once custom turns into a social force, Bacon explained, then its power increases:

> if the Force of *Custome* Simple and Separate, be Great; the Force of *Custome* Copulate, and Conjoyned, and Collegiate, is far Greater. For there Example teacheth; Company comforteth; Emulation quickeneth; Glory raiseth: So as in such Places the Force of

[39]Bacon 2000, 145–150.

[40]Bacon 1985, 121.

[41]Bacon 1996b, 115–116.

[42]Bacon 1996b, 117.

[43]Bacon 1985, 121.

> *Custome* is in his Exaltation. Certainly, the great Multiplication of Vertues upon Humane Nature, resteth upon Societies well Ordained, and Disciplined.[44]

Laws, customs, habits, religious beliefs and opinions were all forces which, in Bacon's view, could be used to master human appetites.

Analogous forces could be found to control the appetites of matter. In his search for possible ways of connecting the realms of nature and man, Bacon concentrated his efforts on the material spirits. He thus explained the influence of the passions of the human soul upon the body in the same way as he described the primordial reactions of the spirits. As already mentioned, Bacon's natural philosophy is characterised by a deliberate overlapping of explanatory levels. He described the motions of matter in terms of animal appetites, while interpreting human emotions as one of the most sophisticated forms of vital reactivity in the material universe.

Here, for instance, is the way Bacon examined the passion of fear in the experimental setting of *Sylva Sylvarum*. The analysis works on two levels: human and natural.

> Fear causeth paleness, trembling, the standing of the hair upright, starting, and skriching. The paleness is caused, for that the blood runneth inward to succour the heart. The trembling is caused, for that through the flight of the spirits inward, the outward parts are destituted, and not sustained. Standing upright of the hair is caused, for that by shutting of the pores of the skin, the hair that lieth aslope must needs rise. Starting is both an apprehension of the thing feared, (and in that kind it is a motion of shrinking,) and likewise an inquisition, in the beginning, what the matter should be, (and in that kind it is a motion of erection); and therefore when a man would listen suddenly to any thing, he starteth; for the starting is an erection of the spirits to attend. Skriching is an appetite of expelling that which suddenly striketh the spirits: for it must be noted that many motions, though they be unprofitable to expel that which hurteth, yet they are offers of nature, and cause motions by consent: as in groaning or crying upon pain.[45]

The important contribution of Bernardino Telesio's natural philosophy to Bacon's theory of spirits has been long recognised.[46] In his *De natura iuxta propria principia* (1565, 1570, 1586), Telesio had provided a systematic account of man's emotional and ethical life based on the natural motions of the spirits and their tendency to self-preservation. Bacon was very keen on adopting this explanatory framework and on applying it to the analysis of both natural and social phenomena. His most accurate analysis of Telesian philosophy can be found in *De principiis atque originibus* (c.1610s). In Telesio's cosmological model, Bacon asserted, conflicts are deemed to be less widespread than they actually are, for Telesio confined them to the outermost borders of the only two spheres of activity, cold (earth) and heat (heaven). There, in the interface separating the two kingdoms – a sort of rind

[44] Bacon 1985, 122.

[45] Bacon 1857–1874, II, 567–568.

[46] Bacon 1996a, 128; 224, 250–258. On Bacon's attitude towards Telesio, see Giachetti Assenza 1980; Margolin 1990; De Mas 1990; Bondì 2001.

(*cortex*) and coating (*incrustatio*) surrounding the inner core of the universe – the hostile forces of nature meet and fight each other. Bacon gave a brief outline of the Telesian cosmos, a cosmos ravaged by wars led by opposing armies:

> after he [Telesio] has sufficiently fortified the innermost parts of both kingdoms, he sets a military campaign in motion, for all tumult, conflict and infernal disorder can be found in those spaces between the outmost parts of the heaven and the innermost parts of earth as happens in empires where the borders are exposed to incursions and aggressions, while the internal provinces enjoy peace and security. Thus these natures [hot and cold] and their concretions have the appetite and faculty of generating and multiplying themselves continually, of spreading out in every direction, of occupying the whole mass of matter, of fighting and invading each other, of ousting and driving away each other from their respective domains and settling in the places that have been conquered; furthermore, they have the appetite and faculty of perceiving and apprehending (*appetitus et facultas percipiendi et prehendendi*) the force and the actions of another nature and their own, and of moving and adjusting themselves according to such perception. All the differences of beings, action and power are derived from this battling (*decertatio*).[47]

Bacon acknowledged the value of Telesio's cosmology. In particular, he praised the way in which the Italian philosopher had characterised the turbulent conflicts pervading nature, which he called the "implacable and deadly war": "one nature desires, strives and tries hard to destroy the other and to impose its exclusive domain on matter." Bacon did not accuse Telesio of having underestimated the seriousness of the war. Like Bacon, he had described the contrast dividing the active natures (heat and cold) in terms of a war where no alliance is possible and no prisoners are taken. "Every generation and every effect in a natural body are the result of a settlement based on victory and predominance (*praedominantia*), and not on an agreement or covenant."[48] Constant conflict is the rule, and if some sort of harmony seems to keep the world together it is only because the unbounded energy of one nature's power is curbed by the similarly unbounded energy of another's. The disposition of the universe, seemingly stable but in fact precarious and always on the point of breaking into complete anarchy, "does not result from the laws that regulate covenants and agreements, but from sheer power" (*impotentia*, in the sense of inability to restrain one's own power);[49] "every increase and decrease in power and action does not derive from a moderation of the force (*intensionis moderamen*) – it being a force that desires (*concupiscit*) the whole (*integrum quiddam*) – but from the blow and resistance of the opposite nature." Telesio, Bacon went on, managed to identify three factors that account for the order and

---

[47]Bacon 1996a, 230–232. The same image is already in *Cogitationes de natura rerum*. See Bacon 1857–1874, III, 34: "Itaque tumultus fere omnis, et conflictus, et perturbatio, in confiniis tantum coeli et terrae locum habere videtur. Ut in rebus civilibus fit; in quibus illud frequenter usu venit, ut duorum regnorum fines continuis incursionibus et violentiis infestentur, dum interiores utriusque regni provinciae secura pace atque alta quiete fruuntur."

[48]Bacon 1996a, 248.

[49]Bacon 1996a, 234. Both Francis Headlam's 1858 translation (Bacon 1857–1874, V, 481) and Rees' and Edwards' rendition of *impotentia* as want or lack of power is a serious misunderstanding (besides contradicting the meaning of the whole passage).

variety of the universe: the "force of heat," the "disposition of matter" and the "type of subjection." These three factors "are related to each other in some sort of a nexus and one is cause to the other" (*nexu quodam inter se implicantur, atque sibi ipsis concausae sunt*).[50] More specifically, with respect to subjection (*subactio*), Bacon distinguished two possible ways in which such action of subjection manifests itself in nature: by rejection (*ejectio*) or by "going over to the enemy" (*versio*). Like two enemy armies, when they are both large and fully arrayed for battle, heat and cold try to conquer each other's position (*Nam entia, veluti acies, loco moventur et impelluntur*); when the quantity of matter is smaller, "then acts of defection (*versio*) follow, for beings are destroyed, and they change their own nature rather than their place."[51]

Bacon's use of this characteristically Telesian framework is particularly evident when he discusses the nature of material and human appetites, as when in *Sylva Sylvarum* he presents his experimental observations conducted on such passions as fear, grief and pain, joy, anger, dislike, shame, pity, wonder and laughing. Compression, contraction and contention are all motions involved in the expression of emotions, based on the alternation of opposing tendencies to resist and relax.[52] In an "experiment solitary" concerning "the flight of the spirits upon odious objects," Bacon accounts for various reactions of fear by referring to the most original appetites of matter. "All objects of the senses which are very offensive," he notices, "do cause the spirits to retire: and upon their flight the parts are (in some degree) destitute: and so there is induced in them a trepidation and horror."[53] The most striking aspect of his argument is the ease with which Bacon switches back and forth from the level of human senses to that of natural sentience. It is not merely that matter is in an unremitting state of trepidation: such a state of physical trepidation seems to be a condition of animal trepidation as well. If this is really the case, what prevents Bacon's view of nature and his form of experimental practice from ending in a form of crass animism and delusion? Is not Bacon's new "magic" after all just a kind of primitive magic?[54]

A plausible answer to this difficult question can be found if we turn our attention to Bacon's ideas regarding longevity. In *Sylva Sylvarum*, we find a series of interesting remarks concerning the possibility of prolonging life, in which Bacon seems to admit forms of direct control over the life of matter. Since Bacon's spirits have a predatory nature, prolongation of life depends on finding ways of producing a "more placid motion of the spirits," which "thereby do less prey and consume the

---

[50] Bacon 1996a, 234.

[51] Bacon 1996a, 248–250.

[52] Bacon 1857–1874, II, 568–571.

[53] Bacon 1857–1874, II, 597.

[54] Bacon had no qualms in seeing his experimental investigations of appetites as a new form of natural magic. "For this writing of our *Sylva Sylvarum* is (to speak properly) non natural history, but a high kind of natural magic. For it is not a description only of nature, but a breaking of nature into great and strange works" (Bacon 1857–1874, II, 378).

juice of the body." Bacon recommends two possible strategies: "either that men's actions be free and voluntary, that nothing be done *invita Minerva*, but *secundum genium*," or "that the actions of men be full of regulation and commands within themselves: for then the victory and performing of the command giveth a good disposition to the spirits." In this last case, Bacon points out the beneficial effects that the idea of accomplishing great results through gradual training may have on the mind of men ("especially if there be a proceeding from degree to degree; for then the sense of victory is the greater"). To extend the course of his life, therefore, man can rely either on fostering the absolute spontaneity of the natural functions or on moulding these functions through rigorous self-control and discipline. He provides an example for each of the two lifestyles: "a country life" and the life of "monks and philosophers."[55] The shift from one level to the other – from the body to the soul, so to speak – is possible because of the ubiquitous character of material cupidity. Therefore man can exercise the power of his will on his body, and a particular disposition of his body can increase the power of his will. The passions of matter are like the passions of human nature in that they are both governed by motions of liberty and continuity, and, even more deeply, by motions of trepidation and rest. Country life (that is, a life centred around the motion of liberty) and philosophical life (a life centred around the motion of resistance) represent the two alternatives available to man in his attempt to control the appetites of matter, from the original motions of matter up to the motions of the will.

4 Conclusion

According to a famous anecdote reported by John Aubrey in his *Brief Lives*, William Harvey, asked about his opinion concerning Bacon's philosophy, came up with the often quoted sentence: Bacon writes of natural philosophy like a Lord Chancellor.[56] To be sure, we should take this statement with a grain of salt. Whether Harvey was being malicious (which is highly likely, as Aubrey himself assumes) or whether he was expressing a sincere compliment, the fact remains that, *malgré* Harvey, this statement can be used as an excellent way of summing up Bacon's agenda and peculiar style of philosophising. Once stripped of the original authorial intention, Harvey's opinion, far from sounding like an insult or a piece of learned gossip, can thus be used as a convenient interpretative tool.

[55] Bacon 1857–1874, II, 437. On the predatory nature of spirits, see Bacon 1857–1874, II, 440. On Bacon's ideas concerning the prolongation of life, see Webster 2002; Serjeantson 2002; Giglioni 2005.

[56] Aubrey 1958, 130: "He had been physitian to the Lord Chancellour Bacon, whom he esteemed much for his witt and style, but would not allow him to be a great Philosopher. Said he to me, *He writes Philosophy like a Lord Chancellor*, speaking in derision; *I have cured him*."

That Bacon writes of natural philosophy like a Lord Chancellor can be read in at least two ways. It can mean that he applies (or projects, Harvey would say) political and legal views onto the understanding of nature. Or, on a less judgemental and more stimulating level, we can say that Bacon writes like a Lord Chancellor in the sense that a bold form of political realism may be the key to understand the inner workings of both nature and human society. As I hope I have shown in this paper, it is not too much to say that, in Bacon's view, *war* and *interest* are the ultimate moving forces in the world of nature, long before they make their appearance in the world of man.

We also may come close to addressing the question of why Bacon decided to devote so much time and energy to the reform of natural philosophy when his main preoccupations were of a legal and political matter. It is well known that the advancement of learning, and especially of learning concerning nature, was an integral part of Bacon's programme of universal reform. One might argue that, precisely because he adhered to a view of nature grounded on ontological realism and on the primacy of appetite over knowledge (that is to say, the activity of nature, and not the knowledge that man has of nature, comes first; appetites, and not men's representations of nature, are the essential thing), Bacon was seeking in nature a solid basis upon which to reorganise the human commonwealth. He offered us a view of the conflicts of nature based on political realism and a view of human society based on materialism. In such a view, it is not always easy to establish what comes first, whether political realism or naturalism. Worst of all, it is difficult to avoid the impression that Bacon falls short of animism when the two levels of analysis – the natural and the political – are so muddled.

A possible key to understanding such a controversial view of nature lies in Bacon's notion of appetitive atomism. As shown in the previous sections, appetites are atomic and self-interested. They have no long-term knowledge concerning the purpose of their actions. The direction of their tendencies depends on the need to find immediate satisfaction for their desires and on matter's overall tendency to defuse the tension that accumulates from conflicting motions. Particles of matter are under constant pressure; they are caged, ready to flee or to attack depending on the circumstances. Bacons' matter is extremely sensitive and delicate. It perceives the slightest touch as a threat. We know that *trepidatio* is its normal condition. Unsettled by actions and reactions, upset by constant motions of assault and resistance, matter is therefore divided within itself, an unstable aggregate of conflicting particles. Bacon's constantly ambivalent attitude towards both particulate and homogenous views of matter may therefore be interpreted as an attempt to devise a view of matter in which conflicts can be controlled and reabsorbed into a broader unitary vision. As an important political figure during both Elizabeth's and James' reigns (Solicitor-General, Attorney-General, Lord Chancellor), a man interested in programmes of legal and state reforms, he knew very well that a centrally organised state must cope with the centrifugal tendencies of unruly and intractable elements;[57]

[57] See Martin 1992.

likewise, matter understood as a unitary continuous substratum may be seen as capable of curbing the turbulent and divisive propensities of numberless material particles. In this case, too, Bacon writes of philosophy as a Lord Chancellor.

Since tensions created by conflicting appetites are more original than states of motion and rest, there is no condition of absolute rest in Bacon's material universe. Indeed, appetite is even more original than life and death (or than any vital and mechanical properties of matter). Natural appetites are more original than vital motions and they are an indication of matter's tendency to remain the same and resist any attempt to change its configuration. Rest prevails over change because the primordial appetite is the appetite which always seeks to re-establish the original condition. Every change is produced in order to resist change. It may sound like a paradox, but life, for Bacon, has a predatory quality (a view that is particularly evident in his account of the vital spirits) because the innermost tendency is the one to restore the original balance: any vital reaction is a reaction on the part of matter against being forced to change.

It does not come as a surprise, therefore, that the list of the basic motions of matter given in the *Novum organum* ends with the "motion of trepidation" (*motus trepidationis*) and the "motion of rest," also called "dread of motion" (*motus decubitus sive exhorrentiae motus*). The two final motions recapitulate the whole account of motions provided in the *Novum organum* and they also represent the metaphysical background for Bacon's apparently haphazard collection of experiments in *Sylva Sylvarum*. In the *Novum organum*, the motion of trepidation is also called the motion of "eternal captivity" (*aeterna captivitas*) and it occurs every time "bodies are placed in situations that are not suitable to their nature, and yet they do not feel completely uncomfortable. As a result, they quiver with trepidation and feel restless, not content with their condition and yet not daring to move further." This motion characterises the state of tension and fluctuation (*status anceps*) between easiness (*commoda*) and uneasiness (*incommoda*), in which "bodies, drawn in different directions, try to free themselves, and again they are rejected, and nevertheless they still keep trying to free themselves." The motion of rest, too, is at the very core of matter, occurring when matter reaches its highest level of density. In this case, bodies abhor motion and their only desire is the desire not to move. They are "teased (*vellicentur*) and provoked to motion in infinite ways," but "they maintain their nature as much as they can." And even when they are forced to move, "they always seem to do this in order to restore their condition of rest and not to move any longer."[58] These two motions, the motion of trepidation and the motion of rest, are both expression of matter's longing for repose and, at the same time, of its ultimate inability to avoid motion and change. Reaching a state of rest is the deepest desire in matter and yet the atomistic constitution of the appetites, their factional and divisive nature, prevents the particles of matter from reaching a state of peaceful

[58] Bacon 2004, 410–412.

coexistence. Matter is constantly at war with itself, for the state of desire is ambivalent to the point of contradiction. It is not by accident that Bacon calls such condition *status anceps*. Here *anceps* is definitely an adjective fraught with meaning: literally, it means "facing in two directions," but also, "critically poised," "wavering," "indecisive," "unsettled." The two directions that constantly destabilise the appetites of matter are, on one hand, the search for ways to put an end to the state of motion and, on the other, the condition of perpetual motion. We can now understand better, I hope, why Bacon decided to refer to the last type of motion he examined in the *Novum organum*, paradoxically, as "motion of dread of motion." Not *horror vacui* (one of the typical bugbears of atomism), but *horror motus* is the real linchpin of Bacon's natural philosophy.

Once Baconian appetites are seen as manifestations of matter's innermost tendency to reduce the tension of the system, then we may venture a possible explanation of the reason why the "families of things" manage to persist despite conflicts and oppositions. If there is a telos in Bacon's universe, it is the dread of motion and the ensuing sense of agitation and alarm. Many of Bacon's descriptions in *Sylva Sylvarum* assume that cohesion and coherence result from counterbalancing appetites. The universe as a whole forcefully exerts its natural equilibrium against a world constantly thrown out of balance. Bacon never tires of alerting the reader that every time he refers to the motions of liberty and union, attraction and gravity, sympathy and antipathy, trepidation and rest, he is not assuming the existence of a harmoniously organised system of nature based on intentional design. While dealing with the tendencies of sympathy and antipathy allegedly inherent in plants, for example, Bacon explains that every plant is in fact "an enemy to any other plant" because they all compete to draw the best juices from the soil.[59] Competition for food, not attraction to the like, is the principle that accounts for a large number of actions in nature. Fulfilment of particular and immediate urges, rather than deferral to long-term planning, is the driving force behind the life of nature. The experimental evidence resulting from *Sylva Sylvarum* confirms the self-adjusting and inevitably precarious character of material reality. The universe persists in its course because it is a steady-state system pervaded by a general condition of uneasiness and trepidation.

It is only by knowing and subduing the appetites of matter that man can master the intractable forces of nature, thereby restoring humankind's original control of its appetites – i.e., the original paragon of cognitive and moral perfection embodied by the pre-lapsarian Adam. In Bacon's philosophy, then, there is a basic correspondence between the motions of the mind and the motions of the bodies, a correspondence which derives from the fact that both orders of reality are rooted in the unstable realm of appetitive drives. Bacon is of the opinion that both human and natural creativity, both *ingenia* and *conatus* are arbitrary forces. Just as a "ruler" (*regula*) is presented in *Novum organum* as the remedy for the bewilderment of

[59]Bacon 1857–1874, II, 494.

knowledge, so custom is the force that disciplines the unruliness of human action. The *New Organon* – the reorganisation of man's mental faculties – and the *New Atlantis* – the reorganisation of the social order – work in parallel. This is the fundamental reason that the chief institution in Bensalem (the utopian political organisation in *New Atlantis*), which is devoted to the study of the appetites in matter (the "secret motions of things"), is also in charge of preserving the political order. Men of science are in power because by knowing the appetites of matter they have a better knowledge of the very roots of human appetites.

References

Aubrey, John. 1958. *Brief Lives*, ed. O. Lawson Dick. London: Warburg and Secker.

Bacon, Francis. 1857–1874. *Works*, eds. J. Spedding, R. L. Ellis and D. D. Heath, 14 vols. London: Longman.

Bacon, Francis. 1985. *The Essayes or Counsels, Civill and Morall*, ed. M. Kiernan. Oxford: Clarendon Press.

Bacon, Francis. 1996a. *Philosophical Studies c. 1611–c.1619*, ed. Graham Rees. Oxford: Clarendon Press.

Bacon, Francis. 1996b. *The Major Works*, ed. B. Vickers. Oxford: Oxford University Press.

Bacon, Francis. 2000. *Advancement of Learning*, ed. M. Kiernan Oxford: Clarendon Press.

Bacon, Francis. 2004. *Instauratio magna Part II: Novum organum and Associated Texts*, ed. G. Rees. Oxford: Clarendon Press.

Bondì, Roberto. 2001. Bacon e la restaurazione di Parmenide. *Rivista di Filosofia* 92: 327–339.

De Mas, Enrico. 1990. Bernardino Telesio e la falsità di Aristotele. Il giudizio di Bacone e di Tobia Adami. In *Atti del Convegno Internazionale di Studi su Bernardino Telesio*, 167–179. Cosenza: Accademia Cosentina.

Giachetti Assenza, Valeria. 1980. Bernardino Telesio: il migliore dei moderni. I riferimenti a Telesio negli scritti di Francesco Bacone. *Rivista di Storia della Filosofia* 35: 41–78.

Giglioni, Guido. 2005. The Hidden Life of Matter: Techniques for Prolonging Life in the Writings of Francis Bacon. In *Francis Bacon and the Refiguring of Early Modern Thought. Essays to Commemorate* The Advancement of Learning *(1605–2005)*, eds. C. Gimelli Martin and J. R. Solomon, 129-144. Ashgate: Aldershot.

Margolin, Jean-Claude. 1990. Bacon, lecteur critique d'Aristote et de Telesio. In *Atti del Convegno Internazionale di Studi su Bernardino Telesio*, 135-166. Cosenza: Accademia Cosentina.

Martin, Julian. 1992. *Francis Bacon, the State and the Reform of Natural Philosophy*. Cambridge: Cambridge University Press.

Serjeantson, Richard. 2002. Natural Knowledge in the *New Atlantis*. In *Francis Bacon's* New Atlantis. *New Interdisciplinary Essays*, ed. Bronwen Price, 82–105. Manchester and New York: Manchester University Press.

Webster, Charles. 2002. *The Great Instauration: Science, Medicine and Reform 1626–1660*. Bern: Peter Lang.

'A Corporall Philosophy': Language and 'Body-Making' in the Work of John Bulwer (1606–1656)

Justin E.H. Smith

Abstract Francis Bacon, in his 1605 work *The Proficience and Advancement of Learning*, analyzes "Notes of Cogitations" into "twoo sortes... [t]he one when the Note hath some Similitude, or Congruitie with the Notion; [t]he other... hauing force onely by Contract or Acception." The latter are either "Hierogliphickes" or "Gestures," and the latter of these "are as Transitorie Hierogliphickes, and are to Hierogliphickes, as Words spoken are to Wordes written, in that they abide not." In some fashion or other, it is the first kind of hieroglyphics that will dominate in the seventeenth-century efforts to develop an ideal, artificial writing system, one that would not be based on mere convention, but would instead serve transparently for producing emblems of the things one wishes to denote. The second variety Bacon identifies, gesture, will in contrast gain little attention. Yet little attention is not none at all. Over the course of the 1640s, the obscure Baconian natural philosopher John Bulwer would develop his predecessor's notion of transitory hieroglyphics into an elaborate system, one that would serve as the starting point for the later sciences of, among other things, sign language and sociolinguistics. Bulwer's theory of gesture reveals an important rift in seventeenth-century debates about the universal character, between those who believe that this can be nothing other than an artificial language, and those who believe that it is precisely artifice that obscures meanings, and that any universally comprehensible system of communication will be perfectly natural as opposed to artificial. In exploring this rift, we are also able to gain access to a curious, if not terribly influential, theory constituting a point of contact between early modern philosophy of language on the one hand and the early modern metaphysics of body on the other.

J.E.H. Smith
Department of Philosophy, Concordia University, Montreal

C.T. Wolfe and O. Gal (eds.), *The Body as Object and Instrument of Knowledge: Embodied Empiricism in Early Modern Science*, Studies in History and Philosophy of Science 25,
DOI 10.1007/978-90-481-3686-5_9, © Springer Science+Business Media B.V. 2010

1 Introduction

Francis Bacon, in his 1605 work *The Proficience and Advancement of Learning*, argued for a 'real character' or artificial language that would be able to communicate intended meanings from one person to another with perfect transparency. This is a project that would be important throughout the rest of the century, with thinkers such as Samuel Hartlib, John Wilkins, and G. W. Leibniz all making significant contributions to the study of artificial and formal languages.

> Notes of Cogitations are of twoo sortes; The one when the Note hath some Similitude, or Congruitie with the Notion; The other Ad Placitum, hauing force onely by Contract or Acception. Of the former sort are Hierogliphickes, and Gestures. For as to Hierogliphickes, (things of Ancient vse, and embraced chiefly by the AEgyptians, one of the most ancient Nations) they are but as continued Impreases and Emblemes. And as for Gestures, they are as Transitorie Hierogliphickes, and are to Hierogliphickes, as Words spoken are to Wordes written, in that they abide not.[1]

In some fashion or other, it is the first kind of 'Hierogliphickes', in the sense Bacon describes here, that will dominate in the seventeenth-century efforts to develop an ideal, artificial writing system, one that would not be based on mere convention, but would instead serve transparently for producing 'Emblemes' of the things one wishes to denote. The second variety Bacon identifies, gesture, will in contrast gain little attention. Yet little attention is not none at all. Over the course of the 1640s, the obscure Baconian natural philosopher John Bulwer would develop his predecessor's notion of transitory hieroglyphics into an elaborate system, one that would indeed serve as the starting point for the later sciences of, among other things, sign language and sociolinguistics.

According to Jeffrey Wollock, Bulwer would entirely ignore Bacon's interest in an ideal language, focusing instead exclusively upon Bacon's characterization of gesture, indeed turning this into the centerpiece of his chirological project. According to Wollock, "this was in part because [Bulwer] retained older views on the inherent ontological harmony between man and the universe, but also because, for Bulwer the physician, the underlying neurophysiological basis of gesture confirmed it as the universal 'language' of humanity."[2] It would be more correct to say, however, that Bulwer does not abandon the search for an ideal language, but indeed believes that he has *already found one* in gesture. In examining why he believes this, we might be able to discern an important rift in seventeenth-century debates about the universal character, between those who believe that this can be nothing other than an artificial language, and those who believe that it is precisely artifice that obscures meanings, and that any universally comprehensible system of communication will be perfectly natural as opposed to artificial. But in considering Bulwer's understanding of the natural, and of the way that nature equips bodies with a sort of mute natural language, we are also able to gain access to a curious, if not

---

[1] Bacon 2000, 120.

[2] Wollock 2002, 227.

terribly influential,[3] theory constituting a point of contact between early modern philosophy of language on the one hand and the early modern metaphysics of body on the other. The best way to draw this connection out, over the course of the following two sections, will be to focus on Bulwer's very different – and at first glance unjustifiably different – judgments about two different ways in which the body is implicated in human activity: as the vehicle of meanings in body language, and as the object of human artifice in tattooing, foot-binding, and other forms of body modification.

2 'The Hand' as Extended Mind

According to William H. Sherman, in contrast with Wollock, Bulwer's work was in fact a continuation of the seventeenth-century, and in part Baconian, project of developing a universal language, "one that would not only allow people from different places to communicate with each other but would, more importantly, recover the integrity of language before the Tower of Babel and even before the Fall itself."[4] In the *Chirologia*, Bulwer explicitly cites Bacon as the inspiration for his own philosophy of gesture, and in the prefatory poem even describes his own work as a sort of completion of what Bacon did not live long enough to do: "Let *Bacons* soul sleep sweet: the time is come / That Gesture shall s[no?]o longer be dumbe."[5] But how does a gesturology amount to the culmination of something left incomplete in Bacon's own work?

Bulwer often invokes the metonymy of 'the hand' to describe body language in general. The hand, he proclaims, "speakes all languages, and as an *universall character of Reason*, is generally understood and known by all Nations, among the formall differences of their Tongue." For Bulwer, the mutual comprehensibility of different human groups by means of gesture proves that this is "the onely speech that is naturall to Man," and that

> it may well be called the *Tongue and generall language of Humane Nature*, which, without teaching, men in all regions of the habitable world doe at the first sight most easily understand. This is evident by that trade and commerce with those salvage Nations who have long injoyèd the late discovered principalities of the West, with whom (although their Language be strange and unknowne) our Merchants barter and exchange their Wares, driving a rich and silent Trade, by signes, whereby many a dumb bargaine without the crafty Brocage of the Tongue, is advantageously made.[6]

---

[3] As Lewis notes, Bulwer was far from being an important player in seventeenth-century discussions of the universal character: "Although it is certain that [Samuel] Hartlib was familiar with his work, and that he passed details of it on to those he knew to have an interest in universal communication, it does not appear that Bulwer's work had much impact on the main body of language projectors" (Lewis 2007, 46).

[4] Sherman 2004.

[5] Bulwer 1644, Preface, no page numbers; also cited in Lewis 2007, 45.

[6] Bulwer 1644, 3–4.

Notwithstanding Bulwer's intense cultural chauvinism, to which we shall be exposed shortly, here he is making an ethnologically subtle point, one repeated frequently in the following centuries, that whatever the otherness of different cultures consists in, it is not an absolute otherness, permitting no foothold in like responses and behaviors under like conditions that may serve as a starting point for exploring differences. Human groups are, in short, both similar *and* different enough to make comparison meaningful. It is, moreover, with respect to the pre- or sublinguistic expression of meanings or intentions that they are similar, while language is the principle source of differentiation and so also of miscomprehension.

Most early modern theories of a primordial Adamic language held that it was not spoken language as such that compromised true meanings in favour of their mere approximation in the sounds that humans agree by convention will stand in for them. But Bulwer appears to want to argue that speaking, whether pre- or post-lapsarian, corrupts meanings precisely because sounds can only ever be conventional. For this reason, the true primordial embodiment of meaning is only to be found in the body itself, which is to say in gesture. As William Diconson writes in the commendatory verse that opens Bulwer's *Chirologia*:

> ...At first sight we learne to read; and then
> By Natures rules to perce and construe Men:
> So commenting upon their Gesture, finde
> In them the truest copie of the Minde.
> The Tongue and Heart th'intention oft divide:
> The *Hand* and Meaning ever are ally'de.[7]

One might think that gesture is itself already artificial, already a sort of technology of the body. Aristotle had said that the hand is "the instrument of instruments,"[8] and Bulwer interprets the hand's unique place among instruments as consisting in the fact that nothing is lost through its mediation. Intention is not divided. If we think of the extended-mind hypothesis as holding that technologies can store and transmit information that initiates in the mind, and keep this information available for future access,[9] then for Bulwer the hand is certainly the paradigm instance of the extended mind. But unlike writing in a natural language, or the recording of spoken language in a magic sponge,[10] in the case of the hand, to speak with the information theorists, the signal does not degrade at all, and it can be transmitted from anyone to anyone, without prior agreement upon a key for decoding the message.

This is not to say that all gestures are universal. In his *Anthropometamorphosis*, treated in detail in the following section, Bulwer distinguishes between the 'native' and the 'naturall':[11] The latter is what comes directly from nature, the former what

[7] Diconson, William. Commendatory verse in Bulwer 1644, a1v.

[8] *De Anima* 3, 8.

[9] As is well known, this hypothesis has been developed at length by Andy Clark and David Chalmers. See Chalmers and Clark 1998.

[10] See Sutton 2007.

[11] Bulwer 1653, Epistle Dedicatory, no page numbers.

may accompany one from birth, but only as a result of one's cultural environment. The 'native' is the 'nationall', and while there are many gestures that may be native in this sense, these are the object of the special domain of study Bulwer calls "Chirethnicalogia, or the Nationall expression of the Hand." Yet there are also many gestures that are in fact natural, in contrast with 'native' in the sense just described. Thus he explains in the *Chirologia* that "[t]o extend out the right Hand by the arme foreright, is the naturall habit wherein we sometimes allure, invite, speak to, cry after, call, or wanre to come, bring into, exhort, give warning, admonish, protect, pacifie, rebuke," and so on.[12] And further on: "To lift up the right Hand to heaven, is the naturall forme and ceremony of an oath, used by those who call God to witness."[13] Even what is sometimes called 'giving the finger' is qualified by Bulwer as 'naturall': "The putting forth of the middle-Finger, the rest drawn into a Fist on each side, which is then called... vulgarly *Higa*, in the ancient tongue, *pugner* ..., is a naturall expression of scorne and contempt. Hence also Martial calls this Finger, *Digitum impudicum*." Bulwer's insight here seems to be that the middle finger functions as a sort of 'natural symbol'[14] in virtue of certain anatomical correspondences that are transparent to any observer. It thus has a meaning that human beings could not have made up, any more than the correspondence between a hieroglyph of a beetle and an actual beetle is established by convention.

In the *Philocophus, or, The Deafe and Dumbe Mans Friend* of 1648, Bulwer extends his theory of gesture from the hand to the face, arguing that the deaf can, by learning how to read lips, attain to a sort of 'Ocular Audition':

> a treasure reserved for these times, which had escaped their privy search, who guided by the illumination of their owne endeavours had in *sudore vultus* ransackt the bosome of nature, wherein wisdome had hid it among other Arts and Sciences which have their foundation in Nature, and neither grow nor encrease but appeare when time and observation unlockt them unto us: Having well scanned this *Magnale naturae*, I found it to be one of the subtlest pieces of Recondit learning, and that it bordered upon other avenewes unto the braine, as Orall and Dentall Audition, of which wee have discovered sufficient ground to raise a new Art upon, directing how to convey intelligible and articulate sounds another way to the braine then by the eare or eye; shewing that a man may heare as well as speake with his mouth.[15]

Why does Bulwer believe that this new science has waited until the present age to reveal itself? The answer would appear to be connected to his Baconian method, his commitment, as he says, to the growth of knowledge by observation, and to the establishment of "Phylosophicall verity" through "unanswerable Demonstration from matter of fact."[16] Bulwer is extremely loyal to the ancient

[12] Bulwer 1644, 43.

[13] Bulwer 1644, 50.

[14] See Douglas 1996.

[15] Bulwer 1648, Preface, no page numbers. Of course, lip-reading could only ever involve 'nationall' signs in the body, since what is being read are not natural gestures, but only the bodily counterpart to spoken, and thus merely conventional, language. For more on Bulwer's contributions to sign language and to the science of deafness, see Wollock 1996; Norman 1942–1943; Rée 1999.

[16] Bulwer 1648: Preface, no page numbers.

Rhetoricians, and seldom makes a claim without extensive corroboration from Cicero, Martial, and/or? Julius Caesar. But his appreciation of ancient authority does not compromise his commitment to the Baconian commitment to the primacy of observed matters of fact.

'The hand', by which Bulwer means the body, is as we have seen both an extension of the mind as well as its most faithful reflection. It conveys in hieroglyphic form the passing states of the mind. Writing constitutes an advance over oral communication because it can be preserved and transmitted across long distances with little degradation of meaning. Even if the gesture is a more faithful representative of the thought, without webcam technology or something comparable it still requires the 'reader' of the gesture to be in the same place and time as the person doing the gesturing. Nor is it clear what advantage the transience of a gesture confers. If the conventionality and arbitrariness of written alphabetic language is something that is overcome by ideogrammatic writing, then it seems that even if what is preferrable about gesture is its naturalness and its non-conventionalness, these are nonetheless features of a certain kind of writing as well, which also has the virtue of being non-transient. Why, in short, should the hand, with its gestures, be preferred to the page, with its permanent inscriptions, so long as these meet the criterion of non-conventionalness that is also met by gesture? In order to adequately answer these questions, we should perhaps first consider Bulwer's views on what we might call the permanent hieroglyphics of the body, that is, bodily modifications, and in particular tattoos.

3 Foolish Bravery

Why should the transient hieroglyphics of the body be preferred to permanent ones? Beyond the mere xenophobia provoking Bulwer to revolt against exotic practices, there is an important lesson to be learned about the project of Baconian natural philosophy, and about the preference within nascent empiricism for direct observation of nature over the authority of writing.

As many historians and sociologists of knowledge have argued,[17] the technology of inscription may have been an important weapon in the arsenal of expanding Western powers in their encounter with other cultures. This technology enabled the transmission of information across far-flung and complex networks. Yet the exclusively Western character of writing has been called into question by recent scholars. Simon Schaffer, in a compelling article on the encounter between British seafarers and Polynesians in the late eighteenth century,[18] has shown the various respects in which Polynesian tattooing fulfilled many of the same network-sustaining purposes as conventional writing did for the global network sustained by the British.

---

[17] See in particular Todorov 1997; Goody 1986.

[18] See Schaffer 2007.

The most widely accepted definition of writing characterizes it as "a system which records sequences of words in durable signs."[19] These sequences of words are generally supposed (with the exception of Jacques Derrida's puzzling tale of the 'primacy of writing') to have their origin and basis in spoken language, and to be in turn 'cashable' by their readers through vocal pronunciation of them. Writing *stricto sensu* is thus distinct from semasiography, which records ideas in the form of pictograms and ideograms, which in turn could be rendered into speech by different linguistic communities in different, mutually incomprehensible ways.

There is considerable scholarly debate as to which early civilizations developed information-encoding systems worthy of the name 'writing', and which in contrast remained at the level of semasiography. It is certain that the tattooing practices described by Schaffer were at most semasiographic, but as his account of the interaction between the British and the Polynesians shows, there was in this instance certainly no perception of the superiority of the writing on paper over 'quasi-writing' on the body. Moreover, if we are correct in characterizing Bulwer's chirology as part of the broader early modern project of developing an ideal universal language, then we certainly cannot take alphabetic and phonetic writing as in any respect superior to semasiography: from Bulwer's point of view ideographic and pictographic writing are as we have seen preferable to phonetic writing, since only the former could be understood by people belonging to separate spoken-language communities. So when it comes to the permanent hieroglyphs of the body, it could not be out of any disdain for hieroglyphics as merely semasiographic that Bulwer would have denounced them.

Curiously, Bulwer himself describes writing, in particular his own writing, as a sort of spasm of the body, and thus, one would suppose, as deriving its meaning from the state of the body of the writer during a text's composition. In the preface to the *Anthropometamorphosis*, dedicated to Thomas Diconson, Bulwer begins:

> The Heroique Disease of Writing hath (as you well know) long since seized on me, this being the Fifth Publique Paroxisme I have had thereof. It hath been ever the humour of my Genius to put me upon untrodden Pathes, and to make up aggregate Bodies of very scarce and wide dispersed Notions.

He describes his latest 'paroxisme' as

> an Enditement framed against most of the Nations under the Sun; whereby they are arraigned at the Tribunall of Nature, as guilty of High-treason, in Abasing, Counterfeiting, Defacing, and Clipping her Coine, instampt with her Image and Superscription on the Body of Man.

In the breathtaking screed that follows, against all forms of bodily modification, set to verse and included as a prefatory poem to the *Anthropometamorphosis*, Bulwer denounces not just tattooing, but also ear-piercing: "What Gallantry is this, wherein th'appears/So Hell-hound like with long out-stretched Eares?/Whose bored Tips torn wide with the fond weight/Of glittering Stones, thy shoulders over-fraight"; lip-piercing: "The neather Lip's bor'd through to yield a vent/To them, who are not with one mouth content"; the sharpening and cosmetic extraction of teeth: "Here thy Teeth are as sharp as Needles fil'd,/There, in a foolish bravery exil'd."

[19] See Trigger 2007, 586.

After moving, in a similar vein, through nipple-piercing, genital mutilation, foot-binding, and so on, Bulwer finally comes to tattooing:

> Thus we most foolishly our life invade,
> For to advance the Body-makers trade.
> Painted with lists, here, naked arms behold,
> Branded and pounc'd with colours manifold,
> Rich tinctur'd Red, Blacke, Tawny, Yellow, White,
> All badges of the gallants gay delight.[20]

Bulwer's preferred examples of tattooing come from exotic nations, in particular from the ethnographic materials to which he had access concerning the customs and appearance of Native Americans. He writes that 'the Virginian women'

> pounce and rase their Faces and whole Bodies with a sharp iron, which makes a stampe in curious knots, and drawes the proportions of Fowles, Fishes, or Beasts; then with painting of sundry live colours they rub it into the stamp, which will never be taken away, because it is dried into the flesh.[21]

He returns to this same practice a few pages later, this time openly paraphrasing John Smith's *Generall Historie of Virginia* of 1624 (see Fig. 1):

> The Virginian women adorne themselves with paintings; some have their Face, Breasts, Hands, and Legs, cunningly embroidered with divers workes, as Beasts, Serpents, artificially wrought into their flesh with black spots.[22]

Fig. 1 Bulwer 1653.

[20] Bulwer 1653, Dedicatory poem, no page numbers.

[21] Bulwer 1653, 252.

[22] Bulwer 1653, 257.

Yet Bulwer certainly does not see this sort of practice as limited to far-away and exotic cultures. He adds that

> Our Ladies here have lately entertained a vaine Custome of spotting their Faces, out of an affectation of a Mole to set off? their beauty, such as Venus had, and it is well if one black patch will serve to make their Faces remarkable; for some fill their Visages full of them, varied into all manner of shapes and figures.[23]

In general, Bulwer does not seem to distinguish between different forms of body 'painting', which include tattooed representations, or simply designs, or even the use of cosmetics to change the hue of the skin (see Fig. 2). Thus he denounces the Virginian women for the same 'bravery' he sees in 'the Ladies of Italy', who, "to seeme fairer than the rest, take a pride to besmeare and paint themselves."[24] He also tends to mention more instances of non-representational tattooing than the rare case of the fish or beast, such as the practice of the 'Egyptian Moores, both men and women," who "for love of each other, distaine their Chins into knots, and flowers of blew, made by the pricking of the skin with needles, and rubbing it over with inke and the juyce of an herb."[25] He complains of all of these 'Nations', the Virginians, the Englishwomen, the Italians, and the Moors alike, "what needlesse paine they put themselves unto to maintaine their cruell bravery! Nay, which is yet stranger, they

Fig. 2 Bulwer 1653

[23] Bulwer 1653, 261.

[24] Bulwer 1653, 260.

[25] Bulwer 1653, 252.

seeme to love this unnaturall and bloudy Gallantry so well, that they hate their own flesh and bloud, whereof they freely sacrifice to their fantasticall imaginations."[26]

Let us return for a moment to the South Sea encounter described by Schaffer. According to him, when the astronomer aboard the *Daedalus*, a certain Gooch, brought out his pen and ink on deck, one of the islanders lay down in front of him. Gooch reports that the man "wished me to tattoo his hip, but I did not understand the task... On seeing me write, [he] deem'd it tattooing."[27] Unlike the astronomer, Bulwer sees no symmetry between his writing and the tattooing of the Virginian women. He does not, again, take their representation of fish upon their faces as significantly different from the simple use of rouge. Yet at the same time, again, he takes the body as itself a sort of 'writing' in the broad sense of a report upon the state of the soul, and he condemns all forms of bodily modification, including tattooing, on the grounds that these irreversibly freeze the body in a condition in which nature did not intend for it to be. Thus bodily modification is a degradation of what is given by nature, and thus a moral degradation of the soul of the person whose body has been modified. Writing in the strict sense as well is in the end a modification of the body, but one that does not lead to its temporary alteration. It is, like gesture, fleeting, but unlike gesture it leaves a permanent record, or a permanent hieroglyphic, outside of the body on the written page. Whether this permanent record is morally blameworthy or not would, one supposes, be determined by the content; yet even in such a case its blameworthiness would be more akin to that of an obscene gesture than that of a self-inflicted bodily deformation.

Interestingly, Bulwer must concede that not every form of bodily modification amounts to an unnatural intervention in the divinely instituted course of nature. Some things were set up in nature in order to guide human beings toward the right sort of conduct, including those excrescences of the body that require regular grooming. While Bulwer believes that for a man to shave his beard would be to unnaturally do away with the "naturall ensigne of Manhood, appearing about the mouth,"[28] he nonetheless will not go so far as to claim that clipping one's nails is an impermissible derailing of a natural process. Instead, he maintains, in all parts "there is an appointed end, a certain commoderation of the quantity of parts to the actions of them, according to the faculties using the Organ in the Body." He maintains that the "continuall increase [of the nails] in man is an Argument of a Divine Nature, a prerogative in which beasts cannot participate, and teacheth us charity to our Bodies."[29] Bulwer goes on to argue that, before the Fall, in the absence of iron tools, Adam must have kept his nails short by biting them. Why should the line be drawn between nails and facial hair? Why is it in keeping with God's appointed ends to curtail growth in the one case, but an abortion of the same ends to do the same in the other? Clearly, in the end Bulwer's conception of what is natural comes

---

[26] Bulwer 1653, 253.

[27] Cited in Schaffer 2007, 91.

[28] Bulwer 1653, "The Introduction," Scene XII.

[29] Bulwer 1653, 297.

out looking like a list of his preferred conventions. What is interesting for our purposes, however, is Bulwer's conviction that there is a proper way to maintain the body and that this way is dictated by nature. The full elaboration of which modification practices are unnatural and which, in contrast, are there in testimony to the divine nature, will amount to nothing less than 'a corporall Philosophy'.

How, now, does Bulwer's gesturology fit into this philosophy? Interestingly, the *Anthropometamorphosis* is introduced with a commendatory poem by Francis Goldsmith, who makes explicit reference to the continuity of the project of the *Chirologia* with that of the more recent work. He writes of Bulwer:

> He, whose first Lecture was on Natures hand,
> Now all her Features hath exactly scan'd...
> Pliny but Natures History us gave;
> Thou, her Great Champion, dost her honour save.[30]

The project of the *Chirologia* was to give an account of how the body reports upon internal states; the *Anthropometamorphosis*, in turn, draws out the explicitly moral implications of the insight that the body is in the end nothing but a report of this sort. The most important implication is that the body must not be made to deviate irreversibly from the ends appointed to it by nature. Naturally, a human being may on occasion think of fish, but there is nothing so naturally ichthyous about a human being that the form of a fish should deserve a place upon the human face.

What Bulwer argues is, in effect, that physiognomy philosophically understood just is the view that the soul makes the body. For Bulwer, 'the hand' is such an adequate representative of the states of the soul, because it is in the end nothing other than the soul's outward sign. If the soul is in harmony with nature, which is to say for Bulwer in harmony with reason, then the body will only ever grow or move naturally and rationally. In traditional physiognomy, the soul makes the body simply as an outward reflection of an inward state. This can be either the reflection of an individual personality, as in the case of a simpleton with droopy eyes and mouth, or of national or ethnic character traits. Bulwer however has the soul shaping the body through directly willed intervention in the ordinary development of the body. Again, this can be either individual or 'national', but what is different about Bulwer's account is that he sees the deformation of the body as an end in itself of morally degenerate people, rather than as an unintended consequence or side effect of moral degeneracy. In this respect, Bulwer moralizes the artificial like none of his contemporaries.

4 Bulwer's Stoic Anthropopoeia

We may now be in a better position to return to the question as to why Bulwer should have picked up and continued the Baconian project of developing a universal character, while at the same time abandoning that part of this project that sought to develop an artificial language for the ideal transmission of meanings between

[30] Bulwer 1653, Dedicatory poem, no page numbers.

people. It appears that, for Bulwer, the ideal universal character would not be an artificial language at all, but the true natural language, that differs from what we would call 'natural languages' (e.g., English, Latin), to the extent that it is transparently meaningful to all human beings, and to the extent that these latter 'natural' are in fact only 'native', and already involve a distancing or disconnection from the immediate expression of meaning of which the body is naturally capable. Bulwer's place in the history of early modern thinking about artificial languages, then, is marked out by his opposition to artificiality *tout court*, and by his conviction that we must distinguish between the project of developing artificial languages, on the one hand, and that of developing a universal character on the other. Whatever is going to be truly universal will be natural, and not artificial.

Bulwer does not entirely disdain 'body-making': in the *Chirologia*, he refers to his own project as an 'anthropopoeia'.[31] Let us recall that the final component of this two-part neologism means both 'making' (as in 'chrysopoiesis' or 'gold-making') as well as 'poetry'. We might use this instance – one of many – of linguistic creativity in Bulwer to highlight an interesting division in early modern thought about artificial languages in particular, and in general about the division between the artificial or technological on the one hand, and the natural on the other. For Bulwer, anthropopoeia should only be the poetry of the human being, but never the making of such a being by way of intervention in the course of nature. The proper human poetry, moreover, is the one dictated directly by nature. Indeed, to present any other poetry as anthropopoeia is already to cross over into poiesis as technological intervention, or, which is the same, as deformation of what is given by nature.

The complete pedigree of Bulwer's strict identification of the natural with the rational, and these in turn with the moral, would be difficult to elaborate in full. But one undeniable component of it is the Baconian rediscovery of the Stoic identification of nature and reason, and the corollary view that the best way to live is the way that is most in keeping with nature's dictates.[32] Part of Bacon's own version of the Stoic philosophy of nature had been the view that, through the new method of observation of nature, an ideal language could be reconstituted, for the first time since the Fall. What Bulwer's version of this idea makes clear, however, is that there is a division among the early modern advocates of an ideal language, separating those who would want to build it up artificially and de novo, from those who believe it is dictated directly by nature and that it is in some way primordial, or something to be *re*discovered. G. W. Leibniz is a prominent representative of the former camp: he believes, for example, that Hebrew is just a natural language among others, as its corrupted verb forms show, and that in part for this reason the attainment of an ideal language will only arrive when we overcome our interest in finding a form of communication that is primordial, pure, and coeval with the creation, and instead seek to create that language ourselves by means of artifice. In sharp contrast, Bulwer's project gives a whole new meaning to the notion of 'natural language', and insists,

[31] Bulwer 1644, 44.

[32] For a classic exposition of Stoic elements in Bacon's writing, even if focused principally on rhetorical matters, see Croll 1989.

against the advocates of an ideal, artificial language, that it is precisely by leaving all artifice behind that one arrives at the truly transparent embodiment of meanings.

As this last phrase suggests, and as Bulwer's *oeuvre* vividly shows, meanings may be thought of as literally embodied, and there may be an important connection between early modern debates about the nature of the ideal language, on the one hand, and on the other the contemporary debates as to the role of the soul, if any, in the body. Again, for Bulwer the physiognomist, the soul makes the body in the most direct way possible: through active, willful intervention, whether of the sort practiced at a 'nationall' level, or as the result of an? individual caprice. The best body is the one had by a moral soul, which is to say the one left to develop as nature intends. Failure to respect this intention can lead even to the loss of a properly human nature. Thus Bulwer relates in the *Anthropometamophosis* that "in discourse I have heard to fall, somewhat in earnest, from the mouth of a Philosopher (one in points of common beliefe (indeed) too scepticall) That man was a meer Artificial creature, and was at first but a kind of Ape or Baboon, who through his industry... by degrees in time had improved his Figure & his Reason up to the perfection of man." Bulwer believes that the 'Philosopher's' opinion, in comparison with those of Plato and Galen, constitutes a symptom of the moral decline of the modern period. For, he thinks, if mutation of humanity over time can occur, it will not, as the Philosopher thinks, take the character of an ascent from beast to man, but rather the reverse, a descent into ape-likeness:

> But by this new History of abused Nature it will appeare a sad truth, that mans indeavours have run so farr from raising himselfe above the pitch of his Originall endowments, that he is muchfallen below himselfe; and in many parts of the world is practically degenerated into the similitude of a Beast.[33]

As with the body, so too with language, the best expression of meaning is the one that respects a human being's 'Originall endowments' by, so to speak, leaving intentionality to nature. This means that speaking – not to mention alphabetic writing – is already, for Bulwer, a denaturing artifice. The transient hieroglyphics of gesture, in contrast, or at least those of them that are not merely 'native', amount to a universal *natural* character, as opposed to a universal *artificial* language, insofar as they convey nature's meanings directly.

It is hard to imagine that Bulwer might have thought that any proposition whatsoever could be expressed by natural gesture. Did he really think that one could debate, say, transubstantiation vs. consubstantiation without resorting to conventional signs? Probably not, but he also probably does not think that such doctrinal disputes, or adherence to doctrine, is necessary for the possession of natural wisdom. In this connection, Bulwer again seems to be in agreement with ancient Stoic, and perhaps Cynic, doctrine, according to which animals are the supreme embodiment of reason. The language of the hand, which is to say gesture, is for Bulwer just the language of animals, which, while lacking hands strictly speaking, nonetheless comport themselves in such a way that their internal states

[33] Bulwer 1653, 455f.

can be read directly off of their bodies. He writes of "the common tongue of Beasts, who by gestures declare their senses, and dumb affections," adding:

> [A]s Montaigne (in that elegant Essay of his, where he in imitation of Plutarch, maintaines that Beasts participate with us in the rationality of thier discourses) shewes, that even they that have no voyce at all, by their reciprocall kindnesse, which we see in them, we easily inferre there are some other meanes of entercommunication: their gestures treat, and their motions discourse.[34]

Bodily modifications of any sort, in turn, including the permanent hieroglyphics stamped into the body by tattooing, cannot convey nature's meanings, even if they can, perhaps, denote fish or beasts, since nature never meant for the face of a human to permanently convey the idea of a fish or a beast. Body modification is for Bulwer an unnatural anthropopoeia or body-making, while the anthropopoeia that is in agreement with nature is the one that does not interfere with the body's natural form, even as it allows the body to express the passing states of the soul through the transitory hieroglyphics of gesture.

5 Conclusion

This is only the beginning of a sketch showing the intersection of some important themes in the work of an obscure and colourful seventeenth-century thinker. In the future, it will perhaps be fruitful to further investigate the relationship between, on the one hand, the emerging ethnography of the early modern period, with its inevitable accounts of the causes and nature of 'racial' difference, and on the other hand theories of language and meaning, which often sought to move back behind 'natural' languages in order to find features of human existence that served as natural vehicles of communication, and which thereby worked in a complementary fashion to the protoethnography of the period, to the extent that both were preoccupied with determining the limits of universalist thinking about the human race. Finally, it will be useful to think about both of these issues in relation to the perhaps more familiar metaphysical questions in the early modern period about the nature and ontology of the body, and about the causal influence of the soul upon the body. As is familiar to historians of early modern philosophy, the idea that the soul somehow makes the body was one widely available option (alongside preestablished harmony, occasionalism, and many other theories), often associated with, but certainly not limited to, Platonist schools of thought, in accounting for the way in which a particular soul has this particular body, with its unique conformation and capacities. As Bulwer's work shows, 'body-making' was not necessarily just an abstract metaphysical process, but could also be carried out through technological intervention in the course of nature: a power enjoyed by human beings but not animals, and one that was and is accompanied by moral concerns as to its proper use.

[34] Bulwer 1644, 5–6.

References

Bacon, Francis. 2000. The *Advancement of Learning*. In *The Oxford Francis Bacon*, Vol. IV, ed. Michael Kiernan. Oxford: Oxford University Press.

Bulwer, John. 1644. *Chirologia: or the Naturall Language of the Hand. Composed of the Speaking Motions, and Discoursing Gestures Thereof. Whereunto Is Added Chironomia: or, the Art of Manuall Rhetoricke. Consisting of the Naturall Expressions, Digested by Art in the Hand, as the Chiefest Instrument of Eloquence, by Historicall Manifesto's, Exemplified Out of the Authentique Registers of Common Life, and Civill Conversation. With Types, or Chyrograms: A Long-Wish'd for Illustration of This Argument.* London: Thomas Harper.

Bulwer, John. 1648. *Philocophus, or, A Deafe and Dumbe Mans Friend. Exhibiting the Philosophicall Verity of That Subtile Art, Which May Inable One with an Observant Eie, to Hear What Any Man Speaks by the Moving of His Lips.* London: Humphrey Moseley.

Bulwer, John. 1653. *Anthropometamorphosis: Man Transform'd: or, the Artificiall Changling Historically Presented, in the Mad and Cruell Gallantry, Foolish Bravery, Ridiculous Beauty, Filthy Fineness, and Loathsome Loveliness of Most Nations, Fashioning and Altering Their Bodies from the Mould Intended by Nature; with Figures of Those Transfigurations. To Which Artificiall and Affected Deformations Are Added, All the Native and Nationall Monstrosities That Have Appeared to Disfigure the Humane Fabrick. With a Vindication of the Regular Beauty and Honesty of Nature. And an Appendix of the Pedigree of the English Gallant.* London: William Hunt.

Clark, Andy and Chalmers, David. 1998. The Extended Mind. *Analysis* 58: 10–23.

Croll, Morris W. 1989. Attic Prose: Lipsius, Montaigne, Bacon. In *Style, Rhetoric, and Rhythm: Essays by Morris W. Croll*, eds. Max Patrick and Robert O. Evans, 167–202. Woodbridge: Oxbow Press.

Douglas, Mary. 1996. *Natural Symbols: Explorations in Cosmology*. New York: Routledge.

Goody, Jack. 1986. *The Logic of Writing and the Organisation of Society*. Cambridge: Cambridge University Press.

Lewis, Rhodri. 2007. *Language, Mind and Nature: Artificial Languages in England from Bacon to Locke*. Cambridge: Cambridge University Press.

Norman, H. J. 1942–1943. John Bulwer (fl. 1654), 'The Chirosopher', Pioneer in the Treatment of the Deaf and Dumb and in Psychology. *Proceedings of the Royal Society of Medicine* 36: 589–602.

Rée, Jonathan. 1999. *I See a Voice: Deafness, Language and the Senses: A Philosophical History.* New York: Metropolitan Books.

Schaffer, Simon. 2007. 'On Seeing Me Write': Inscription Devices in the South Seas. *Representations* 97: 90–122.

Sherman, William H. 2004. Toward a History of the Manicule. Presented at the conference, Book-Trade Consumers: Owners, Annotators, and the Signs of Reading. London: Birbeck College.

Sutton, John. 2007. Spongy Brains and Material Memories. In *Environment and Embodiment in Early Modern England,* eds. Mary Floyd-Wilson and Garrett Sullivan, 14–34. New York: Palgrave Macmillan.

Todorov, Tzvetan. 1997. *The Conquest of America: The Question of the Other*. Norman: University of Oklahoma Press.

Trigger, Bruce G. 2007. *Understanding Early Civilizations*. Cambridge: Cambridge University Press.

Wollock, Jeffrey. 1996. John Bulwer's (1606–1656) Place in the History of the Deaf. *Historiographica Linguistica* 23: 1–46.

Wollock, Jeffrey. 2002. John Bulwer (1606–1656) and the Significance of Gesture in 17th-Century Theories of Language and Cognition. *Gesture* 2: 227–258.

Memory and Empirical Information: Samuel Hartlib, John Beale and Robert Boyle

Richard Yeo*

Much memory, or memory of many things, is called *Experience*.
Thomas Hobbes, *Leviathan*[1]

Abstract Robert Boyle and John Beale had connections with Samuel Hartlib and his correspondence circle. The position of these three figures can be taken as an 'empirical' one in the sense that they favoured 'particulars' over 'systems'. But differences emerge if we consider their attitudes towards the role of memory in Baconian natural histories. Hartlib's call for empirical particulars coexisted with an expectation that information could be reduced and arranged to aid both memory and thinking. As one model, William Petty promoted John Pell's reductions of mathematical knowledge. Beale's letters to Boyle (in the 1660s) urged systematic ordering of empirical data in the service of memory and hypotheses. Although Boyle did believe that a disciplined individual memory could embody multifarious experiences, he resisted Beale's advice. What we accept as Boyle's 'empirical' attitude was not so much a distinctive commitment to gathering matters of fact – something also professed by Hartlib and Beale – but a refusal to condense and arrange material in the way they demanded. Beale's promotion of memory techniques that relied on highly structured arrangements of units seems to have aggravated Boyle's existing suspicion of premature systems.

---

*I am grateful to the Australian Research Council for funding the larger project to which this chapter belongs. For their advice and comments, I thank Peter Anstey, Mary Louise Yeo, an anonymous referee, and participants at a workshop at Sydney University convened by Charles Wolfe. I also thank Jacky Hodgson and her staff in Special Collections, University of Sheffield Library, for access to the original Hartlib Papers.

[1] Hobbes 1651, 16.

R. Yeo
Centre for Public Culture and Ideas, Griffith University, Brisbane

C.T. Wolfe and O. Gal (eds.), *The Body as Object and Instrument of Knowledge: Embodied Empiricism in Early Modern Science*, Studies in History and Philosophy of Science 25,
DOI 10.1007/978-90-481-3686-5_10,

1 Introduction

John Passmore once remarked that if William Whewell, the nineteenth-century historian and philosopher of science, had not lived, J. S. Mill would have had to invent him. Mill himself admitted that he needed Whewell's strong statement of an idealist philosophy of science before he was able to express his own empiricist account.[2] Their debate made epistemology the basis of an argument between rival twins – empiricism and idealism – that spread across domains from science to morality, from aesthetics to politics. In early modern England, the term empiricism (or, more usually, variants such as 'empirick' or 'empirical') did not stand for a well-defined epistemological position delineated against a clear alternative. For the purposes of this paper, I want to consider the ways in which it is helpful to speak of an 'empirical stance' (rather than a doctrine of 'empiricism') as characterising those who favoured 'particulars' over 'systems'.[3] In his *Experimental Philosophy* (1663–1664), Henry Power ridiculed excessive attachment to systems as entirely at odds with the spirit of the new science:

> Me-thinks, I see how all the old Rubbish must be thrown away, and the rotten Buildings be overthrown, and carried away with so powerful an Inundation. These are the days that must lay a new Foundation of a more magnificent Philosophy, never to be overthrown: that will Empirically and Sensibly canvas the *Phaenomena* of Nature.[4]

Of course, as Power's title indicated, experimental inquiry was crucial to his position; but I think it is instructive to consider the word 'empirical' as describing the quest for information derived from books, testimony, and observations as well as experiments.[5] This broader set of referents captures the sense in which Robert Boyle defined "Experience" to include "all those ways of Information, whereby we attain any Knowledge that we do not owe to abstracted *Reason*."[6] As he elaborated, this "*Experience*" was wider than "*Personal Experience*"; it encompassed "the knowledge we have of any matter of Fact, which, without owing to Ratiocination, either we acquire by the Immediate Testimony of our Own Senses and other Faculties, or accrews to us by the Communicated Testimony of Others."[7] He championed "the modern Virtuosi" over members of other philosophical sects, arguing that it was the builders of Systems who "make little use of Experience; contenting themselves for the most part to employ but few and obvious Experiments,

---

[2]Passmore 1970, 19 cited in Yeo 1993, 178. See also Snyder 2006. For information on historical figures mentioned in this chapter, see the *Oxford Dictionary of National Biography*.

[3]See van Fraassen 2002.

[4]Power 1663–1664, 184 and 192; my emphasis. Here, and in all subsequent quotations from primary sources, I preserve the original spelling and punctuation.

[5]Peter Anstey (2005) contrasts 'experimental' and 'speculative' philosophy. I think the other kinds of empirical data that I mention here need to be added to the 'experimental' side of this contrast.

[6]Boyle 1690, 54; also 4–5. I leave aside the question of the various meanings of the word information (and 'informations') in this period.

[7]*Ibid.*, 57; see also 53–7; this quotation also in Boyle 1999–2000, 11, 306.

and vulgar Traditions, usually Uncertain, and oftentimes False."[8] Boyle portrayed the Christian *virtuoso* as one who carefully collected and examined empirical information from diverse sources:

> That, when, in this Discourse, I speak of an Experimental Philosopher, or Virtuoso; I do not mean, either, on this hand, a Libertine, tho' Ingenious; or a Sensualist, though Curious; or, on that hand, a mere Empirick, or some vulgar Chymist, that looks upon nothing as Experimental, wherein Chymistry, Mechanicks, &c are not employ'd; and who too often makes Experiments, without making Reflection on them, as having it more in his aim to Produce Effects, than to Discover Truths. But the Person I here mean, is such a one, as by attentively looking about him, gathers Experience, not from his own Tryals alone, but from divers other *matters of fact*, which he heedfully observes, though he had no share in the effecting them.[9]

What is the link – implied in the title of my chapter – between this 'empirical' stance and the role of memory? In promoting sensory evidence of particulars against false universals or premature theoretical systems, Boyle confronted the problem Francis Bacon had diagnosed: how to manage the plethora of empirical particulars demanded by the new philosophy?[10] As early as 1657, in "A Proemial Essay," Boyle expressed suspicion of 'systems' and 'superstructures' not founded on observation or experiment.[11] But one virtue of systems was their ability to order and condense material. This is the admission in a suggestive passage from *The General History of the Air* (1692). When canvassing the "Peripatetic Doctrine about the Limits and Temperaments of the three Regions, into which they divide the Air." Boyle cautioned that "it becomes a *Naturalist* to consider, not so much how easy a Doctrine is, by reason of its Concinity [internal harmony], to be *remembred* or *supposed*, as how strongly "tis to be *proved*."[12] He thus conceded that certain features of leading doctrines might make them easy to remember, but insisted that this was not a voucher for their truth. Although harmony and order might be seductive *aide-mémoires*, he preferred an honest mass of empirical particulars, even if not yet methodized and unable to be remembered. Why did he need to issue such a warning?[13]

---

[8]Boyle 1690, 5.

[9]*Ibid.*, 52. This passage shows that Boyle had to distinguish his position from that of the 'empiricks', who also professed opposition to general systems. See Galen 1985; also Cook 1990.

[10]For the tension created by Bacon's linkage of *Memoria* and *Historia*, see Yeo 2007.

[11]Boyle 1661, 1–36, at 6–9; also the 2nd edn. of 1669 in Boyle 1999–2000, 2, 9–34. See Anstey 2000, 4–7 for the caveat that Boyle was not opposed to systematisation as such. On this, see Boyle 1674.

[12]Boyle 1692, 150 for Title XIX, 'Of the Heat and Coldness of the Air'; also in Boyle 1999–2000, 12, 100. The material for this work was gathered from the 1660s onwards. The Title XIX cited here appeared as no. 18 in a copy of a list of thirty-one Titles dating from 1682; this is in Bodleian Locke MS c. 42, part 1, 16–17. Apart from this clue, I have not attempted to establish the likely date of this specific Title. See the editors' discussion in Boyle 1999–2000, 12, xii–xxi.

[13]In this instance, Boyle was not referring to grand systems, but to a doctrine about the nature of air and some of its effects in climatic regions.

I think that one answer lies in Boyle's lifelong preoccupation with the proper role of memory in the custody and organization of empirical information and ideas. Some sensibilities about memory are highlighted in the letters Boyle received from John Beale between 1663 and 1666.[14] These reveal interestingly *different* approaches to the question of how far collaborative Baconian natural histories (collections of medical, chemical and other data) should rely on individual memory, either natural or trained. Beale was an avid contributor of empirical information (especially, but not only, about fruit trees) to the Royal Society; but he claimed that such information must be arranged to assist memory and thinking. Boyle tried to extend his own empirical collections by pooling data gathered by others, and thus confronted the sheer mass of information potentially entailed by Baconian natural histories. He also developed his own memory discipline, but cautioned against the artificial classification that mnemonic arts usually relied upon.

2 Samuel Hartlib and His Circle

Before coming to Boyle's own approach and his exchange with Beale, we need to consider the Prussian émigré, Samuel Hartlib, the 'great intelligencer' with whom they were both linked. Beale had corresponded with Hartlib from the spring of 1656; he continued to write to him, almost on a weekly basis, until Hartlib's death in 1662.[15] Members of Hartlib's correspondence network were among the young Boyle's earliest intellectual contacts.[16] His first letter to Hartlib dates from early 1647, and from early 1648 there are frequent references to him in Hartlib's diary (his Ephemerides) as the source of reports, recipes (or 'receipts') and observations.[17] Boyle's first publication, "An Invitation to Free and Generous Communication" (1655), appeared in a volume orchestrated by Hartlib; its message was that 'empiricks' should share their secret knowledge, such as cures for the stone.[18] Boyle was sympathetic to the general tenor of Hartlib's approach: in a letter to John Mallett, he punned on

[14]Beale was elected FRS in January 1663. Maddison (1958) lists 27 letters from Beale to Boyle between these dates; in Boyle 2001 there are 30 letters, two of which are no longer extant.

[15]See Greengrass et al. 1994.

[16]See Stubbs 1982, 477–85 and 464 for suggestion that Beale's correspondence amounts to about 400 letters. There is no mention of Beale in More 1944.

[17]Boyle to Hartlib, [early 1647], in Boyle 2001, 1, 51; and Hartlib, Ephemerides, January–June 1648, HP31/22/1A. Hereafter, all citations from the Hartlib Papers (Hartlib 2002) are given in this format. For Boyle's early links with Hartlib's circle, including Beale, William Petty, Benjamin Worsely, William Brereton and John Worthington, see Maddison 1969, 61–63, 68, 71 and 95. See also O'Brien 1965; Shapin 1994, 137, 144, 175.

[18]Boyle 1655, 113–50; also in Boyle 1999–2000, 1, 1–12. This piece was written circa 1647–48. See Rowbottom 1950.

"The Legacy of Husbandry" (a Hartlib publication), affirming that "for my part I make no doubt that the Husbandry of Knowledge will be dayly improv'd too: (though by throwing downe of Enclosures) & all Parts of Philosophy, be both better cultivated & more fruitfull."[19]

Hartlib pressed the need for what we might call empirical information, or what he and his correspondents referred to as 'particulars' gained from observations and experiments. It is difficult to find a precise definition of this notion, and I leave aside the question of how such 'particulars' relate to Boyle's notion of 'matters of fact'; but Hartlib invariably used this term in the context of rejecting excessive deductive methods. For example, in 1640 he noted in his diary that "Those abstract Axioms which are made in Philosophy breed but a slavish assent in men. For they must beleeve only that they are true not knowing out of what particulars they come to bee so … It is a very hard matter to come once to know throughly that a thing is certain and true. Therfore wee must labour principally in the Historical part of all things."[20] Hartlib professed that "The more new particulars wee meet withal the more knowledge will bee enlarged."[21] This assertion was pitched against both the scholastic pedagogy of the 'schools' or universities and, less aggressively, against modern philosophical systems, including that of René Descartes.[22]

In a letter of 13 September 1630, Hartlib told John Dury [or Durie] that he was sending something from "my by-collections," so named because he happily rejected "ordered Systemes."[23] Hartlib favoured Bacon's "Aphorismes as the onliest way for deliverie of Knowledge … For discourse of Illustration must bee cut of." He suspected "the shew of a Totall … Whereas the Aphorismes representing a knowledge broken doe invite men to inquire further."[24] Writing five years later he declared that: "the only way to write for the encreasing of Learning is to write Truths by way of Aphorismes. Systematical Method is like a bag or sack which come bound up."[25] Like Bacon, Hartlib viewed aphorisms not as pithy compressions of authoritative

[19]Boyle to John Mallett, 2[3] March 1652, Boyle 2001, 1, 133. For the context of Hartlib's interests, see Bennett and Mandelbrote 1998, 33–42 and 157–68.

[20]Hartlib, Ephemerides, August 1640, HP30/4/53B. Hartlib's stance matches the spirit of Bacon's plea, in his *Novum organum,* for 'a greater abundance of experiments' and a 'store of particulars' as a foundation for further inquiry. See Bacon 1963, 4, part I, aphorism nos. 100 and 103, 95–96; also in Bacon 2004, 159–61.

[21]Hartlib, Ephemerides, January 1640, HP30/4/42A.

[22]See Hartlib, Ephemerides, HP30/4/3A: 'Cartes in his apriori philosophizing' against 'the sonder a posterior path of Junguis'.

[23]Hartlib to Dury [or Durie], 13 September 1630, HP7/12/1B.

[24]See also Webster 1970, 76–77.

[25]Hartlib, Ephemerides, 1635, HP29/3/13A.

notions, but as spurs to further investigation and collection of particulars.[26] But his support of Bacon was not unreserved. As Stephen Clucas has indicated, Hartlib was consciously eclectic, aiming to combine the best of insights from a range of authors; he sometimes expressed preference for Giacomo Aconzio (or Acontius), Joachim Jungius or Joachim Hüber over Bacon.[27]

In Hartlib's estimate, even Bacon had underestimated the scale of effort and time dictated by his programme. The question was how to gather, order and store information, especially when much of it was at first necessarily disconnected. Although Hartlib sponsored Jan Comenius' visit to England in 1641, he complained that the Czech reformer "play[ed] so much upon … [the point that] all things are reducible to certaine maine heades and principles from whence all other particulars can bee deduced." Hartlib responded that the collection of "a great Copia Rerum" must take precedence over the quest for an "Ars Universalis."[28] Referring to the ideal of a "Pansophia," he cautioned that first "wee must labor to get more particularia. Else wee shall repeate the selfe-same Notions by several expressions and advance not a whit further. Wheras when wee must write of particulars wee must always bring some new matter or other … wee can not say the same things of a dog and a peacock, but must needs bring new notions of them."[29] The endemic problem here was that "Every body will flie presently to abstractions and generals but they are loath to meddle with the gathering of the Experimental History because it is more troublesome then the other, 2. because a man cannot here compendiat as in the other but must bee as large as the things require themselves."[30]

How did Hartlib propose to manage and process this information? Two features of his thinking (and that of some close associates such as the physician, William Petty and the mathematician, John Pell) are relevant here: first, the emphasis on the role of memory in storing experience of empirical particulars, such as observations and experiments; second, the importance given to selecting and commonplacing material found in books.

Hartlib maintained that the proper collection of 'particulars' required the cultivation of both observation and memory. He believed that individuals should build up their experiences of particulars in memory, beginning in childhood and continuing until at least the age of twenty-five. Writing in his Ephemerides for 1639, he asserted that "wisdome of Arts, Sciences and Inventions will never bee enlarged till Men furnish themselves in their yonger yeares with a World of all manner of

---

[26] On Bacon's view of aphorisms, see Clucas 1997.

[27] Clucas 1994, 58–63 and Clucas 2009. See Hartlib, Ephemerides, 1639, HP30/4/22A on Acontius; and HP30/4/24B criticizing Comenius; and Hartlib to Boyle, 8 or 9 May 1654, praising Jungius, Boyle 2001, 1, 172–73.

[28] Hartlib, Ephemerides, 1640, HP30/4/42B and HP30/4/39A.

[29] Hartlib, Ephemerides, 1640, HP30/4/43A.

[30] Hartlib, Ephemerides, 1640, part 2, HP 30/4/49B; this echoes aphorism no. 19 in Bacon's *Novum Organum*, Bacon 1963, 4, 50; also in Bacon 2004, 71.

particulars til 16–25, or 30, or more. Afterwards when they are come to maturity of judgment, they will bee able to some purpose to exercise their reason and philosophat upon them."[31] Castigating traditional pedagogy, he alleged that children were taught "the most abstractive things and that in a most verbal way." Instead, there was no reason "why Children should very soone learne to write, … that from their very infancy they may bee taught how to make observations and how to write into their Ephemerides of whatever they shall see and heare which afterwards being reduced into Loci Communes when they are of judgment, would make them learned in the whole Encyclopaedia or Pansophia before they are aware of."[32] This preoccupation appeared in his correspondence with Beale, who agreed that from an early age each individual should build up a cognitive structure into which later, and novel, material can be embedded. He told Hartlib that the categories and themes internalised from "your Child-hoode" will be the "best & truest Topiques for our future improvement of Memory during life."[33] Two years later, Beale repeated this advice to Boyle, asserting that the strength of memory could be prolonged if an early mental structure has been established, if "the studdes or Nayles were engrafted in our childehood."[34] I return to the significance of this below.

The second of these commitments – reading and summarizing books – appears to clash with Bacon's rhetorical contrast of books with nature. However, it was mainly Thomas Sprat and other polemicists who exaggerated this; Bacon himself said that the best books needed to be examined and new ones written.[35] Hartlib contended that the classical texts could still be read with profit:

> Now these particulars are to bee learned by Sensual observation in conversation as likewise by reading of Histories where there is nothing required then Memory and faith to beleeve it. Those wee see that from their yonger years have read all manner of Classick Authors exceed all your pedantick Systematik's in true knowledge, because those Authors containe a World of realitys in them which their other bookes doe not.[36]

The aim was to extract useful knowledge from all extant books, both ancient and modern, by commonplacing material under identifiable Heads. This method was constantly recommended. Hartlib and his colleagues believed that orderly reduction and summary of knowledge would show up duplication and repetition, thus allowing an economical compression. Then the collection of new material could proceed, following up gaps revealed by this accounting process.[37]

---

[31] Hartlib, Ephemerides, 1639, HP30/4/22A.

[32] Hartlib, Ephemerides, 1650, HP28/1/53B.

[33] Copy in scribal hand of Beale to Hartlib, 2 December 1661, HP67/22/13B.

[34] Beale to Boyle, 29 September 1663, Boyle 2001, 2, 134.

[35] See Bacon's *De Augmentis*, Bacon 1963, 4, 290: '… the true remedy is not to destroy the old books, but to make more good ones'.

[36] Hartlib, Ephemerides, 1639, HP30/4/22B.

[37] For the application of this approach to information of various kinds, and the function of Hartlib's 'Office of Publike Addresse', see Hartlib 1647. See Yeo 2007, 11–17 for collective commonplacing.

In this thinking there was a juxtaposition of two seemingly incompatible assumptions: a call for collections of copious information *and* the conviction that memory could master its essential elements. In one sense, this combination was not unusual, since it underpinned standard Renaissance rhetorical techniques. The *copia* of tropes and quotations had to be counterbalanced by *brevitas* – hence the use of pithy aphorisms and parables that put an idea in a nutshell. However, the Baconian imperative to collect 'particulars' could result in *copia* of data without the advantages of *brevitas*. Nevertheless, Hartlib did not despair that such a mass of particulars would overwhelm memory. The benefit of ordered information was that it functioned as a prompt for the recollection of related material; it lightened the load on memory by offering external support. Hartlib decided as early as 1639 that the putative merit of artificial memory as a technique for recalling and juggling long lists of names or textual passages (*copia verborum*) was pointless. Instead, he averred that: "A Rational Reminiscentia to remember things apposite when wee would have them is far better than a bare Memoria of confused things at random which the Localists are able to doe. But that is the best Memory that helpes most the judgement."[38] The contrast here signals Aristotle's distinction between *memoria* (memory) and *reminiscentia* (recollection), the latter being a deliberate search for something stored in memory.[39] Hartlib recognized that the 'judgement' (or understanding) was best served if recollection delivered salient material quickly from memory; in turn, attention, judgement and choice exercised while material was being committed to memory enhanced efficient recollection. As he put it: "that which is to be learned may be comprehended by the least labor et in the shortest time, et may be easiest kept in memorie."[40] In his *True & Readie Way to learne the Latine* (1654), Hartlib recommended this cooperation of mental faculties: "For man's memory, imagination, and reason hath this peculiar, that the more Things it knowes, the more it can still further receive."[41] On this basis, we can appreciate the faith in careful commonplacing of information under Heads: by forcing selection of material and allocation to categories, this method created a set of prompts for recollection.

In William Petty's *Advice ... to Hartlib* (1647) we find a programme deriving from these assumptions. Petty suggested a set of eight desiderata, including the

[38] Hartlib, Ephemerides, 1639, part 1, HP30/4/4B. See also Hartlib, 'Notes on Preaching, from Perkins, in Hartlib's hand', April–August 1655, HP27/8/12B: 'Artificial Memorie is not to bee used at all. Because it is vaine or impious, in the inventing of Images, and also it is burdensome in the threefold apprehension of places, images and the thing to bee apprehended of, and so it dulleth wit and memorie'.

[39] See Sorabji 1972, 52–60. For the relationship between memory and recollection in medieval thought, see Carruthers 1990, 61–64 and *passim*; and Lewis 2009b for Bacon's position.

[40] Hartlib, undated, 'Tract on Logic in Hartlib's hand', HP24/6/12A.

[41] Hartlib 1654, 33. Hartlib agreed with Bacon's stress on the 'exercise' of the intellectual powers as crucial for improving memory. See Bacon 1657, 226 and 229–31; also in Bacon 1963, vol. 7 95–103, at 97.

following: that "all the Real and Experimental Learning" should be "sifted and collected"; the "appointment of able Readers" suitably instructed "with certain and well-limited Directions"; and finally, "Out of all these Books one Book, or great Work, may be made, though consisting of many Volumes." In show-casing a model for this approach, Petty cited John Pell's *An idea of mathematicks* (1638).[42] As he explained, "for the more explicit understanding of our Meaning herein, we refer to Mr. *Pell's* most excellent Idea [of mathematics] thereof written to Master *Hartlib*."[43] Pell sought to epitomise mathematical knowledge in a form that would allow the user to "lay them [all necessary axioms and principles] up in their heads, so as to need no booke at all."[44] Significantly, the abstract level of the subject matter did not deter Petty from affirming that it contained lessons for *other* forms of inquiry. This assumption requires notice because there is a substantive difference between reducing already established knowledge into its key elements and gradually developing propositions, themes and concepts from disaggregated 'particulars'.[45] Indeed, Hartlib championed the latter task and underscored the different levels of knowledge. In 1639 he made this entry in his diary: "Two great Faults have beene committed in our philosophy the one is that it hase not beene made truly universal. 2. that it hase not begun a particularibus."[46] In the following year he noted, as the view of Joachim Hubner, that both Descartes and Jungius "make always in knowledge an 1. Historical and 2. scientifical Part and to distinguish betweene those two accurately… If wee know how to universalize every particular truth wee shall store ourselves with a world of new notions continually." But this would only happen if the proper sequence were followed: empirical (including experimental) data gathered under 'Historia' should be arranged in ways that helped reason to perceive patterns and relationships. This method would facilitate the progress from *historia* to *scientia*.[47]

On this basis, it is possible to appreciate that Pell's *Idea* was an aspirational document; it was more ambitious than a preliminary commonplacing of material under Heads. For Pell, once various domains of knowledge had been systematized it would be possible to crystallize the essentials, shedding all unnecessary information. Then, as he said, an individual might carry these fundamental propositions,

[42]Pell's *Idea* was printed as a folio broadsheet in 1638. See Wallis 1967, 141–45. The English version appeared as an addition to Dury [or Durie] 1650 and again in the second edition of 1651. I cite from the former. Robert Hooke published the Latin version, together with comments from Marin Mersenne and René Descartes, in the *Philosophical collections*; see Pell 1681/1682.

[43]Petty 1647, 5.

[44]Pell 1650, 45. On Pell, see Malcolm 2005, 65–76 for the 'Idea' and its distribution via intermediaries.

[45]For a highpoint of the typical Renaissance desire to reduce knowledge in books to common topics, see Nelles 2009 on Conrad Gessner's *Pandectarum sive partitionum universalium* (1548–1549).

[46]Hartlib, Ephemerides, 1639, part 3, HP30/4/25B.

[47]Hartlib, Ephemerides, 1640, part 2, HP30/4/45B; and part 3, HP30/4/53B.

concepts, and axioms in a pandect or pocket booke, or even "laie them up in their heads, as to need *no Book at all*"; a mathematician "utterly destitute of bookes and instruments" might solve any problem "exactly as if he had a complete *Library* by him."[48] Pell admitted that such a goal "will perhaps seem utterly impossible to most," but indicated that it all rested on the appropriate use of external aids, such as his pandects and summaries, acting in consort to "fortifie the imagination, to prompt the memory, to regulate our reason."[49]

3 Improving Memory, Enlarging Experience: Boyle's Early Writings

What was Boyle doing at this time, that is, during the 1640s and 1650s? In short, he was preoccupied with memory in a way that seemed to follow Hartlib's recommendation about the individual building up experiences in youth. As several scholars (for example, Hunter, Harwood and Principe) have indicated, Boyle began as an author of works on moral edification, such as *Seraphic Love* (1659) and *Occasional Reflections* (1665).[50] These were composed in the late 1640s, together with unpublished manuscripts such as "The Aretology" (1645–1647), "The Dayly Reflection" (1646), and "The Doctrine of Thinking" (late 1640s).[51] Parts of the last two surfaced in *Occasional Reflections*. As Michael Hunter has put it, these writings dealt with the "pursuit of moral balance, self control and piety."[52] They did so via a consideration of the reading of romances for moral lessons and the proper direction of thoughts in meditation.[53] Since Boyle was only about eighteen when the earliest of these was written, it might be objected that we are dealing with juvenilia; but the themes treated appeared in the later published works.[54] In fact, there is a chronological overlap between these moral writings and his earliest 'scientific' publications: *New Experiments Physico-Mechanicall* (1660), *The Sceptical Chemist* (1661), and

---

[48] Pell 1650, 45 and 40.

[49] *Ibid.*, 44–5. Beale used these words in his writings about the art of memory copied by Pell in 1663. See BL Add MS 4384, fols 64–117; dated 'June xi.1663', at fol. 64^r: 'And in pursuance of Mr Pell's Idea of Mathematics, viz: That men may consider what meanes may be used to fortify the imagination, to prompt the Memory, or regulate our Reason, and what effects may be produced by the uniting of these meanes, and the constant exercise of them.'

[50] Only about half of Boyle's writings concern natural philosophy or natural history. See Hunter 1995; reprinted in Hunter 2000, chapter 2.

[51] Boyle 1991. The title 'Doctrine of Thinking' is supplied by Harwood in Boyle 1991, 185, n. 1 for the RS, MS 197, fols 4–43. See also Principe 1995a, 63–64.

[52] Hunter 1994, Sixteen.

[53] See Principe 1995b, 379; and Principe 1994.

[54] For Boyle's sense of his 'Ethics' (that is, his 'Aretology' begun in 1645) as part of his studies, see Boyle to Isaac Marcombes, 22 October 1646, Boyle 2001, 1, 37. Marcombes was his tutor during his time in Geneva from 1639. Boyle returned to England in 1644.

Some considerations touching the Usefulness of Natural Philosophy (1663; composed from late 1650s) appeared before *Occasional Reflections* (1665).[55]

I am interested in the strong role that Boyle assigned to memory in the cultivation of a virtuous self, and how this relates to his concern with enlarging individual experience, both moral and empirical. What kind of 'experience' did Boyle recommend and seek? In these early writings he meant the incidents, behaviour and thoughts that touched upon his moral self. Thus the purpose of *Occasional Reflections* was to "make the little Accidents of his Life, and the very flowers of his Garden, read him Lectures of Ethicks or Divinity."[56] In his autobiography, "An account of Philaretus" (written during 1648 and 1649), Boyle said that as a young student he was "a passionate Friend to Reading" and confident in his own memory: "what time he [Boyle] could spare from a Schollar's taskes (which his retentive Memory made him not find uneasy) he would usually employ so greedily in Reading."[57] This assessment was backed by those who knew him at Eton: Robert Carew (a member of the staff reporting to Boyle's father) said that he possessed "the rarest memory that I ever knew."[58] In writings from the late 1640s, Boyle assumed that memory could be trained via practice, not necessarily by classical memory techniques (mainly using sequences of places, or *loci*, combined with vivid images), but rather by drawing out the circumstances and consequences of observations and ideas, through careful attention, repetition, and recollection. His emphasis was on choosing and selecting what to commit to memory. This was one basis for his attack on "those memorys now so much in Fashion, which are stuff't with almost nothing else then what deserves to be excluded." His target here was the exercise of mnemonic techniques without proper regard to the content of what was stored. Thus Boyle derided those memories that keep "nought but Strawes, Dust, Feather, and such lighter Trash."[59]

Boyle discussed memory in relation to the practices of reading and thinking, or meditation, by which he meant a carefully directed train of thoughts. In "The Doctrine of Thinking" he explained how he "set my Thoughts awork upon in those shreds of Time … to recall to mind any thing I have almost forgotten, or repeate

[55]See 'Another Advertisement' in *Seraphic Love* (1659), Boyle 1999–2000, 1, 60 for the explanation by a friend that Boyle had not dropped his scientific work, but would continue to publish 'those Experimentall Essay's and other Physiologicall Writings, which he is known to have, lying by him'.

[56]Boyle, 'A Discourse touching Occasional Meditations', in Boyle 1665, 1–80, at 4; also in Boyle 1999–2000, 5, 22.

[57]Boyle, 'An account of Philaretus during his Minority', in Maddison 1969, 2–45, at 15; also in Boyle 1772, 1, xii–xxvi; and in Hunter 1994, 1–22. Boyle's amanuensis, Robin Bacon, said he had a good memory. See BL Add MSS 4229, fol. 66, cited in Boyle 1991, 194, n. 21. But compare Boyle to Hartlib, 8 April 1647 in Boyle 2001, 1, 56: 'the treacherousness of my memory'.

[58]Maddison 1969, 11 citing a letter of [December] 1635.

[59]Boyle, 'The Dayly Reflection', BP 7, fols 269–87, printed in Boyle 1991, 203–35, at 232–33. I use this edition, edited by John T. Harwood, for all subsequent page references.

any thing I desire to retaine more firmly in my Memory."[60] He gave a list of six exhortations to himself about how best to exercise his mind. The pervading theme was that active reflection was necessary to guard against "Intervening Fancys" and the proclivity of the mind to wander.[61] This watchfulness is also apparent in the "Dayly Reflection" in which he advised that we should "recall orderly to mind" whatever we have read or observed or thought "the foregoing Day. The memory may be commanded to make this Restitution either in the order their Seniority/ Priority … gives them, or that which their Considerablenesse assignes them in your Esteem."[62]

Boyle affirmed that experience could be magnified and extended by intensifying observation. The central purpose of his "Dayly Reflection" was to dispense experience from "the Lawes of Time" and to improve the "Judgment." He contended that "what men Commonly stile Experience is nothing else but a certin Dexterity of conduct, resulting from the Remembrance and Consideration of [the] Occasions suitably circumstanced."[63] But the retention of experiences in memory depended on active engagement with the world. Experiences must be examined and analysed so that "our Reflections on what we have observ'd, improves it into consequences new Axioms and Uses." Such a practice helped memorization because "the Repetition of what we learn greatly contributes to secure our Acquists from (the Danger of Oblivion)."[64] In this way, Boyle promised, an individual can gather more experience than his age might seem to allow, because "Experience Consists, not in the multitude of yeares, but in that of Observations, Experience is the result, not of yeares but of Observations."[65] He elaborated on this theme in *Occasional Reflections*, claiming that by such meditations "a man comes to discover a multitude of particulars even in obvious things." He added that "this exercise of the mind must prove a compendious way to Experience, and make it attainable without grey-hairs; for that, we know, consists not in the multitude of years, but of observations, from Numbers and variety of which it results."[66] Boyle applied this conceit in a letter to Hartlib when describing John Hall, the poet and pamphleteer, as having "September in his judgment, whilst we can scarce find April upon his chin."[67]

Cultivated in this fashion, memory supported thinking by providing a stock of experiences and by retaining previously forged links between them. The discipline

---

[60]Boyle, 'The Doctrine of Thinking', Boyle 1991, 194.

[61]*Ibid.*, 195–201, at 200–01.

[62]Boyle, 'The Dayly Reflection', Boyle 1991, 222.

[63]*Ibid.*, 208.

[64]*Ibid.*, 207.

[65]*Ibid.*, 208.

[66]Boyle 1665, 29 and 30–1.

[67]Boyle to Hartlib, 8 May 1647, Boyle 2001, 1, 59. See also de Montaigne 1987, 1263 and 1225, who professes to enjoy all things 'twice as much as others', and reports Socrates' view that if a man lived long enough and experienced many illnesses he would not need the art of medicine.

Boyle prescribed – careful selection of materials, repeated reflection on key themes, and rehearsal of a skeletal direction of meditation etc – promised to expand the experience stored in memory and facilitate its retrieval. Boyle aimed to enrich his experiences by building up associations to the places, times and circumstances in which these occurred. He argued that by taking "notice of the properties and circumstances of most things that Occur to him," and by relating them by "Resemblance or Dissimilitude"[68] to each other, a person can not only "Revive the Memory" of good thoughts, but make "almost the whole World a great *Conclave Mnemonicum*, and a well furnished *Promptuary*, for the service of Piety and Vertue."[69] In this way, the individual is made self-sufficient, a walking library:

> Besides, whereas Men are wont, for the most part, when they would Study hard, to repair to their Libraries, or to Stationers Shops; the Occasional Reflector [i.e. a person following Boyle's method] has his Library always with him, and his Books lying always open before him, and the World it self, and the Actions of the Men that live in it, and an almost infinite Variety of other Occurrences being capable of proving Objects of his Contemplation; he can turn his eyes no whither, where he may not perceive somewhat or other to suggest him a Reflection.[70]

When Boyle developed his interest in natural philosophy, he continued to find moral lessons in the study of nature and regarded the 'experimental life' as a morally worthy one. Indeed, he forecast that the habits acquired in daily meditations might be applied not only to moral topics, but to "Oeconomical, Political, or Physical matters."[71] In the manuscript "Of the Study of the Book of Nature," Boyle extrapolated what he said about the effects of gathering moral and devotional thoughts: "the Study of Nature is the Noblest Memoria Localis of a Christian, & that he may turne the whole world into a *Conclave Mnemonicum*."[72]

Does this notion of a self-sufficient memory equate with John Pell's plans for mathematics? There is some resemblance, but in these early writings Boyle was not concerned with the organization of specific bodies of knowledge. Explaining the term 'Mnemonicum' in a marginal note, he wrote: "So they call a certain Room, Artificially furnish'd with Pictures or other Images of things, whereby to help the Memory."[73] Yet he said nothing about the crucial complementary device in the classical technique – the ordered sequence or chain (*catena*) of places (*loci*) that allowed the

---

[68] Boyle 1665, 41; Boyle 1999–2000, 5, 37.

[69] Boyle 1665, 'An introductory Preface', sig. b2^r; also in Boyle 1999–2000, 5, 19; cited in Harwood, 'Introduction', in Boyle 1991, lxiv.

[70] Boyle 1665, 15; also in Boyle 1999–2000, 5, 26.

[71] Boyle 1665, 25.

[72] Boyle, 'Of the Study of the Book of Nature, For the first Section of my Treatise of Occasional Reflections ', 1650s, BP 8, fols 123–39, at fol. 137^r in Boyle 1999–2000, 13, 168; and editors' comments, vol. 1, xxxi. See Boyle to Lady Katherine Ranelagh, 31 August 1649, Boyle 2001, 1, 82–83 for mention of this work; and Principe 1995b, 393 on the continuity here between Boyle's moral and scientific attitudes.

[73] Boyle 1665, sig. b2^r; in Boyle 1999–2000, 5, 19.

recollection of one thing after another. Although he did refer to having a "Modell" or plan for the direction of his meditations, there is nothing indicating the use of a mnemonic structure in the early accounts of his memory practice.[74] Boyle's reports of his memory of ideas and experiences were often instances of *remembering the experience* of learning something, of seeing, doing, or touching certain things.[75] In the terms suggested by Julia Annas, this can usefully be classed as "personal memory," as opposed to "non-personal memory" of a fact, a date, or a theorem – one that "lacks those features of [personal] memory which make the past acquiring of knowledge part of what is remembered."[76] In any case, Boyle showed little interest in putting his knowledge into the kind of condensed and codified form that Pell imagined. Boyle's "Library" and "Mnemonicum" were personal, deriving from intense practices of observation and meditation that seem closer to the tradition of Protestant poetics than to the Baconian legacy.[77]

4 Advice to Boyle

Boyle's position can be set in sharper relief by examining two pieces of advice from Petty and Beale, both members of Hartlib's circle. Writing in April 1653 in the role of a physician, Petty counselled Boyle about the dangers of "your continual reading." He joked that

> like a Quacksalver [an 'empirick'] I might tell you how it weakens the brain, how that weakness causeth defluxions, and how those defluxions hurt the lungs and the like. But I had rather tell you that although you read 12 hours *per diem* or more, that you shall really profit by no more of what you read, then by what you remember, nor by what you remember, but by so much as you understand & digest, nor by that, but by so much as is new unto you, and pertinently set down.[78]

Petty was asking Boyle to give more weight to reason than to memory. This is the point of his rhetorical questions: "what a stock of experience have you already

---

[74]Boyle, 'Doctrine of Thinking', in Boyle 1991, 198–99.

[75]Compare Petrarch's account of a friend's amazing power of recollection, apparently achieved by 'personalizing bits of information'; Carruthers 1990, 61.

[76]Annas 1992, 307–10 especially 308. Boyle accepted the standard view that memory was a 'corporeal faculty', and therefore wondered how 'a multitude of various things' can be stored and found again. See Boyle, *Christian Virtuoso. The Second Part*, in Boyle 1999–2000, 12, 463. With many of his contemporaries, Boyle did not believe that there was a satisfactory account of the relationship between classical mnemonic practices and contemporary views of the physical workings of memory.

[77]Lewalski 1979, especially 151–52 and 161–62. See also Fisch 1953.

[78]Petty to Boyle, 15 April 1653, Boyle 2001, 1, 142. For the link between melancholy and excessive study, see Shapin 1994, 153–56.

in most things? What a faculty have you of making every thing you see an argument of some usefull conclusion or other? How much are you practiced in the method of cleere and scientifical reasoning?"[79] John Harwood cites these remarks as praise of what he calls Boyle's '*method*' of thinking; but I suggest that Petty was chiding Boyle for trying to remember without appropriate abbreviation and rational ordering of ideas.[80] Petty believed that Boyle was overloading his memory with *copia* that he had not thus reduced. This message was put more insistently by Beale.

By 1663, when Beale began advising Boyle about various ways of improving his memory, he was addressing someone who might listen. What transpired, I think, is that Beale urged Boyle to provide, in effect, a version of Pell's *Idea* for natural history, including experimental work. This required a systematic arrangement of Boyle's ideas and results which would, in turn, serve as a mnemonic framework. Even though both men espoused the collection of empirical information, their exchange reveals a clash about the way natural history and natural philosophy should be done. Beale stressed early systematic ordering of empirical information because this aided memory; Boyle accepted the value of Heads for the collection and storage of particulars, but his existing antipathy towards hasty systems seems to have been inflamed by Beale's focus on memory training.[81] It may be that this was a delicate point for Boyle, since his references to a personal mnemonicum (published in 1665, during his exchange with Beale) suggest that he had confronted the problem of how memory could be improved without acceding to premature systems, more generally.

In 1658 Hartlib told Boyle about the Somerset virtuoso, John Beale, saying that there was no one in the world to match his universal knowledge and Baconian enthusiasm.[82] When introducing himself to Hartlib, Beale had declared himself "very willing to bee an incendiary to inflame with the Love of profitable knowledge ... You will finde mee a diligent collector, and fayre interpreter of other mens notions, but noe greater plagiary."[83] Beale first wrote to Boyle on 23 February 1663, prefacing his letter with a long Latin ode to Boyle's accomplishments. Beale revealed that he had acquired "the Nic Name of Erasmus Junior," and he threatened weekly letters.[84] The next day he sent a long letter titled 'The Mnemonicalls' (or mnemonics), and

[79] Petty to Boyle, 15 April 1653, Boyle 2001, 1, 142–43.

[80] Harwood, 'Introduction', in Boyle 1991, xli.

[81] For Boyle's use of Heads, see Hunter 2007a.

[82] Hartlib to Boyle, 27 April 1658, Boyle 2001, 1, 268. Maddison 1969, 14 (in notes) says Beale 'was a near contemporary' of Boyle at Eton, 'though a trifle earlier'. This is understated, since Beale enrolled at Eton in 1622 and Boyle did not do so until October 1635. Beale was twenty years older than Boyle. For Boyle's possible reference to Beale as 'a particular friend, (a great virtuoso of the Royall Society)', see Hunter 2000, 230, n. 27 citing RS, MS 187, fols 32^{v}–34 reprinted in Boyle 1999–2000, 11, lxvii.

[83] Beale to Hartlib, undated, HP25/6/1A and 4B; but some of these words also appear (crossed through) in a fragment dated 18 March 1656, HP52/4A.

[84] Beale to Boyle, 25 February 1663, Boyle 2001, 2, 62–68, especially 66; and Beale to Boyle, 28 September 1663, *ibid.*, 127.

confided that before going to Eton he had "in secrete corners, conceald from other eyes" read and memorized key texts. Then "afterwards in Cambridge proceeding in the same order, & diligence with their Logicians, philosophers, & Schoolmen, I could at last learne them by hearte faster than I could read them."[85] They continued to exchange letters until Beale's death in 1682 but, frustratingly, we no longer have Boyle's side of the correspondence.[86]

There were two main preoccupations in Beale's letters to Boyle from February 1663 to August 1666: the art of memory, and the best arrangement of Boyle's publications.[87] These were linked since, as all adepts of memory training agreed, orderly arrangement was a reliable aid. Beale announced his own scheme for an art of memory in the next extant letter of 29 September 1663 (it is hard to believe that he waited seven months without writing again). In what is actually an essay of about 8000 words, Beale outlined a series of lessons about memory drawn from a wide range of sources: classical mnemonic techniques, implications of the Cartesian theory of brain processes, and his own observations about the habits of individuals endowed with both naturally powerful, and expertly trained, memories.[88] He offered Boyle a range of hints on memory improvement, some drawn from anecdotes about his mother's astonishing memory, others from his own experience. Some of these were proverbial, as seen in these extracts: "whatever is not offered to the Memory upon very easy Termes, is not duely tendred"; "The more we acquire, & the more often wee visite & imploy Memory, the firmer & stronger wee find it."[89]

Beale's own memory art was a development of the classical technique. Instead of visual images, usually contrived and chosen by each individual, he proposed a grid of symbolic characters that had universal pretensions.[90] He announced to Hartlib that he set out "To devise Millions of Millions of Characters, each one soe apparently differing from each other throughout the whole immense variety, That the eye at first glance shall discerne, & distinguish the difference."[91] Beale was confident that these characters would be "retained in the minde, & in a moment producible in fit order for any kind of occasion." Moreover, he promised: "All this, the reading, writing, use & practise, soe ye Learner bee willing, & of ordinary capacity & skill in Clerkship, I undertake to teach with ease, due vacancyes & refreshmen^{ts}

[85] Beale to Boyle, 29 September 1663, *ibid.*, 128–42, at 129–30.

[86] See the editors' 'Introduction' to Boyle 2001, 1, xii.

[87] Of course there were other topics, including religious ones. See Wojcik 1997, 22–23.

[88] Beale to Boyle, 29 September 1663, Boyle 2001, 2, 128–42.

[89] *Ibid.*, 132, 135.

[90] Beale had discussed the art of memory and the prospect of 'an Universall character' with Hartlib on 9 January 1658, HP31/1/61A–63A, at 61B; he had circulated comments on Caleb Morley's memory treatise on 23 December 1656, HP31/1/7A–8B and 2 December 1661, HP67/22/13A–14B. See also Lewis 2005.

[91] Copy in scribal hand of Beale to Hartlib, 4 October 1661, HP67/22/11B; see also the version copied by John Pell, titled 'The Mnemonicall Probleme', BL Add MS 4384, fols 64–117 (dated 'June xi.1663') at fol. 64^{r}.

in one weeke."[92] This optimism was apparent in his assurance that the characters could be taken in "as by one glance or blink of the eyes."[93]

Beale acknowledged that the classical art of memory had fallen out of favour, mentioning the "prejudice most learned men have against all discourses of Artificiall Memory." For this reason, he was not seeking, via Boyle, "to engage the Royal Society in it" – although he had suggested precisely this about his *own* scheme in what he sent to Hartlib.[94] His main message was that natural memory was not harmed by mnemonic techniques, and that a constitutionally weak memory could be enhanced by various methods.[95] Indeed, he added a warning for Boyle: "And I know I could once boaste of a naturall memory beyond beliefe, but I found it true, that ... The first grey hayre would signify the decay of it."[96] Without the appropriate habits, Beale predicted, each of us would live the fate of the "Merchant Whose wares are not placed in order, Or He seldome in his owne shoppe or Warehouses. His store confounds him, When the Fayre comes."[97] The insinuation was that for someone trying to manage a large array of material, such habits were crucial. Beale added an incentive that must have resonated with Boyle: namely, that his method was one "which heapes up, digests, & firmly retaines whole Libraryes."[98]

From 18 April 1666, Beale turned his attention to the best publishing format and sequence for Boyle's works. Much of what he suggested was influenced by his preoccupation with memory. For example, printed material could benefit from special typography, coloured ink and visual symbols, like those in medieval manuscripts. In his second letter of 25 February 1663, he referred to "the beautifying letters" which marked "the fronts of Chapters, & Sections" in old manuscripts, and stressed what a "vaste ayde might be to the Memory in the very printing of bookes that are worthy to be learned in the reading."[99] In terms of the physical format of Boyle's works, Beale advised that it was best that they "were abroad in Quarto,

---

[92] Copy in scribal hand of Beale to Hartlib, 4 October 1661, HP67/22/12A; another copy at HP71/6/1A–2B. See also BL Add MS 4384, fols 64–117, at fol. 64ʳ.

[93] Beale to Boyle, 29 September 1663, Boyle 2001, 2, 141. A similar confidence underlay Beale's belief that an artificial language of the kind proposed by John Wilkins could be quickly learnt, that 'the reall Character may be easily taught in few dayes' (Beale to Boyle, 23 June 1682, Boyle 2001, 5, 301). For a detailed account of Beale's views, see Lewis 2005; and more generally, Lewis 2007, especially chapter 5.

[94] Beale to Boyle, 25 February1665, Boyle 2001, 2, 69.

[95] This point is developed in 'Notes upon Mr Hartlib's Accompt of Mr Morleys Art of Memory', BL Add MS 4384, fol. 65ʳ.

[96] Beale to Boyle, 25 February 1663, Boyle 2001, 2, 69–70. See Oldenburg 1965–1986, 1, 320–21 for the editors' bad opinion of Beale: 'He suffered from total recall and confident reliance upon an unreliable memory'.

[97] Beale to Boyle, 29 September 1663, Boyle 2001, 2, 132.

[98] Beale to Boyle, 2 October 1663, *ibid.*, 145.

[99] Beale to Boyle, 25 February 1663, *ibid.*, 70. In a subsequent letter Beale himself added pointing-hand signs (manicules) in the margins; see 10 August 1666, Boyle 2001, 3, 198–200. For discussion of the manicule, see Sherman 2008, 29–40. See also Carruthers 1990, 107–09 on 'notae'.

& rather in thinner Tomes, than in thicker." Thin volumes were desirable because "every man may sorte them in a Methode more agreeable to his owne humour & concernments."[100] He reinforced this strategy in a subsequent letter, reminding Boyle that the Jesuits had already pioneered it with success. These zealots had realized the power of short pamphlets, or even single pages, which could win hearts and minds more quickly than larger tomes. In this way, he observed, the "Jesuits, doe infatuate the world, as well by their shorte manualls, as by their endlesse volumnes … [for] by their single sheetes they catch him that runneth by." These brief pieces, like "shorte daggers" used against an enemy, did the job quickly, allowing one to move on to the next task. Beale urged that all proponents of the new philosophy, especially Boyle, should do likewise.[101]

Apart from having Boyle's readers in mind, Beale wanted to make it easier for him to manage his own unpublished manuscripts. He may have known about Boyle's workdiaries, containing entries of "Promiscuous Experiments," for he suggested that once these reached 100 experiments they should be published: "That assoone as they amounted to a Century, they deserv[d] to be abroad. For thus you may empty your deskes often; & be lesse overwhelmed with your owne abundance."[102] Nevertheless, in Beale's view, the density of empirical information did not entail a diminished role for memory. Having acknowledged the particularity and scale of empirical data, he reiterated the point about not publishing "your Pandects, or promiscuous Experiments … more than a Century at a Time," now adding that "they may easily overwhelme an ordinary Industry, & confound Memory."[103] In the same letter it is clear that this stress on brevity was meant to facilitate the rehearsal of information: "Tis impossible that we should keepe *our Memories firme for our own improvements*, if by thiese strong impressions, renewalls, ruminations, & inculcations they should not be fortifyed."[104]

What might Boyle have made of all this? Was he spooked by Beale's tales of decaying memory capacity? Did he feel that his own memory was becoming impotent in the face of empirical particulars, as Bacon had warned? Unfortunately, Boyle's replies are not extant; but we can detect something of what he might have said from

[100] Beale to Boyle, 13 July 1666, Boyle 2001, 3, 187.

[101] Beale to Boyle, 10 August 1666, *ibid.*, 200. The editors of Boyle's *Works* note that after the peak of his publications in 1666 (that is, after his correspondence with Beale) he changed to books resembling 'short essays' (Boyle 1999–2000, 1, xxxvii). I cannot say whether this indicates Beale's influence.

[102] Beale to Boyle, 18 April 1666, Boyle 2001, 3, 139–40. Boyle's workdiary entitled 'Promiscuous Experiments, Observations, and Notes' (no. 21 in BP 27, 5–159) dates from the 'late 1660s', according to Hunter 2007b, 414. Boyle made entries in these workdiaries in sets of 100, thus 'centuries'. Beale had earlier alluded to the 'Sylva of promiscuous Experiments, Upon which you may discharge such of your papers & informations'. Beale to Boyle, 28 September 1663, in Boyle 2001, 2, 127.

[103] Beale to Boyle, 10 August 1666, Boyle 2001, 3, 208.

[104] Beale to Boyle, 10 August 1666, *ibid.*, 205; italics in original. Beale made similar points in a letter to an unidentified recipient of 2 December 1661, HP67/22/13A–14B.

the conciliatory mood of two letters from Beale in 1666. There is a sign in the letter of 13 July 1666 that Beale recognized (perhaps because Boyle raised it) that this preoccupation with brevity, order and memory might not be appropriate for the new empirical Baconian sciences. Beale acknowledged that some heavily systematised subjects (he refers to these as "loosely notionall" but does not cite examples), might be "skim'd over, as with the glance of the eye" and understood, and remembered. He accepted that this was not the case with the material of "Experimentall Philosophy," in which the various observations and experiments needed to be continually revisited and revised: "But thiese [new sciences] doe require a frequent & assiduous reviewe, & a kind in incubation, as for innumerable applications, for remoter discoveryes, & for seasonable inventions upon all imaginable occasions." Nevertheless, his compulsion to look for brevity and compression is seen in a marginal addition suggesting "Howe you may make 100 experiments serve for 1,000 uses, & escape all oppositions."[105]

In the letter of 10 August 1666 (about 5,000 words), Beale acknowledged Boyle's suspicion of premature systems: "I do not forget that you have renderd sufficient reasons against the praesumptuous affectation of Methodes, & hasty Systemes." He did not relent, however, drawing attention to Boyle's "Concessions in *your Proœmiall Essay pag. 5. & 6* That there is a Usefullnes & a Seasone for Systems. And certainely when store off good materialls are collected, … It will have more usefullnesse, ornament, & strength, if skillfully Ordered into a fit building, than in a confused heape."[106] His plea to Boyle was "*to drawe foorth your Experiments & Observations into Hypothesis,* such as they *doe fayrely beare,* & unite them as far as *they give mutuall strength, & light, & assistance.*"[107] Beale maintained that Boyle owed it to the world to publish a systematic presentation of the new philosophy, one that might compete with the traditional doctrines.

Beale's advocacy of a nomenclature of plants offers an example of his conviction that multifarious data could be reduced to a simple, and memorisable, form. The background is a letter from Cyprian Kinner (an associate of Comenius) to Hartlib of 27 June 1647. Addressing the difficulty of remembering the huge number of details about the qualities of plants, Kinner asked: "For how few of the most experienced Botanists are there who know all the virtues and names of every plant, when the Authors are so at variance with one another about every single one of them; and how few of them can by the common way of learning impress them on the memory and retain them in it?" Kinner thought this was impossible, and called for a new botanical terminology that captured the qualities and powers of plants, herbs etc. Combinations of consonants and vowels in syllables would indicate these

[105]Beale to Boyle, 13 July 1666, Boyle 2001, 3, 187–88. For Boyle's own remark on 'how vast a Disparity there is betwixt experimentall & notionall Learning', see Boyle to John Evelyn, 23 May 1657, Boyle 2001, 1, 214.

[106]Beale to Boyle, 10 August 1666, Boyle 2001, 3, 198–210, at 198. Here Beale muted Boyle's resistance to systems. Compare note 11 above.

[107]Beale to Boyle, 10 August 1666, *ibid.*, 200. See also editors' comments in Boyle, 1999–2000, 1, lxxxiv.

common and differentiated natures, so that "if someone can at least memorise that kind of technical word in this way, he will have fully in his grasp all the virtue of the whole plant denoted by that word, its usage and its common nomenclature, in a wondrous, easy and pleasant small compass."[108]

On 11 October 1665, Beale mentioned to Boyle a manuscript he received from Hartlib called *De Herbis sine Duce cognoscendis* (A way of recognizing plants without a guide). He gave this to William Brereton to pass on to Boyle.[109] Beale referred to it again on 28 April 1666, expanding the title to say that it enabled one to distinguish "all plants by their affinityes & differences, in their rootes, stemms, branches, blades, stature, color, leaves blossoms, fruites, seedes &c." Beale wanted it made more public because he saw it as a model "allso for the collections of those infinite varietyes into fewe heades"; we might then "cleare our apprehensions of the nature of gravity, & levity, It may give us some satisfaction concerning the Systeme of the World." Beale called it the "*Cribrum divinum*" ("divine sieve") by which the essential natures of things in the world are sorted.[110]

We do not know whether Boyle commented on this, having received it from Brereton; in any case, Beale had to remind him of it in a letter 13 July 1666. Now he asked Boyle to have the pamphlet translated into English and disseminated. Importantly, he thought this way of reducing and codifying information could work just as well for Boyle's own experimental natural history: "to Exemplify what you have written of colors, & other qualityes in that Generall Physiology."[111] Here Beale seemed to endorse the view that there are simple natures, or primitive forms, which are few in number and underlie all complex phenomena.[112] One implication of this outlook was that once these forms had been identified, results would follow quickly.[113] This is certainly what Beale professed in a letter to Hartlib: "Tis my great joy that Mr. B[oyle] is so far engaged to give us the rest of his notes and following experiments. In these he hath obliged all the intelligent inhabitants of this world, and hath given us hope, that we shall *shortly complete humane sciences*."[114]

---

[108]Cyprian Kinner to Hartlib, 27 June 1647, HP1/33/12A–14B, at 13A–B; this is an English translation by W. J. Hitchens of the Latin original; also cited in DeMott 1957, 7. See DeMott 1957 and Lewis 2007, 55–56 for the connection with artificial languages.

[109]Beale to Boyle, 11 October 1665, Boyle 2001, 2, 554. For an earlier reference to Jungius in this connection, see Hartlib to Boyle, 8 or 9 May 1654, Boyle 2001, 1, 172–73.

[110]Beale to Boyle, 28 April 1666, Boyle 2001, 3, 159; see also 18 April 1666, 138 for '*Cribo divino*'.

[111]Beale to Boyle, *ibid.*, 192.

[112]On this conviction in Bacon, see Zagorin 1998, 104–105; Gaukroger 2001, 138–48.

[113]This assumption fits with the Puritan millenarian expectation, shared by Hartlib's circle, that a quick installation of recovered prelapsarian knowledge could be achieved. See Webster 1975; Harrison 2007, chapter 5.

[114]My emphasis. Hartlib copied this to John Worthington, 26 August 1661; printed in Worthington 1847–1886, 1, 365–76, at 369; original letter in BL Add MS 32498; a copy in Add MS 6271, fols 12r–13v, at fol. 13r.

5 Conclusion

On the basis of this exchange between Beale and Boyle, we might conclude that their quest for information from observation and experiment signified a shared *empirical* stance. But these letters also reveal a tension among the Baconians, despite agreement about the collection of particulars and the rejection of premature systems. Pell and Petty, and especially Beale, sought to condense knowledge and information into an ordered structure that facilitated learning and recollection. To various degrees, they hoped that essentials, radicals, or simple natures, could be discerned after current information was sifted and arranged; and that this ordering would provide a sound basis for the integration of new discoveries. They aimed for an agreed set of abbreviations in various subjects and, potentially, a classification of the world that supported a shared mnemonic system.

Boyle's early writings show some affinity with Hartlib's emphasis on individual memory as a store of embodied experiences. Boyle wanted to thicken and deepen his own experiences so that these would be more securely impressed in his memory. He aimed to select experiences (from observations and experiments) and draw out inferences; but he was far less interested in placing the facts and ideas thus acquired in a sequence within some mnemonic grid. What we accept as Boyle's "empirical" attitude was not so much a greater commitment to gathering matters of fact – also professed by Hartlib and Beale – but a refusal to condense and arrange material in the way they demanded. Beale's promotion of memory techniques that relied on highly structured arrangements of units seems to have aggravated Boyle's existing suspicion of premature systems.

During 1666, the final year of this particular correspondence between Boyle and Beale, there were several important discussions within the Royal Society about the proper management of information. These included Robert Hooke's unpublished papers entitled "Lectures of things requisite to a Ntral History" and "A general scheme, or idea of the present state of natural philosophy"; and Boyle's own "General Heads" of April 1666 and his "Designe about Natural History" of June 1666 – all of which concern the best methods of gathering, storing and analysing both old and new information.[115] These documents stressed the need for a long-term view. Oldenburg underlined the scale envisaged when he told Boyle that "Mr Hook has also ready … A Method for writing a Naturall History, which, I think, cutts out work enough for all Naturalists in the World."[116] Yet we now know that it was precisely in this context that Hooke speculated about the possibility of enhancing Baconian "Natural Inquiry" by seeking a higher level of abstraction, one that might

[115]For Hooke's paper, see RS, Classified Papers, vol. 20, no. 50a, fols 99–109; Boyle 1666; Boyle to Oldenburg, 13 June 1666, Boyle 2001, 3, 170–75; and the recent edition of this letter and associated manuscripts in Hunter and Anstey 2008.

[116]Oldenburg to Boyle, 27 January 1666 in Boyle 2001, 3, 46. He was probably referring to Hooke's 'General Scheme', composed about this time.

"not improperly be call'd a Philosophical Algebra, or an Art of directing the Mind in the search after Philosophical Truths." He promised that this would be "a vast Help to the Understanding and Memory, as in Geometrical Algebra, the expressing of many and very perplex Quantities by a few obvious and plain Symbols."[117] This kind of expectation is not usually associated with Boyle; but in a manuscript from this time there is indeed the suggestion that "We may also give Symbolical marks to our *Data*, and other Particulars and by adding, subtracting &c. in a way suitable to the nature of this Physical Algebra, we may frame new Propositions, whence will oftentimes result new Truths."[118]

This is a reminder that the optimism of some members of the Hartlib circle about reducing data for ease of memory, recollection and thinking was not wholly alien to some key figures in the Royal Society.[119] One difference in emphasis, however, was that Boyle stressed the time and collective effort required to reach this stage. The study of nature, he noted to himself, was "a subject so vast & comprehensive, [it] will afford exercise to ye Curiosity & Industry of more than one Writer, perhaps more than one Age."[120] In the meantime he insisted that the focus must be on the collection of copious particulars, complete with their circumstances, even to the point of including some that "seem mean trivial and as to immediate use barren."[121] We can see that Boyle did exactly this in his notebooks and workdiaries that recorded reading, observations and experiments.[122] In the quantity and detail of these notes, there is possibly some trace of his preoccupation with cultivating a personal memory of thick observations and reflections, as expressed in his notion of a mnemonicum. However, Boyle did not want memory aids to sanction premature systematizing of empirical observations and experiments; such a choice might lead the mind away from the world. Thus when he warned about the attraction of systems or doctrines in *The General History of the Air* (1692), it is quite likely that Boyle was thinking about his earlier exchanges with John Beale.

Abbreviations

HP Samuel Hartlib Papers, Sheffield University
BL British Library
BP Robert Boyle Papers, Royal Society of London
RS Royal Society papers

[117]Hooke 1705, 6–7 and 64. See Yeo 2007, 26–31.

[118]Boyle, BP 9, fols 72–3; see transcription in Hunter and Anstey 2008, p.7.

[119]Lewis 2009a, 355–58.

[120]Boyle, RS, MS 198, circa 1680, fol. 104^{r}.

[121]Boyle, RS, MS 189, 1689–90, fols 27^{v}–28^{r}.

[122]In contrast, Beale thought notebooks weakened the memory; see Yeo 2007, 1–2. For Boyle's management of his papers and notes, see Knight 2003; Hunter 2007b; and Yeo 2010.

References

Annas, Julia. 1992. Aristotle on Memory and Self. In *Essays on Aristotle's De Anima*, eds. M. C. Nussbaum and A. O. Rorty. Oxford: Oxford University Press, 297–311.

Anstey, Peter. 2000. *The Philosophy of Robert Boyle*. London: Routledge.

Anstey, Peter. 2005. Experimental Versus Speculative Natural Philosophy. In *The Science of Nature in the Seventeenth Century: Patterns of Change in Early Modern Natural Philosophy*, eds. P. Anstey and J. Schuster. Dordrecht: Springer, 215–42.

Bacon, Francis. 1657. A Letter, and Discourse, to Sir Henry Savill, Touching Helps, for the Intellectual Powers. In *Resuscitatio, or, Bringing into Publick Light Severall Pieces, of the Works,* ed. W. Rawley. London: Printed by Sarah Griffin for W. Lee, 225–31.

Bacon, Francis. 1963. *The Works*, 14 vols., eds. J. Spedding, R. L. Ellis and D. Heath. Stuttgart: F. Frommann Verlag G. Holzboog. Reprint of 1857–1874 London edition.

Bacon, Francis. 2004. *The Instauratio Magna Part II: Novum Organum and Associated Texts*, eds. G. Rees and M. Wakely. Oxford: Clarendon Press.

Bennett, Jim and Mandelbrote, Scott, eds. 1998. *The Garden, the Ark, the Tower, the Temple: Biblical Metaphors of Knowledge in Early Modern Europe*. Oxford: Museum of the History of Science.

Boyle, Robert. 1655. An Invitation to Free and Generous Communication. In *Chymical, Medicinal, and Chyrurgical Addresses: Made to Samuel Hartlib*. London: Giles Calvert, 113–50.

Boyle, Robert. 1661. *Certain Physiological Essays*. London: H. Herringman.

Boyle, Robert. 1665. *Occasional Reflections upon Several Subjects. Whereto Is Premis'd a Discourse About Such kind of Thoughts*. London: W. Wilson for Henry Herringman.

Boyle, Robert. 1666. General Heads for a Natural History of the Country. *Philosophical Transactions*, 1, 186–9.

Boyle, Robert. 1674. About the Excellency and Grounds of the Mechanical Hypothesis. In *The Excellency of Theology, Compared with Natural Philosophy*. London: H. Herringman.

Boyle, Robert. 1690. *Christian Virtuoso ... The First Part*. London: Edward Jones.

Boyle, Robert. 1692. The *General History of the Air.* London: printed for Awnsham and John Churchill.

Boyle, Robert. 1772. *The Works of the Honourable Robert Boyle*, 6 vols., ed. T. Birch. London: J. and F. Rivington et al. Reprinted 1965. Hildesheim: Georg Olms.

Boyle, Robert. 1991. *Early Essays and Ethics of Robert Boyle,* ed. and annotated and with an Introduction by John T. Harwood. Carbondale: Southern Illinois University Press.

Boyle, Robert. 1999–2000. *The Works of Robert Boyle*, 14 vols., eds. M. Hunter and E. B. Davis. London: Pickering & Chatto.

Boyle, Robert. 2001. *The Correspondence of Robert Boyle*, 6 vols., eds. M. Hunter, A. Clericuzio and L. M. Principe. London: Pickering & Chatto.

Carruthers, Mary. 1990. *The Book of Memory: A Study of Memory in Medieval Culture.* Cambridge: Cambridge University Press.

Clucas, Stephen. 1994. In Search of "The True Logick": Methodological Eclecticism Among the "Baconian Reformers". *Samuel Hartlib and Universal Reformation: Studies in Intellectual Communication*, eds. M. Greengrass, M. Leslie and T. Raylor. Cambridge: Cambridge University Press, 51–74.

Clucas, Stephen. 1997. "A Knowledge Broken": Francis Bacon's Aphoristic Style and the Crisis of Scholastic and Humanist Knowledge–Systems. In *English Renaissance Prose,* ed. N. Rhodes. Tempe, Arizona: Medieval and Renaissance Texts and Studies, 147–72.

Clucas, Stephen. 2009. *Scientia* and *Inductio scientifica* in the *Logica Hamburgensis* of Joachim Jungius. In *Scientia in Early Modern Philosophy. Seventeenth-Century Thinkers on Demonstrative Knowledge from First Principles,* eds. T. Sorell, G. A. J. Rogers and J. Kraye, *Studies in History and Philosophy of Science*, vol. 24. Dordrecht: Springer, pp. 53–70.

Cook, Harold. J. 1990. The new philosophy and medicine. In *Reappraisals of the Scientific Revolution*, eds. D. C. Lindberg and R. S. Westman. Cambridge: Cambridge University Press, 397–436.

de Montaigne, Michel. 1987. On Experience. In *The Complete Essays*, Bk III:13, ed. M. A. Screech. London: Penguin Books, 1207–69.

DeMott, Benjamin. 1957. Science versus Mnemonics: Notes on John Ray and on John Wilkins' *Essay toward a real character, and a Philosophical Language*. *Isis*, 48, 3–12.

Dury [or Durie], John. 1650. *The Reformed Librarie-Keeper, with a Supplement to the Reformed School ... Wherunto is Added 1. An Idea of Mathematicks*. London: W. Du-Gard.

Fisch, Harold. 1953. The Scientist as Priest: A Note on Robert Boyle's Natural Theology. *Isis*, 43, 252–65.

Galen. 1985. An Outline of Empiricism in *Three Treatises on the Nature of Science*, eds. R. Walzer and M. Frede, introduction by M. Frede. Indianapolis: Hackett Publishing, 21–45.

Gaukroger, Stephen. 2001. *Francis Bacon and the Transformation of Early-Modern Philosophy*. Cambridge: Cambridge University Press.

Greengrass, Mark, Leslie, Michael and Raylor, Timothy eds. 1994. *Samuel Hartlib and Universal Reformation: Studies in Intellectual Communication*. Cambridge: Cambridge University Press.

Harrison, Peter. 2007. *The Fall of Man and the Foundations of Science*. Cambridge: Cambridge University Press.

Hartlib, Samuel. 1647. *A Brief Discourse Concerning the Accomplishment of our Reformation*. London.

Hartlib, Samuel. 1654. *True & Readie Way to Learne the Latine*. London: Printed by R. and W. Leybourn.

Hartlib, Samuel. 2002. *Hartlib Papers: A Complete Text and Image Database of the Papers of Samuel Hartlib c. 1600–1662.*, 2nd edn. Sheffield: HROnline, Humanities Research Institute. Held in Sheffield University Library, Sheffield, UK.

Hobbes, Thomas. 1651. *Leviathan*, London: Reprinted in 1996, ed. R. Tuck. Cambridge: Cambridge University Press.

Hooke, Robert. 1705. A General Scheme, or Idea of the Present State of Natural Philosophy. In *The Posthumous Works of Robert Hooke*, ed. R. Waller. London: Printed by S. Smith and B. Walford, 1–70.

Hunter, Michael. 1994. *Robert Boyle by Himself and His Friends*. London: William Pickering.

Hunter, Michael. 1995. How Boyle Became a Scientist. *History of Science*, 33, 59–103.

Hunter, Michael. 2000. *Robert Boyle 1627–91*: *Scrupulosity and Science*. Woodbridge: Boydell Press.

Hunter, Michael. 2007a. Robert Boyle and the Early Royal Society: a Reciprocal Exchange in the Making of Baconian Science. *British Journal for the History of Science*, 40, 1–23.

Hunter, Michael. 2007b. *The Boyle Papers: Understanding the manuscripts of Robert Boyle*. Aldershot: Ashgate.

Hunter, Michael and Anstey, Peter, eds. 2008. *The Text of Boyle's 'Designe About Natural History'*. London: Robert Boyle Project, Occasional Papers no. 3.

Knight, Harriet. 2003. Organising Natural Knowledge in the Seventeenth Century: the Works of Robert Boyle, unpublished PhD Thesis, University of London.

Lewalski, Barbara Kiefer. 1979. *Protestant Poetics and the Seventeenth-Century Religious Lyric*. Princeton: Princeton University Press.

Lewis, Rhodri. 2005. "*The Best Mnemonicall Expedient*": John Beale's art of memory and its uses. *The Seventeenth Century*, 20, 113–44.

Lewis, Rhodri. 2007. *Language, Mind and Nature: Artificial Languages in England from Bacon to Locke*. Cambridge: Cambridge University Press.

Lewis, Rhodri. 2009a. Hooke's Two Buckets: Memory, Mnemotechnique and Knowledge in the Early Royal Society. In *Ars Reminiscendi: Mind and Memory in Renaissance Culture*, eds. D. Beecher, and G. Williams. Toronto: Centre for Reformation and Renaissance Studies Publications, 339–63.

Lewis, Rhodri. 2009b. A Kind of Sagacity: Francis Bacon, the *Ars Memoriae* and the Pursuit of Natural Knowledge. *Intellectual History Review*, 19, 155–75.

Maddison, Robert Edwin Witton. 1958. A Tentative Index of the Correspondence of the Honourable Robert Boyle, F.R.S. *Notes and Records of the Royal Society*, 13, 128–201.
Maddison, Robert Edwin Witton. 1969. *The Life of the Honourable Robert Boyle F.R.S.* London: Taylor and Francis.
Malcolm, Noel. 2005. The Life of John Pell. In N. Malcolm and J. Stedall, *John Pell 1611–1685 and His Correspondence with Sir Charles Cavendish: The Mental World of an Early Modern Mathematician.* Oxford: Oxford University Press, 12–244.
More, Louis Trenchard. 1944. *The Life and Works of the Honourable Robert Boyle.* New York: Oxford University Press.
Nelles, Paul. 2009. Reading and Memory in the Universal Library: Conrad Gessner and the Renaissance Book. In *Ars Reminiscendi: Mind and Memory in Renaissance Culture*, eds. D. Beecher, and G. Williams. Toronto: Centre for Reformation and Renaissance Studies Publications, 147–69.
O'Brien, John J. 1965. Samuel Hartlib's Influence on Robert Boyle's Scientific Development. *Annals of Science*, 21, 257–76.
Oldenburg, Henry. 1965–1986. *The Correspondence of Henry Oldenburg*, 13 vols., eds. A. R. Hall and M. B. Hall. Madison: University of Wisconsin Press.
Passmore, John. 1970. *A Hundred Years of Philosophy*. Harmondsworth: Pelican.
Pell, John. 1650. An Idea of Mathematics, written by Mr Joh. Pell to Samuel Hartlib. In *The Reformed Librarie-Keeper ... wherunto is added 1. An Idea of Mathematicks*, J. Dury [or Durie], 33–46. London: W. Du-Gard.
Pell, John. 1681/1682. An Idea of Mathematicks, long since written by Dr. John Pell. In *Philosophical collections*, ed. Robert Hooke, no. 5. London: Printed for R. Chiswel, 127–34.
Petty, William. 1647. *The Advice of M. P. to Mr. Samuel Hartlib for the Advancement of some Particular Parts of Learning*. London.
Power, Henry. 1663–1664. *Experimental Philosophy.* London: T. Roycroft.
Principe, Lawrence. 1994. Style and Thought of the Early Boyle: Discovery of the 1648 Manuscript of *Seraphic Love*. *Isis*, 85, 247–60.
Principe, Lawrence. 1995a. Newly Discovered Boyle Documents in the Royal Society Archive: Alchemical Tracts and His Student Notebook. *Notes and Records of the Royal Society of London*, 49, 57–70.
Principe, Lawrence. 1995b. Virtuous Romance and Romantic Virtuoso: The Shaping of Robert Boyle's Literary Style. *Journal of the History of Ideas*, 56, 377–97.
Rowbottom, Margaret E. 1950. The earliest published writing of Robert Boyle. *Annals of Science*, 6, 376–89.
Shapin, Steven. 1994. *A Social History of Truth: Civility and Science in Seventeenth-Century England.* Chicago: The University of Chicago Press.
Sherman, William Howard. 2008. *Used Books: Marking Readers in Renaissance England.* Philadelphia: University of Pennsylvania Press.
Snyder, Laura. 2006. *Reforming Philosophy: a Victorian Debate on Science and Society.* Chicago: University of Chicago Press.
Sorabji, Richard. 1972. *Aristotle on Memory*. London: Duckworth.
Stubbs, Mayling. 1982. John Beale, Philosophical Gardener of Herefordshire. Part I. Prelude to the Royal Society 1608–1663. *Annals of Science*, 39, 463–89.
van Fraassen, Baas C. 2002. *The Empirical Stance*. New Haven: Yale University Press.
Wallis, Peter John. 1967. An Early Mathematical Manifesto. John Pell's "*Idea of mathematics*". *The Durham Research Review*, 18 April, 139–48.
Webster, Charles, ed. 1970. *Samuel Hartlib and the Advancement of Learning*. Cambridge: Cambridge University Press.
Webster, Charles. 1975. *The Great Instauration: Science, Medicine, and Reform 1626–1660.* London: Duckworth.
Wojcik, Jan W. 1997. *Robert Boyle and the Limits of Reason.* Cambridge: Cambridge University Press.

Worthington, John. 1847–1886. *The Diary and Correspondence of Dr. John Worthington,* 3 vols., ed. J. Crossley. Manchester: Printed for the Chetham Society.

Yeo, Richard. 1993. *Defining Science: William Whewell, Natural Knowledge, and Public Debate in Early Victorian Britain.* Cambridge: Cambridge University Press.

Yeo, Richard. 2007. Between Memory and Paperbooks: Baconianism and Natural History in Seventeenth-Century England. *History of Science*, 45, 1–46.

Yeo, Richard. 2010. forthcoming. Loose Notes and Capacious Memory: Robert Boyle's Note-taking and its Rationale. *Intellectual History Review*.

Zagorin, Perez F. 1998. *Francis Bacon.* Princeton: Princeton University Press.

Lamarck on Feelings: From Worms to Humans

Snait B. Gissis

Abstract Lamarck's theory of '*sentimens*' (feelings) is presented. Within this framework, what he called '*sentiment intérieur-sentiment d'existence*' ('inner feeling') and his conception of 'attention' are analyzed. It is argued that Lamarck's innovative stance stemmed from his evolutionary perspective on "mental events" and his conception of 'self' that derived from it. The 'mental events' of empiricism – subsumed under the experiencing of 'feelings' – were assumed by Lamarck to be direct consequences of physiological processes. The transition from experience to feeling which Lamarck posited could not, in a strict sense, be justified within the sensationalist framework. He argued that the capacity of experiencing was dependent on bodily structures that were emergent. The resulting experiencing and the resulting 'feelings' were also an emergent category, embodied, yet endowed with a qualitative surplus. Organisms, as well as their components, were assumed to be in constant interaction with their specific environment. Consequently, the feelings-events in the body, through the processes of the body as integral parts of the body, bore the specificities of this or that particular body, and were embodied in this double sense. Lamarck's assumption of a gradually emergent *sentiment intérieur* was intended to resolve the empiricist dead-end on identity-and-continuity of the self. The evolutionary perspective which assumed both growing complexification, openness to environment and interaction with it, compelled Lamarck to look for a unifying cohering foundation for the individualized entities which populated his theory. This foundation had to cater to the demands of 'science' as well as those of 'experience'. Thus he produced the first version of an evolutionary 'self', and a new conception of 'internality'.

This is an elaboration of a paper presented at the February 2009 University of Sydney workshop on "Embodied Empiricism."

S.B. Gissis
Cohn Institute for the History and Philosophy of Science and Ideas, Tel Aviv University.

C.T. Wolfe and O. Gal (eds.), *The Body as Object and Instrument of Knowledge: Embodied Empiricism in Early Modern Science*, Studies in History and Philosophy of Science 25,
DOI 10.1007/978-90-481-3686-5_11, © Springer Science+Business Media B.V. 2010

1 Introduction

Lamarck[1] was a transitional figure, who was deeply conscious of the contemporaneous attempts[2] to explain the macroscopic features of the inanimate world in terms of the motions of, and forces between, the constituent microscopic entities. He ascribed to the observed phenomena involving living forms on earth the attribute of "transformism"[3] and attempted to give a systematic and all encompassing "microscopic" explanation to them. Given that he belonged to the rapidly dwindling community of naturalists for whom the elementary entities grounding physics and chemistry were still part of earlier systems, e.g. the caloric and electric fluids, he did not reduce his mechanisms and processes of life and that of experiencing to the elementary units of the new contemporaneous physics and chemistry. Yet his endeavour was in a deep sense analogous to that of the 'new science', with a marked emphasis: he was acutely aware of both the complexity and the time directedness characterizing the phenomena of life. He considered organization and self organization their principal features, and thus complexification as an inherent property of life. He consequently was deeply cognizant of the enormity of the task to provide such explanations.[4]

[1]Lamarck has been re-instituted as an important contributor to "evolutionism" in the early nineteenth century, and in particular, as an important influence on Geoffroy Saint Hilaire, Frédéric Cuvier, Grant, Chambers, Spencer, Darwin, Baden Powell, and others who concerned themselves with the history of life on earth during the first half of the nineteenth century in France and in Great Britain. Corsi 1997, 2005 has corroborated this by making clear the role Lamarck and his students played in making the history of life on earth a central issue. He and Barsanti 1995, 1997a, have also pointed to the extensive considerations of the subject matter instigated or influenced by Lamarck or his auditors outside of France and Great Britain, the latter two countries being the locations usually addressed by scholars in the field until the last decades of the twentieth century.

[2]For example, Lavoisier, Laplace and others. See Fox 1990, Gillispie 2004.

[3]A later term, not his own.

[4]Lamarck saw the phenomena of the environment as falling under physical explanation, though subdivided among chemistry, geology, hydrogeology, meteorology etc. He did not reduce the phenomena of life to physico-chemical explanation but assumed that their understanding was a necessary though not a sufficient, condition for understanding living organisms. He believed in a physics of the four elements, with a major role ascribed to fire/heat (in its three assumed states). The tendency of these elements to return to their simple state was used as an explanation of the decay of their compounds, both organic and inorganic. He assumed (mostly until around 1799–1800) that the existence of "delicate fluids" accounted for phenomena such as heat and magnetism, and served as an element in the explanation of gravity. His positions in his numerous chemical–physical–geological publications both before and after 1795 on some of the major issues debated at the time (e.g. the geology of earth, the new chemistry of Lavoisier and its implications, the explanation of respiration, Spallanzani's explanation of digestion, …) had been such that they were shunned by those at the head of the science teaching institutions and at the forefront of the contemporaneous science. See Cuvier 1810, 1815/1989, 1860. Likewise, his views on the formation of the earth's surface through constant changes, most importantly on the location of the seas and the impact of the atmosphere, estranged him from those scientists with whom he had much in common concerning chemistry. The same goes for his assumption regarding the constant dying out of organisms as the source of inanimate materials of earth. Lamarck's views on chemistry and geology and the explanation of what he termed" physical phenomena," including their revision after the turn of the century, have been expounded in detail in a number of studies, notably in Barsanti 1997b,

One can notice changes in Lamarck's publications written during the last years of the eighteenth century. Foremost among them was the distinction he introduced between living and non living entities in which the concept of organization played a subtle role. At first, it merely distinguished between living and non living, and somewhat later it drew lines between different kinds of living entities. Instead of the simply 'graduated nuanced differences' of the early botanical period, one now had differences in accordance with the degree of complexity of the organization.[5] The latter was to be gauged according to faculties, functions and modes of activity, and later according to the existence of specific organs shaped by them.[6] In order to discuss his novel conceptualization, one has to include both the function played by the directionality of time and the role of the environment in the shaping of a living entity and its life span. Incidentally, Lamarck's conceptual grid coincided in interesting ways with the 'organizational' discourse and rhetoric of both the revolutionary period and the early Thermidorean post-Jacobin regime. It was also very much in line with the part of the 'progress discourse' that compared the socio-organizational forms of various societies encountered in colonial journeys and exploratory voyages, and constructed hierarchies on the basis of the complexity of social institutions and the functions they fulfilled in those societies.[7]

A comment is in order here regarding a significant departure by Lamarck from the practice of Buffon, and that of many naturalists of his time.[8] Numerous contemporaries, particularly the *idéologues*, and emphatically so Cabanis and physicians in general, used the human individual, usually termed 'man', as the starting point of their discussion of living nature. Even though they looked upon 'man' as part of nature, they devoted long and protracted discussions to man's special, at times even separate, status within nature, and to the particular conceptual tools needed to analyze his unique faculties.[9] Not so Lamarck. Lamarck instead started with a panorama of living nature. He felt it incumbent upon him to provide explanations for experiencing, feeling and intellection across the branching scale of the living.

Burkhardt 1977, 1995a, Corsi 1988, Gohau 1997a, b, Jordanova 1984, and Roger 1979 and most recently in Gohau 2006, Gohau 1977a. See also the sympathetic eulogy by Geoffroy Saint Hilaire in Latreille and Geoffroy 1829.

[5]For example, already in *Biologie* and in the *Discours* of 1801.

[6]Note that this was the opposite way from Cuvier's view on the relationship between functions, structures and activity (behavior). Cuvier 1799, 1805.

[7]See Gissis 2009.

[8]But not so with regard to at least two of his predecessors, Charles Bonnet and Benoît de Maîllet. Furthermore, most current histories of eighteenth century diverse traditions of natural history assume a certain measure of continuity throughout that century. Lamarck (1744–1829) spent most of his adult life within that framework. However, there had been marked changes within that framework towards the last two decades of the eighteenth century and during the first decade of the nineteenth century. I would suggest that Lamarck, in his new capacities within the Museum National d'Histoire Naturelle, was struggling rather with the newer cluster of problems, which arose during the last decades of the eighteenth century and the very beginning of the nineteenth century, transforming and deploying earlier solutions to serve the new ones.

[9]In this sense, some of the members of the *Société des observateurs de l'homme* were within the Buffonian tradition; see also Blanckaert 1992.

When he finally discussed man,[10] he deemed it necessary to emphasize that all a naturalist – i.e. he, Lamarck – could do was to apply uniformly the laws of physics, i.e. his Lamarckian chemistry, geology and physics. For Lamarck, the faculty of reason itself was looked upon as depending causally – physically, functionally – on matter, matter being the sole entity the naturalist could observe, investigate and explicate. Though Man had been mentioned earlier, it was only in the *Philosophie Zoologique* of 1809 that Lamarck started detailed discussion of humans.

During the latter part of the eighteenth century the term "sensibility" was used in a number of fields – within philosophy, natural history, the sciences, as well as in culture, art, morals and politics, and also in medicine and in literature. Thus the term (in the culture within which it was used) was polysemic. Almost all the historians of the last third of the eighteenth century either used the term, or explicated it within their specific work or did both. Its primary signification related to the senses and to the significant status assigned to the ability to use the senses – that is, to have sensations – as a venue of the world of phenomena and of meaning. This over and above that of reflection on one hand, and above the 'response' of sentiment to the sensation on the other. Towards the end of the century sensibility was conceived as the locus of interactions between mind and body. Thus sensory experience and sentiment constituted the foundations of the self, as well as of both subjectivity and sociability. Though originating in Locke's sensationalist epistemology, Rousseau, Helvetius, Condillac, Buffon and Diderot can be viewed as some of principal proponents of "sensibility." 'Sensibility' was translated into the moral-political idiom during the revolutionary decade.[11] I view 'sensation' and 'feeling' and their derivatives in Lamarck's writings as an outgrowth of a predominant epistemology and its transposition to a scheme of nature in which matter and modes of organization relevant to it served as the major formative assumptions. To give an account on how the world was apprehended by living organisms/organized bodies within such a framework meant that it was related to their specific bodily functioning and their specific bodily structures. That is to say 'sensations' and 'feelings' had to be embodied in order to be; they could not be related to semi-external sense organs whose outputs would then become located and worked out in a disembodied mind, psyche, or non descriptive mediating organ.

I shall discuss aspects of Lamarck's treatment of sensations and feelings during the period that started with his 1800 opening lecture course and ended with his 1820 *Système Analytique*[12] with the following questions in mind: "In what ways was the presupposition of evolution significant?" and "In what ways did Lamarck's

---

[10] See below.

[11] See the by now classic essay, Moravia 1978. Within the limits of the present paper I cannot provide even a semblance of a detailed contextualized narrative for that history.

[12] The following writings of Lamarck will be used: Lamarck 1800, 1800/1944, 1801, 1802, 1809, 1812, 1815–1822, 1817, 1820, 1907, 1933, 1972. See Pietro Corsi's website (http://www.lamarck.net).

particular models that dealt with sensation and feeling reflect his evolutionary entwining of behaviour and inheritance?"[13]

More specifically, the main emphasis will be on Lamarck's work from 1806 onwards, when his transformist perspective had been worked out in more detail and was assumed in everything he said. My presentation is mostly an internalist one, in that the main body of texts used will be Lamarck's. However, in my attempt to describe, analyze and conceptualize,[14] I shall also resort to the writings of earlier naturalists and physicians and also those of some of his contemporaries. My discussion is sectioned into two unequal parts:

1. '*Sentimens*' – 'Feelings' – discussed through an evolutionary sieve,
2. '*Sentiment intérieur*' – Inner feeling – the more traditional functions of the 'soul' and the emergent evolving 'self', with a short note on 'Attention'.[15]

Lamarck's methodological reflections will be briefly commented upon in both parts, and related to specific issues. 'Habit', 'need', 'use', 'instinct', will be discussed only in relation to the *sentiment intérieur*. The higher mental functions will not be discussed. Likewise, I shall not discuss Lamarck's general view of human nature, individual and social, except in so far as it is relevant to support my argument on the emergent evolving 'self'.

2 Sentimens

Lamarck's discussion of feelings braided together environment, living body (i.e. an organism), nerves-nervous system, behaviour as components and attempted to draw a systematic evolutionary account of the interactions – the "*rapports*" – among them. Loosely speaking, it was conducted within the framework of sensationalism, later also called associationism,[16] and I would argue that the events and the processes at its focus were considered as being neither solely physical nor solely mental. However – time-wise and space-wise – these events and processes did happen and

[13] In this connection let me note that in order not to use the term 'evolution' anachronistically, I use it only to depict Lamarck's *transformism* or his 'march of nature'. I do use the term 'evolutionary' to refer to processes related to the dynamics of Lamarck's transformism.

[14] I have used as secondary sources primarily the following: Duchesneau 1982, Dixon 1988, Farber 1981, Gross 1979, Hannaway and La Berge 1998, Jacyna 1987, Lesch 1984, Moravia 1972, Reil 2005, Rey 2000, Roger 1993, Sloan 1990, 1995, Smith 1973, Spary 1996, Steinke 2005, Williams 2002; and with a different perspective Reddy 2001.

[15] Wherever I use terms and categories used by the subjects of this study, I have tried to put them in quotes or in italics when not in English. I also took the liberty of changing these into researchers' – i.e. contemporary – terms and categories where deemed useful e.g. physiological-in modern usage, rather than physical, even though in Lamarck's time 'physiology' was deemed to be more closely related to medicine than to zoology and natural history.

[16] O'Neale 1996, Riskin 2002, Richards 1979.

did take place inside the organism, and were to be detected through the reactions and the actions of its various subsystems and parts. Having adopted an empirical stance Lamarck was not unique in trying to discuss the phenomena of the traditional category of mind within the empiricist range of available models.[17] In Lamarck's case, being empirical meant using the materials in natural history cabinets,[18] making use of the diverse findings (whether experimentally or observationally empirical) of other contemporary or earlier eighteenth century naturalists and zoologists, as well as those of anatomists and physicians, though he carefully distinguished his positions from all these practitioners.[19] However, his discussion of mind was anchored in the discourse of 'faculties' that had been firmly established in the planning of the *Encyclopédie*, but became somewhat attenuated in its entries. Thus, though presented evolutionarily, Lamarck's discussion of the faculties was devoid of historical context-dependence. In fact they were treated as "natural kinds." When tracing the development and crystallization of Lamarck's position on feelings, I emphasize the more general framework of changes in his discussion of feelings, emotions, sensibility and movement during the last decades of the eighteenth and the first of the nineteenth century.[20] These changes consisted in a more focused discussion of the kind of organic *system* that was deemed capable of feeling and of movements, i.e. not any organism or any of its parts, nor of the relationships among them, e.g. nerves and muscles. These changes can be characterized as exhibiting

[17]In Lamarck's later writings the scientific status of the work and the conclusions of others were dealt with in two ways: (a) In the discussion of the multilayered use of imagination (not only in his later works but also in the dictionary item by that name) (b) In the distinction made between the immutable laws of nature and their representation in the laws of science. For example, "Aussi, jamais les principes des sciences, quelles quelles soient, et qui sont tous des résultats de nos jugemens, ne pourront égaler en certitude les observation bien faites et les faits constatés qui y ont donné lieu" and this is because "Les objects et les faits observés appartient à la nature" while scientific laws bear the mark of human means. These appear in his *Apperçu des connaissances humaines* 1810–1814, Lamarck 1972, 37. See also Daston 2005, 2008, Daston and Pomata 2003, Farber 1981, Pickstone 2007, Sloan 1995, Terrall 2004.

[18]These were extremely rich at the *Muséum d'Histoire Naturelle* (MHN), in part by virtue of confiscations during the revolutionary decade and the Napoléonic wars. Lamarck took an active part in the sorting, classifications and organization of these cabinets. He did venture once to do actual field work – beyond cabinet-work – when he collected materials for the *Flore Françoise.*

[19]From the late 1790s on Lamarck added philosophy to his professional self-definition. In his *Système* of 1801, he succinctly identified himself as: "naturaliste observateur et philosophe" (359). In almost every opening of his lecture courses he insisted on the need to exercise the imagination in order to encompass the vast variety and diversity of living nature, and to contemplate the meaning of the impact of environment on living organisms. In 1806 he actually suggested to his audience the following thought experiment (repeated almost verbatim in later texts): Suppose that the whole of nature were laid out for a spectator to contemplate as if it were a collection, could one then classify nature in a way that would cut it really at its joints? For example, "... il est nécessaire avant tout d'embrasser par l'imagination le vaste ensemble des productions de la nature ..." in the introductory lessons in 1800, An 9 et An 10, Lamarck 1972, 10. On this see also Gohau 2006.

[20]The distinction between zoological and medical questions, assumptions and interests had begun to diverge during this period. Lamarck's treatment of these issues seems to be on the finely drawn line between these two.

a. A growing differentiation between the more generalized notions of irritability and sensibility,
b. A differentiation within specific organic systems related to their functions e.g. the various parts of the nervous system,
c. The rethinking of the relations between feeling and thinking,
d. The questioning as to which are to be considered the basic functions of an organism.

In Lamarck's earlier "transformist" writings (1800–1806) the influence of the (constantly changing) environment on living organisms was considered to be direct.[21] He practiced classifying the non-vegetable living organisms in accordance with their functioning. In the early writings in which transformism was announced, the three functions – and thus the three organic structures or organic components, deemed essential for a non-vegetable living organism – were sensibility, digestion and movement (voluntary). This in fact indicated that Lamarck at the time was using a generally accepted definition. Cuvier somewhat later used a variation of it in which functions were modelled on those of vertebrates,[22] and he became severely criticized for it by Lamarck in his late writings. The division among the invertebrates was based directly on the presence or absence of specialized organic structures for respiration, movement of body fluids and feeling *(sentiment)*, rather than on functions. If we follow Lamarck in his works we find that in some sense, he tended to posit the invertebrates as "model organisms." One can call them model organisms because they were selected as exemplars of widely observed features, were accessible, were perceived as typical, and thus they served as an index to the group of instances, and to the set of *problématiques*. He applied fine distinctions and divisions among them in a mode analogous to distinctions and divisions in living nature at large. The same was true for the relationships between functions and structures.

Operating within the Hallerian legacy,[23] in his first article in which a transformist ascending scale appeared, – the *Prodrome de l'histoire naturelle des animaux invertébrés*, which he presumably edited in 1799–1800 – Lamarck was starting to distinguish between sensibility and irritability,[24] and to relate them to different levels of complexity of structure, and thus to different stages in the order of organisms. Though the separate mechanisms and supporting organic structures of sensibility and irritability were not as yet specified, it was not the case that both stretched over the whole animal scale. Furthermore, individuation meant animalization (and vice versa). Plants were not endowed with animalization. Thus combining irritability with

---

[21] See Burkhardt 1977, 1995a, Conry 1994, Corsi 1988, Herbert 2005, Hodge 1995, Jordanova 1984, 1989, Spary 2000.

[22] Lamarck looked upon the distinction between invertebrates and vertebrates as one of his important original contributions. It effectively changed both the boundaries of, and the perspective on, the realm of animals he had been nominated to investigate at the *Muséum d'histoire naturelle*.

[23] For example, see Duchesneau 1982, Steinke 2005, Boury 2008.

[24] Due to space constraints I cannot provide an adequate contextualized narrative of the rich history of these terms, central to understanding eighteenth century debates on the nature of the living.

lower, simpler – one would say more *elementary* in an *evolutionary* sense – forms of organized animal life (in this case, the polyps which had the most basic form of animalization), signified establishing the connection between order of appearance in evolutionary time, and scantiness of functions.[25]

In the other writings from this period two features stand out that illustrate the gradual crystallization of Lamarck's views:

a. The explicative discussion of functions, their impact on organic organization – on structures – was conducted either as if it did not matter which way one started on the scale; or that the order of explanation was from the more complex to the simple, even though in most of these early writings at some earlier or later stage in the work there was a statement as to the 'right order of nature'.[26]
b. The distinction between sensibility and irritability became sharper; and the comparison between elementary forms and more complex ones of the nervous system became a heuristic feature of the distinguishing tool among organisms.[27]

In Lamarck's *Système des animaux sans vertèbres* (1801) irritability was still considered a derivation or modification of sensibility, which had appeared in the very rudimentary forms of living bodies as "ébauches de l'animalisation" – preliminary forms of animalization – in spite of the fact that these organisms have: "Neither brain, nor an elongated cord nor nerves." "In sum, all the points of their bodies have in them, no doubt, this mode of the faculty of sensing which constitutes irritability."[28]

But in 1806, in the introductory lecture to his course on invertebrates the distinctions and differentiations were clear cut. Feelings and sensibility at large were completely dependent on the existence of any vestige of nerves, and muscular movement at large depended on it too. Irritability became totally separated from sensibility. It could be found in the most elementary forms of organized life, and in them bodily movements were caused from the outside. Thus instead of the definition of living organism used earlier, Lamarck suggested a new one based on the most general feature found in the lowest forms of organization, rather than on the most sophisticated ones, i.e. feelings and voluntary movement.

---

[25] In his *Prodrome d'une nouvelle classification des coquilles* 1799, (Lamarck 1972, 4).

[26] ".... et qu'il est plus convenable de procéder du connu à l'inconnu, je vais prendre l'ordre en sens inverse de celui de la nature ..." Lamarck 1802, 14.

[27] For example, in his *Recherches sur l'organisation* (in book I under *Les mammaux*) Lamarck states "remarquez que vers cette extrémité de l'échelle animale, tous les organes essentials sont isolés ou ont des foyers isolés en des lieux particuliers. Vous verrez bientôt que le contraire a parfaitement lieu vers l'autre extrémité de la même echelle" (Lamarck 1802, I 16). Note though that in the *Ancien discours de mon cours* of 1806 the notion of a natural order was becoming more stable and the changes, mutations etc were supposed to take place within that order and its immutable laws: "un ordre afin qui ne permet dans ses masses aucun arbitraire de notre part, et qui doit offrir à ses deux extrêmités les corps vivans les plus dissemblables ou les plus éloignés sous tous les rapports" (Lamarck 1972, 195).

[28] "ni cerveau, ni moelle longitudinale, ni nerfs ... Enfin, tous les points de leur corps ont sans doute en eux-mêmes cette modification de la faculté de sentir, qui constitue l'irritabilité." Ibid, 358.

The evolutionary perspective thus became the foundational one, and thus, the perspective through which one classified both vertebrates and invertebrates, and distinguished between flora and fauna. The predominant function that served as a tool of that classification was the presence or absence of nervous activity, signalled by the presence or absence of any form of a specialized system of nerves:

> "irritability in all or in some parts, is the most general characteristic of animals: it is even more so than the faculty of voluntary movement and than the faculty of sensing" "... irritability, a faculty which pertains generally and exclusively to animals...."[29]

It is remarkable that already in the appendix to the *Recherches* (1802) Lamarck used the same explanatory mechanism for feelings, which is then repeated and developed in detail in his *Philosophie Zoologique*. In contradistinction to Cabanis in his *Rapports du physique et du moral de l'homme* (both series of the *Rapports*[30]) and in line with the functioning of passivity,[31] the basic tenet of sensationalism, for Lamarck the active agent in sensations, in feelings and in volition is not the nervous system nor any part of it – they are the loci of the events and processes! – but a particular fluid: "le fluide nerveux." The presence or absence of a bodily system which enables the activity of the various body fluids, such as blood, was considered a distinguishing feature, marking the whole order of invertebrates, and used later on for internal classification within that order. In the general eighteenth century debate on the choice of bodily structures and mechanisms to explain sensations, feelings etc., there was quite a broad choice of models. Lamarck, in line with his geological and physical views, chose a hydraulic model, i.e. of various active fluids[32] as transferring agents and body structures that would go with it – pipes, tubes, canals, and other equivalent spaces. These structures functioned as linkages between inside and outside, as well as connections between various parts of the inside. More generally, they enabled the *rapports* between the environment and the body, since for Lamarck the organism *interacted*, within certain limits and constraints, with the environment, and was not just acted upon by it. It is within specific parts of these structures that the transformation into "information" of inner and outer cues was being carried out; this occurred by being translated into "behaviour" i.e. modes of experiencing and acting – feelings, movement, acting with the various capacities which Lamarck subsumed under intelligence. Lamarck's use of 'fluids' provided him with an occasion to elaborate on the proper and improper modes of using what would be termed in the 1930s 'theoretical entities' within an empiricist

---

[29]"l'irritabilité dans toutes ou dans certaines parties, est le caractère le plus général des animaux; elle l'est même plus que la faculté des mouvemens volontaires, et que la faculté de sentir" (Lamarck 1806, 126). "... l'irritabilité, faculté qui appartient généralement et exclusivement aux animaux ..." (ibid. 127).

[30]Cabanis 1805; see also Staum 1980, 1996.

[31]See Condillac 1746, 1755.

[32]Not only Haller, much earlier, but other naturalists as well, particularly in the 1780s and 1790s used 'fluids', endowing them with a variety of qualities and powers. For Lamarck the power of life is expressed by the movement of fluids in the living body, whether through special systems or without them, as was the case in the most elementary organisms; see also Smith 1976.

framework (i.e. entities that could not be directly experienced or observed, but that formed part of the structure of a scientific theory).With him too such use was justified by inferring from visible effects to – temporarily – invisible causes.

Even before the detailed discussion found in the *Philosophie Zoologique* Lamarck had established a gauge for the level of complexity of an organism. Its attribution to the organism was to be determined by the number and complexity of specialized sub systems of the organism, the extent and diversity of its functioning, its "behaviour," and the spectrum of its experiencing and acting, i.e. its' "faculties" and their range. In Lamarck's work the correlations of these attributes established a continuity between what earlier had been deemed the Cartesian influenced, divided realms of the mental and the physical, a continuity that seemed self-evident within the contemporaneous idéologue-sensationalist framework.

Lamarck's position on the nature of physico-mental/psycho-physical events becomes clearer in the writings from 1809 to 1820. Taking into account repetitions, similarities and differences, it is nonetheless striking how important the characterization of model 'basic units' (sensations/sentiments) became: they came to be considered individualized rather than generic. They were explicated in physiological (in Lamarck, 'physical') terms and presented as events and processes of the material organism, yet were understood as the discursive equivalents of the traditionally mental, all the more so for taking place solely inside the body. Thus I contend that the "basic units" were neither solely physiological nor solely mental or psychological (and the same applied to any process that recombined them – it was a hybrid one too). Lamarck intentionally resorted to the method of analysis characteristic of the empiricist-sensational associationist schools, namely, *to choose basic units which were assumed to be the most elementary, primitive, while all other elements were regarded as combinations and compoundings of such units.* In contradistinction to his predecessors, the evolutionary mechanism he had adopted operated on these basic units. This allowed him to point to a near zero point, and to adumbrate a gradual compounding across the ascending scale of organisms, and to exemplify an increasing *complexification*. Furthermore, because organisms, as well as their subsystems, were in constant interaction with their specific environment, the feelings-events in the body, through the processes of the body, and as integral components of the body, bore the specificities of this or that particular body, and thus were embodied in this double sense.

This is the gist of my argument.

As indicated above, during the period from 1800 to1806 there was in Lamarck a sharpening of the distinction between the Hallerian sensibility and irritability, and the adoption of a hydraulic model of the nervous system and of nervous fluids. In the years that followed, up to 1820, Lamarck's similarity to and distinction from some of his contemporaries, most conspicuously Cabanis, turned on the explanations proffered for sensations and feelings, instinct and inner-feeling or *sentiment intérieur*.

There has been a subdued but continuous disagreement on where to locate him in the naturalist-medical field of his generation. For Richards (1979, 1987), Lamarck and Cabanis were in the same general camp. Corsi (1988), in his masterful reconstruction of the contemporaneous scientific field tends to disagree, but points to actual and possible allies and missed opportunities for cooperation and support. By contrast, Rey (2000), who looked upon the period from the point of view of medical traditions and vitalist positions, had no place at all for Lamarck in her account. Vidal (2006) and Hatfield (1995) took part in this debate by asking whether psychology was newly constituted in that period, and agreed that it was in fact re-constituted, shunning its former dualist ancestry.

The *Philosophie Zoologique* was Lamarck's comprehensive undertaking to present an extremely detailed theory of the animal world, and his endeavour to tie together evolution, generation and development, genealogical/heredity, systematics and the (non-human and human) psychophysical field. Because of space limitations, I shall only attempt a brief analysis of the main arguments advanced and of the mechanisms posited in the *Philosophie Zoologique* regarding some of these matters. Two main issues are of relevance to my discussion here: feelings, *sentiment intérieur*. The principal components of Lamarck's explanatory model of the 'basic units' and their compoundings were the nervous system at large, certain specific spaces within it – *centres, 'foyers'* – particularly the brain, and the (hydraulic model of) nervous fluids. Based on the kinds of experiencing and movement available to groups of animals,[33] experiencing and movement that were found or were absent in the various organisms along the evolutionary scale, the animal realm was divided into four groups in Lamarck's writings up to 1810 (e.g. Lamarck 1809, 48); and into three groups from 1812 on (e.g. Lamarck 1815–1822, 267–269).[34] Given that the nervous fluids were the active agent in the explanatory model, in the 1809 writings[35] there were also two degrees of their mode of activity which mapped unto two modes of growing complexity and sophistication of the physico-mental apparatus. Furthermore, humans were distinguished by highly specialized activities. Though these could theoretically be found in organisms of lower complexity they stood out in humans: attention, imagination, understanding, thinking and recalling (memory). More than before, in the *Philosophie Zoologique* the evolutionary perspective was conspicuous, as the question of animal–human continuity was brought up directly and indirectly in almost every issue discussed. Two causes of change in organisms were available in the explanatory model of the earlier writings: through direct impact of the 'environment-circumstances' on organisms and their action–interaction with it, and through the inner tendency to complexification. In the *Philosophie Zoologique* the order of importance between these two was reversed, so that the tendency to complexification or the "life force" came to be seen as the predominant mechanism

---

[33] Which in turn depended on the degree of sophistication of their nervous system and its parts.

[34] Burkhardt 1995a argues that although Lamarck's theoretical position was that structures were dependent on functions, yet in actual practice he derived function from structure. His work was less like that of a contemporaneous ethologist and more like that of a comparative (animal behaviour). Lamarck himself, though, viewed his practice differently. In his later writings he emphasized the insights culled by any zoologist who could observe the functioning of organs, and who presumably would also infer from it within the faculties discourse in a way which was not open to an anatomist. Lamarck 1972 (a ms. of 1816), 41–42.

[35] *Philosophie Zoologique* and opening lecture for the 1809 course, Lamarck 1972, 201–214, or in his text *l'avancement de la zoologie* from 1808 to 1810. Ibid, 215–226.

for progress in the organic world,[36] while 'circumstances' were counted as secondary to it. Circumstances impacted on organisms indirectly via the mechanism of use and disuse, with the formation of habits inherited inter-generationally. As Lamarck assumed that functions determined structures, it meant that change of habit could bring about new organic forms. The circle of causality actually ran from circumstances or environment to needs and pressure for adaptation, to exercise of faculties, formation of habits which were instrumental in the rise of new structures through intergenerational transfer. Thus some version of a middle way position was forged in the debate on innate versus external capacities, so central in the eighteenth century, both between Cartesians, Leibnizians and sensualists on the one hand, and in the wide range of sensationalist-associationist field, on the other. In the *Philosophie Zoologique* Lamarck delineated multi-levelled system for the conveyance, transfer and maintenance of external and internal data, transformed into various embodied informational modes which had specific ways of activation attached to them. The basic contemporary distinction between within/outside the body was somewhat blunted by the very fact that everything could be processed, could become an event only inside the body. Thus excitation, perception, sensation, feeling, (and some muscular movement) would be one 'organic' mode termed sensibility; excitation, contraction would be another 'organic' mode termed irritability. These two modes were considered separate, differing, as mentioned earlier, in their extension over the animal world and independent of one another. Only the first was looked upon as depending on any specialized organic sub system, the nervous system. But in both modes it was the movement of various fluids within the body which indicated its being a living organism.

Lamarck provided a different view by virtue of his evolutionizing perspective. There is a persistent effort by Lamarck throughout the *Philosophie Zoologique* to distinguish his positions from those of numerous naturalists, even though these were rarely mentioned by name. Thus the book can be read as a series of exercises in 'dialogues manqués'. I shall not take this path here, because the focus of my paper lies elsewhere, but note that only two of his interlocutors – Cabanis and his disciple Richerand – were explicitly mentioned by name time and again. Their positions, particularly those of Cabanis, were perceived by Lamarck, and perhaps by some contemporaries, to be close enough to his own so as to call for explicit criticism.[37] Here Lamarck's effort to differentiate himself and disconnect himself from their positions had been both more difficult and more nuanced. This is directly related to my claim that Lamarck's original contribution lies in his new perspective. In numerous places in the *Philosophie Zoologique* Lamarck argued that method-wise and content-wise his perspective would be more generative, with higher explanatory power than Cabanis'. For example: in the introduction to Part II of the *Philosophie Zoologique*, after a laudatory introduction on Cabanis's findings concerning Man, he chided him, stating:

---

[36] e.g in *Philosophie Zoologique* II/vi: "Life ... tends incessantly by its very nature to a higher organisation, to the creation of special organs, to the isolation of these organs and their functions, and to the division and multiplications of its own centres of activity" (Lamarck 1984 (1809), 239).

[37] For example, Corsi 1988, particularly 122 and ch. 6; Staum 1980, particularly 182–189.

> "We must not confine ourselves to seeking the proofs of it by an examination of the highly complicated organisation of man and the more perfect animals. Proof will be obtained more easily by studying the diverse progress of complexity of organisation from the most imperfect animals up to those whose organisation is most complex … it is the simplest of all organisations that we should open our inquiry" … And "(examining complex organs of higher animals tends to ignore the fact that) … it does not follow that these same organs are essential to the existence of life in all living bodies whatsoever".[38]

Here and elsewhere Lamarck differentiated between the order imposed by instruction[39] and the order of nature. The order of the investigation ought to follow that of nature, rather than the ploys that make it more accessible to encumbered human understanding.[40] It is only by adopting the elementary forms as the object of description and analysis, that *general* and more specific *features* could be derived, *necessary and sufficient conditions* for life established, and *real relationships* such as between functions and structures, organism and circumstances, the various component parts of the organism determined,[41] and *causality* attributed. These seemed to him to be the desired results characteristic of a scientific endeavour. Furthermore, implied in the injunction to start with the elementary forms was the impossibility of either universalizing any stage of the evolutionary progress, or making predictions from one stage to the next one. This in contrast to a "post factum" description. In that sense Lamarck's injunction provided the possibility of regarding the evolutionary path as containing "emergent" offshoots, and particularly so for feelings and for intelligence.

According to Lamarck the feasibility of ascribing 'feelings' to individuals in the animal world was rather limited.[42] Among the invertebrates – Lamarck's area of specialization at the MHN – the lower taxa were considered not to have even the rudimentary organic structure deemed a necessary condition for the expression of "feelings." The latter require as a minimum a primitive, undeveloped, rudimentary nervous system as "The nervous system constitutes the special organ of feeling when it is composed of a single center of communication and of nerves terminating it."[43] As Lamarck started his analysis from the elementary forms of living organisms, the addition of further functions – whose extension were more

---

[38] Lamarck 1809, 185.

[39] For example, Lamarck 1984 [1809], 8.

[40] In an unpublished text written perhaps between 1816 and 1817 this was expressed most concisely: "cet ordre est le seul que soit naturel, instructif pour nous, favourable à nos études de la nature: et qui puisse, en outre, nous faire connoître la marche de cette dernière, ces moyens et les lois qui régissent ses opérations à leur egard. Des rapports qui doivent être employés dans la distribution et la classification des animaux" (Lamarck 1972, 248). He went on to emphasize that this order should be implemented by both the zoologists and the botanists. Ibid, 270.

[41] In one of the rare semi autobiographical paragraphs in an opening course lecture Lamarck stated (1816) that only gradually did he realize what a rare glimpse into the working of nature the investigation of invertebrates had offered him: "… il existoit réellement un ordre à reconnoitre, relativement à la composition de chaque sorte d'organisation animale …" Lamarck 1972, 29.

[42] For some discussions of animal behaviour and evolution at that time see Burkhardt 1981, 1995b.

[43] Lamarck 1809, 274.

bounded and constrained – and of structures – whose specialization became more sophisticated – turned his narrative into a kind of division of labour. The narrative could be characterized as being one of accumulation, and at the same time one of complexification. "Feeling" could be defined from the point of view of the discourse on organic functions, and also from that on faculties. Lamarck was clearly juggling to make the two overlap so as to constitute "feelings" as a basic physico-mental unit. "Feeling, or the capacity for experiencing sensations"[44];

> Feeling is not a faculty … feeling is only an effect; that is to say the result of an organic act and not a faculty inherent in any of the substances, which enter into the composition of a body that can experience it …. none of our humours and none of our organs, not even our nerves, have the faculty of feeling. It is only by an illusion that we attribute the singular effect which we call sensation or feeling to a definite part of our body … But the very remarkable effect called sensation, or, when more intense, pain, is the product of the function of a very special system of organs, the activity of which is dependent on the circumstances which provoke it.[45]

In those parts of the *Philosophie Zoologique*, where feelings were discussed, Lamarck elaborated on the above trying to differentiate his position from those of Cabanis and of the phrenologist Gall:[46] "… it is our entire being that feels or rather undergoes a general effect … it is a commotion throughout the entire sensitive system …[47]

In the third part of the *Philosophie Zoologique* Lamarck presented a detailed description of the general and specific movements and paths of fluids through variously differentiated nervous systems and their specialized sub parts and spaces. He repeatedly dissociated these processes from any assumption about non-physiological substrates. Thus the conventionally 'mental events' of empiricism – subsumed under the experiencing of feelings – were presented as direct consequences of physiological processes, with Lamarck refusing to give in to the demands of accounting for their being otherwise perceived ("an illusion," "une hallucination" etc.). It seems to me that he was aware of the fact that the shift he was asking his students and readers to perform was neither quantitative nor obvious. Furthermore, it was one which, in a serious sense, could not be justified within the sensationalist framework. He was using the evolutionary mechanism both rhetorically and substantially as the lever which would perform that transition. If the capacity of experiencing was dependent on emerging bodily structures which did not develop by simple quantitative steps, then the resulting experiencing, the resulting feelings, were also an emergent category, embodied yet endowed with a qualitative surplus which pointed beyond.[48]

---

[44]Lamarck 1809, II ix, 266.

[45]Ibid., II, ix and 273.

[46]See also *Gall's system* in Lamarck 1933.

[47]Lamarck 1809, III, 321.

[48]A note is in order here to point out that Lamarck's conception of 'material' was not a reductionist one in the nineteenth century sense. One has simply to recall that he did not subscribe to the 'new science' at large, and specifically to Lavoisier's chemistry. See also footnote 4. Note, though, that his later writings were much more zoological both in the explication of mechanisms and in the examples used. On a somewhat analogous problem see Kaitaro 2008.

Between 1809 and until his last book in 1820, the *Système analytique des connaissances positives de l'homme*, Lamarck published one major work, the *Histoire naturelle des animaux sans vertèbres* (1815), as well as numerous entries in a natural history dictionary (1817)[49] which were later incorporated, together with other materials, into the *Système analytique,* written when he was becoming completely blind. There were also many lecture notes, introductory lectures to the courses he had taught, and other articles and materials that had remained unpublished. There are certain interesting differences in the explication of some major notions between the *Philosophie Zoologique* and these later works. There was also a sharpening of Lamarck's insistence on the correctness of certain methodological approaches; for example, those concerning the proper order of investigation and exposition and those relating to norms of classification. I shall here only relate certain aspects of Lamarck's description and analysis of feelings, namely, those that may further clarify his unique perspective as compared with his contemporaries. In most of the shorter later writings Lamarck did not go directly from a discussion of irritability to a general discussion of feelings, but instead he interposed a discussion of a specific feeling variously named "an obscure feeling," "an intimate feeling," "a feeling of existence," and an "inner feeling." Thus in the text also called *The Harvard MS* (1810–1814) where a program for writing on human knowledge was being laid out, the order of exposition was such that irritability came first, then feeling with interior feeling as one of its sub sections, then ideas etc. In the subsequent editing of this work by Lamarck, the arrangement of the intended exposition was changed to irritability, and within the feelings section the only subsection to appear dealt with interior feeling. In another version only a discussion of the interior feeling remained. The emphasis in the later writings was on the "progressive development" of the nervous system whose complexified aspect I suggest was to be looked upon as organic emergence. Its 'effects' were observable on the ascending scale of organisms, and in particular, in the capacities for feeling in the more "complex" organisms, in which, incidentally, "feeling" would not be manifested when the supporting sub system was damaged.[50] This is almost self evident once we recall that the *Histoire naturelle des animaux sans vertèbres* (1815) was considered at the time a highly significant contribution to systematics. It seems to me that Lamarck had endeavoured to turn it into an exposition of "evolutionary systematics" – with a heavy emphasis on "evolutionary" – because he probably realized that the book would be of interest to a much wider audience than his *Philosophie Zoologique.* Indeed, numerous unsold copies of the latter were found on the auction list of his library after his death.[51] The same applied, and all the more so, to the items Lamarck prepared for the Deterville new dictionary. There all the non specialist items written

[49] *Nouveau Dictionnaire d'histoire naturelle …*, Deterville, Paris, 1817–1819, Lamarck's articles are from 1817. Numerous dictionaries were published in those years, reflecting the changing state of the respective fields.

[50] For example, see *Discours d'ouverture* 1816 in Lamarck 1972, 40–41; such quasi incidental comments indicate the extent to which Lamarck was well versed in the medical debates of his time.

[51] Lamarck 1830.

by Lamarck (i.e. not the ones on conchology and on meteorology) were followed by parallel entries by Virey.[52]

The manner in which evolutionary matters were dealt within the 1815 primarily classificatory work and in the subsequent publications from 1815 till 1822 is of interest. The principal thrust of the general part of the 1815 work was presented in the form of "first principles" in which Lamarck tried to tie together method, model, and a rhetoric adequate to the task, so as to distinguish his position clearly from those of vitalists, idéologues, phrenologists,[53] proponents of the "new science" and the new comparative anatomy. And with respect to "feelings" – first and foremost a re-affirmation of a material-organic interpretation of 'feelings' and this within his framework of relations between functions and structures.[54]

Lamarck summarized his position in a short, condensed manner stating that the faculties were separate in their organic structures, but cohered, and were part of an ascending, interrelated, tangled web of relationships within the same organism. The fundamental relationship was that of functions to specialized organs. The novelty here was the emphasis on the fact that this framework worked adequately, fittingly, all along the evolutionary path. Thus for a lower animal, devoid for example of 'feeling' (advanced faculties such as intelligence and judgement) "will be for them superfluous faculties, of which they have no use" – having them would probably be "damaging/dangerous for animals so delicate".[55]

---

[52]See Virey 1817–1822, Corsi 1987, Benichou and Blanckaert 1992, Blanckaert 2002.

[53]See Destutt de Tracy 1817–1818, *La Décade* 1794–1804, Goetz 1986, 1993, Moravia 1980, and Young 1970.

[54]The meaning of this discussion is enhanced by putting it side by side with the list of "zoological questions" found in the *Harvard MS*, which is believed to have been written around the same time. These two writings seem, indeed, to be carrying a question-answer relationship, particularly on the matters mentioned above. Lamarck 1933, 185–189.

[55]Lamarck 1815–1822, Introduction, 23–24. "seraient pour eux des facultés superflues, et dont ils ne feraient aucun usage," "nuisible à des animaux si délicats" (on the division of invertebrates). In his later writings Lamarck found some difficulties with his former presentation of the division of invertebrates, because according to it one of the two lowest classes by the criteria of the nervous system, namely 'worms', was found to be higher up by the criteria of the stage of development of other systems, creating an unexplained "jump" in an otherwise continuously and gradually ascending series. Thus in a late writing, he decided to divide the invertebrates into two branches rather than his usual unilinear ascending one, as e.g. in 'Des rapports' 1816–1817 (Lamarck 1972, 276–281). S. J. Gould argued that it was a later reflection on an earlier comment by Cuvier on the division of that realm (Gould 1999a, b). Note, this is contrary to Darwin's branching tree of descent with a single trunk. Lamarck's conceptual picture of the series was, indeed, similar and he had uniform ordering rules which applied to all living organisms (Daudin 1926). But in order to account for transformation and non-extinction of species he had added spontaneous generation of most elementary living entities. These would then, by a slow process stretching over long chunks of time, become 'more perfected' organisms, in the various animal realms. But it was primarily the various deviations caused by 'circumstances' that added branches to the earlier picture of the evolutionary tree, thus undermining the simple coherence of the 'echelle' image. In *Histoire des animaux sans vertèbres* one finds a marked emphasis on the adequate fit between organisms and their subsystems and faculties and their 'needs'. Yet in the supplement to that work, Lamarck admitted that the actual order, or 'the order of productions', as he named it, was differently constructed.

Or in a long elaboration to prove that this fit in all its details on various levels, and in relation to differing "faculties" supported by the nervous systems: "… there is always a perfect relationship between the needs, the faculties which satisfy them, and the organs that supply these faculties."[56] In these later writings there was a renewed emphasis on the relationship between individuality and life, which had been a feature mainly of Lamarck's earlier writings. The progressive development of more complex organic sub systems which supported more advanced functions was meticulously presented in the *Histoire* as a result of interactive adaptations to ever newly posited needs, thus attenuating the role of 'inner inclination' to perfection when compared to its exposition in the *Philosophie Zoologique*. This is most conspicuous in analyzing the progressive development and diversification of the nervous system. Given the different context in which this work was composed, Lamarck did not try to distinguish himself from either Cabanis or Richerand.[57] And when discussing the need, already well articulated in the *Philosophie Zoologique*, to separate irritability and sensibility, and to distinguish between muscular movement based on a nervous system and muscular irritability, he criticized Haller in a direct manner that had been absent in the *Philosophie Zoologique* . In general, one can say that critical comments were much more frequent and direct in *Histoire*, and the former tortuously detailed formulation of correlative relationships tended to be collapsed into shorter ones e.g. "Surely, there are none but nerves which would be the true organs of sentiment/feeling."[58]

In his last work, the *Système analytique,* Lamarck followed the same pattern as in all the other major ones: the first chapters dealt with the world at large, including the non living, concept of nature etc. The marked difference in this last work was the perspective from which these were discussed, namely: "What were the sources of knowledge available to humans? Under what constraints were these sources?, and "Were these compatible with his (Lamarck) view of the mechanisms of perceiving and thinking available to human beings?" There is hardly any difference in content or theoretical devices between the other late writings and the *Système*, as far as the explication of 'feelings' was concerned. Both analysis and description were much more succinct and tended to deal only with superior evolutionary forms – i.e.

[56]" … Qu'il y a toujours partout un rapport parfait entre les besoins, les facultés d'y satisfaire, et les organes qui donnent ces facultés" (ibid, 252). This had an underlining pointing at Cuvier's conception of 'nature', as the rhetorical thrust of the introduction was intended to undermine the latter's definition for 'animal'. Lamarck's own exposition of the characteristics of a 'living body' harked back to the project "Biologie" he delineated in 1802 but never carried out. Barsanti 1995, 1997a, Corsi 2006, and from a very different angle, Rey 1994. Corsi argued that it happened for socio-political reasons. In accordance with the 'adequate and fit' framework Lamarck offered detailed and elaborate explications for the appearance of novel forms of organic adaptation in the lower invertebrates without the need for an explanatory mechanism involving 'feelings'.

[57]Cabanis's work mentioned earlier and see Richerand, 1804.

[58]"Assurément, il n'ya que des nerfs qui soient les vrais organes du sentiment …" (Lamarck 1815–1822, I, 240).

humans. Those that were taken from the dictionary items formed in some ways an interconnected chain within the book.[59]

3 "Sentiment Intérieur"

I shall now attempt to analyze a special form of feelings in which, once again, the evolutionary perspective was posited as innovative, even when somewhat similar notions appeared elsewhere. I shall discuss the notion of *sentiment intérieur* – "inner feeling" which Lamarck characterized thus:

> every sensing being, i.e. a being which is endowed with the faculty of sensing, and nowhere else but the animal-realm that beings of that sort exist, possesses a sentiment intérieur, which it enjoys without discerning it, that gives it (the being) a very obscure notion of its existence, or put differently, constitutes in it the feeling of its being, and in this way allows [donne lieu] for that "I" so familiar to us, because we have the capacity to pay attention to it. This intimate feeling of existence, in one word this particular 'I' ['moi'] has been well known to us, as I have said.; but it seems to me that the sentiment intérieur that allows for it [the 'moi'], constituting a power which on the one hand is susceptible to being moved by [i.e. responsive to] any felt need, and on the other hand is capable of immediately causing action, has not been recognized by anyone before me.[60]

This short paragraph encapsulates what I consider one of Lamarck's most original ways of applying the evolutionary grid to contemporaneous naturalist, philosophical and cultural issues.

The *sentiment intérieur* was, on the one hand, a feeling i.e. falls under the explanatory mechanisms of feelings, and was emergent in the same sense that feelings within Lamarck's theory were. Yet on the other hand, it differed in remarkable ways from them, which tended to put into question it being 'a feeling'.

In the earlier period i.e. from 1800–1806 this *sentiment,* in the form of *sentiment d'existence*, is mentioned only once, as a component of the events that 'happen' in that part of the nervous system related to feelings "... In the feeling of existence of an individual organism, and in the need which the organism experiences instantaneously when it thinks."[61] I believe that it indicates that Lamarck was aware of the

---

[59] Direct and indirect criticism of these can be found in some related and parallel items by Virey 1817–1822 see also Corsi 1987.

[60] Lamarck 1817, "Instinct," 332, original emphases. "Tout être sensible, c'est-à-dire, doué de la faculté de sentir, et ce n'est que dans le règne animal qu'il en existe de cette sorte, possède un sentiment intérieur, dont il jouit sans le discerner, qui lui donne une notion très-obscure de son existence, ou autrement, qui constitue en lui le sentiment de son être, et qui y donne lieu à ce moi si connu de nous, parce que nous avons le pouvoir d'y donner de l'attention. Ce sentiment intime d'existence, en un mot, ce moi en question nous étoit bien connu, comme je viens de le dire; mais le sentiment intérieur qui y donne lieu, constituant une puissance, d'une part, susceptible d'être émue par tout besoin senti, et de l'autre, capable de faire agir immédiatement, ne me paroît avoir été reconnu par personne avant moi."

[61] Lamarck 1802, 168. "... dans le sentiment d'existence de l'individu, et dans le besoin qu'il éprouve dans l'instant où il pense."

possible criticism of the rudimentary theory of feelings he was constructing at the time, but was vaguely indicating that whatever solution that would be found had to remain within the bounds of the conceptual scheme delineated.

Let me draw a very rough sketch of the thorny field concerning 'self' and 'feelings': During the eighteenth century at least three major traditions of dealing with the enduring self identity of individuals and with the explanatory mechanisms for perception and intellection, battled, with discrete components of each getting intertwined with others in a wide array of variations, none of which I shall attempt to trace here.[62] These were the theological seventeenth century positions on the soul, either Catholic or Protestant, the Cartesian two substances varieties and the sensualist range of solutions – or rather non-solutions – within their epistemological grid. Beside these one should mention the Spinozist influence, which according to some scholars colored debates at least until the mid-eighteenth century.[63] A plethora of programs and disciplinary frameworks had developed in the spaces between these three. By the end of that century sensualist-empiricist approach seemed to be predominant, particularly in the fields of medical physiology and (empirical) natural history. Naturalists as well as physiologists struggled with the various ways of reading Newton, and of applying mechanistic models to what was gradually perceived as complex time-dependent processes in nature (and in society). These seemed to them to resist mathematization and quantification. How to achieve a fit under these terms between on the one hand the (macro) phenomena perceived as those that demanded an explanation – e.g. life, generation-heredity-development, perception and intellection, cohesive unity and functional diversity in the organism – and the micro mechanisms proffered for that purpose, on the other, became a difficult task.

The positions that resonated with Lamarck, as far as one can judge from his muted interlocutors, were probably the physician-vitalist one, the more severe monist materialism such as d'Holbach's, and the mixed options delineated by some members of the Idéologues group and by Diderot. They had in common the shunning of introspection as a privileged venue to the 'self', and the attempt to construct physical–physiological explanatory mechanisms, preferably non-reductive ones.

Concurrently, this was a period of profound economic, social and cultural changes, which some historians view as the period of the emergence of the "self" both in high and low culture and as a scientific object.[64] Most thinkers during

[62] I have intentionally tried to avoid using general terms such as materialism, vitalism, or mechanism in my analysis, because I think that within the context of the passage from the eighteenth to nineteenth century they are not helpful in trying to understand the particular case I discuss. One obvious reason for that may be that their meaning was changing. But see e.g. somewhat indirectly Blanckaert 2002, and more directly Bourdier 1971, Corsi 1982, Dixon 1988, Hagner 1992, Sloan 1990, Smith 1995, Thomson 2001.

[63] See Israel 2001, 2006, and the debates these volumes sparked.

[64] Goldstein 2000, 2005, Wahrman 2004; Habermas vs. Koselleck and the debate on the public sphere in the eighteenth century. See also Staum 1996, Stedman and Gareth 2002.

the second half of the eighteenth century, particularly in France, looked upon the individual as the basic unit of description and analysis.[65] Thus the 'self' was problematized in both the 'social' and the 'natural' discourses of the time. When Lamarck called the 'self' a feeling, he was using an option already there in the repertoire of models of enduring cohesive identity; once again his innovativeness lay in the evolutionary grid applied to this notion. Let me elaborate and illustrate. Lamarck discussed that notion in the *Philosophie Zoologique* and in fact this was the most detailed working out of its meaning in use. Even though Lamarck used both names – *sentiment intérieur* and *sentiment d'existence,*[66] the former was much more frequent, and when using the latter it was often accompanied by further adjectives such as – the intimate sense of … the obscure sense of … the obscure notion of …, and from 1817 on also '*conscience de*' which appeared in a couple of items of the dictionary. In these cases the distinction between conscious and non conscious animals was significant to his argument, and the same goes for the distinction between humans and animals in the *Système*. *Sentiment intérieur* appeared already in the preface which summarized the *Philosophie Zoologique*. It was defined as a feeling 'aroused by needs' and a causal agent in both movements and actions related to needs. The assumption of special organs for specialized functions meant that there was a particular space within Lamarck's construction of the nervous system, where took place the processes of a need being felt through 'emotion',[67] and the processes resulting in the formation of an action and/or a reaction. Being activated by a need meant, within the Lamarckian explanation, that the *sentiment intérieur* played an important role in the formation of habits, which, when becoming an acquired trait, were passed on inter-generationally. This 'feeling' of one's existence was there continually. This "feeling" pervaded the whole organism as it gathered feelings produced by the very activities of living and was thus based on the interconnectedness of the nervous system. This interconnectedness served to bind together processes. It replaced what earlier in the century had been considered two substances, or alternatively, a bodily substance and some non-bodily substance ruling and activating it. Thus Lamarck provided a substitute to a substantive substrate of the functional unity of the organism, i.e. for him the organism functioned reactively and efficaciously. It meant a stable system which was both self referring and a referral system, referring to the ever present needs of the living organism as a cohesive whole in relation to its inner working and in its distributed functioning. This was in relation

[65] Note, though, that some (e.g. Ferguson and even Condorcet in his late writings) argued that one could not imagine, hypothesize, describe or analyze discrete individuals in isolation: they had to be considered already as socialized, already within the context of some society. They all thought that an explanation for the fact that sociability was a universal human trait had to be provided. From the second half of the eighteenth century onwards the dynamic character of being human became emphasized as the foremost feature at both the individual and the collective level, and became a key notion in eighteenth century writings on the mechanism for culture or for civility to evolve.

[66] See Jordanova 1981.

[67] For Lamarck 'emotion' had to do with having a power to affect movement.

to needs, particularly survival and adaptational needs, arising in its interaction with the environment, as embodied in this generalized continuous state. Lamarck looked upon the nervous system as having distinct and separate parts for sensing and for movement. The uniqueness of the *sentiment intérieur* was, for him, its ability to be both a transmitter, as indicated above, and an initiator of movement.

Viewing the *sentiment intérieur* through the evolutionary grid meant that it was emergent, absent in very low invertebrates and appearing in crude form in invertebrates from insects up. Being 'a feeling' it could not exist without a nervous system plus the fluids mechanism which would support it, and both its degree of sophistication and efficacy in bodily movement – i.e. of constraint, compulsion, control and choice – depended on the complexity of this supportive system. It meant that vertebrates in general, mammals and humans in particular, had a more privileged position in relation to it. The assumption of an emergent, gradual *sentiment intérieur* was intended to resolve the empiricist cul-de-sac of identity-and-continuity of the self[68] on many levels. The *sentiment intérieur* provided the organism with a coherence arising from discrete components, encompassing the whole organized living-body-system, endowing it with an ever newly constituted unity-in diversity, completely embodied and yet accounting for the transition to emergent graduated experiencing, which already at its elementary level was regulating and monitoring.[69] Moreover, because the *sentiment intérieur* was constituted as both receptor and initiator it was constituted as a necessary condition for higher functioning such as a variety of modes of thought of which volitional acts were a derivation.

Lamarck discussed some of the enabling conditions for the higher mental functions in the *Philosophie Zoologique*, in the *Histoire*, in the dictionary items, in the *Système analytique* based on them, as well as a couple of writings from the second decade of the nineteenth century.[70] Let me briefly mention one such condition – 'attention' – which illustrated the growing complexity of the nervous system. Whereas the *sentiment intérieur* was wide ranging, '*attention*' was posited as a narrow and specialized function closely and directly related to it. '*Attention*'[71] was a term which acquired, besides its more generalized use during that period, also

---

[68] Which Goldstein 2005 calls horizontal fragmentation of the self.

[69] Another important facet of the *sentiment intérieur* comes through Lamarck's discussion of instinct and Lamarck's position within the contemporaneous controversy on that, but this is beyond the present paper.

[70] For example, Discours d'ouverture de 1816, Apperçu analytique des connaissances humaines.

[71] As he says in the *Philosophie Zoologique* : "Ainsi, pour que les traits ou l' image de l' objet qui a causé la sensation puissent parvenir dans l' organe de l' entendement et être imprimés sur quelque partie de cet organe, il faut, premièrement, que l' acte qu' on nomme attention, prépare l' organe à en recevoir l'impression, ou que ce même acte ouvre la voie qui peut faire arriver le produit de cette sensation à l' organe sur lequel peuvent s'imprimer les traits de l' objet qui y a donné lieu: *et pour qu' une idée quelconque puisse parvenir ou être rappelée à la conscience, il faut, à l' aide encore de l' attention, que le fluide nerveux en rapporte les traits au sentiment intérieur de l'individu, ce qui alors lui rend cette idée présente ou sensible ...*" (Lamarck 1809, 375–376, my emphasis). Or in 'Idée' (*Nouveau Dictionnaire*) : "Cet acte ... qui seul ... donne le pouvoir d'exécuter toute autre opération de l'intelligence." Lamarck 1817, 83.

a specialized signification related to the practice of naturalists during the eighteenth century, namely, a willed selection of observation as well as that of rapt concentration and absorption with a corresponding physiological mechanism to enable 'the soul' to do that.[72] Lamarck lifted '*attention*' from its more accepted use, and from its use among the Idéologues and particularly Cabanis, in order to deploy it for a differing purpose. Contrary to Cabanis, he assumed a unitary organic individual, and denied the possibility of a scattered or divided 'moi'.[73] As mentioned earlier, the rudimentary sense of existence did not entail in any way either thinking or judging, capacities ordinarily characterizing human singularity at that time. '*Attention*' was explicated as one of the necessary preconditions for enabling the physiological arrangements which catered for the formation of ideas out of impressions, and thereby also for memorization.[74] One could view it as a mechanism for fixing the object of sensation, prolonging its duration in the relevant receptor parts of the nervous system, thereby allowing for a focusing pause. That activity of making the pause possible could be seen as a time-component in the spatial-physiological description of thinking, of the formation of ideas. It allowed for gathering-and-enhancing in order to enable a move from one order of mental functioning to the next-qualitatively higher, i.e. more complex one, possible only in specific spaces within the nervous system, such as available only to the 'highest' among the four groups of organisms endowed with *sentiment intérieur.*

To sum up: Lamarck started from a third person description in order to convey the meaning of a first person experience.[75] Damasio's table of self[76] delineates an evolutionary path from proto self to a core self (with consciousness), which can help in grasping what Lamarck was attempting to do. There were naturalists who suggested inner feelings before Lamarck, most conspicuously Buffon and Cabanis[77]

[72]Daston 2004, 117, 2007, 240–41.

[73]Cabanis discussed the possibility of partial, plural, 'moi', related to differing nervous centers, but he declared that one could say very little about them "puisque toutes nos sensations de moi rapportent exclusivement au centre général et que nos moyens d'acquérir des notions exactes" depend on grasping the circumstances "dans leur enchaînement", ibid, 503.

[74]This is where Cabanis' and Lamarck's positions are similar: Cabanis dealt only with humans, and wrote profusely on 'pain' and its important role in medicine. Thus '*attention*' was posited as a necessary condition not only for the very reception and transmission of an impression, but also for the production of pain. These matters are discussed in the first three memoirs and in the tenth as well.

[75]Taylor 1989.

[76]Damasio 1999, 199, Damasio and Damasio 2006; see also Edelman 2006.

[77]But note that though the term was used by numerous authors in France, particularly in the second half of the eighteenth century, it was rather polysemic. Its varying meanings, though non overlapping, related primarily to humans, e.g. to the moral, to that which is beyond reason etc. and when related to animals (e.g. Buffon 1749–1788, 15 cxxix, see http://www.buffon.cnrs.fr.) the comparison emphasized the gap, the conspicuous difference between the two. Cabanis used only 'sentiment d'existence', and related it to issues of the will, for example being a physician, when he discussed the changes brought about by different living regimes, particularly nutrition, that could "en un mot, à donner un plus grand sentiment d'existence" Cabanis 1805, 386. Or when he discussed sympathy, sentiment of self and will. (ibid, 541). For Cabanis "le moi réside exclusivement dans la volonté," and even the foetus has a sense of that (ibid, 31).

(who restricted it to an inner feeling in humans only), but it was Lamarck's evolutionary perspective on 'feelings' in general and on 'inner feeling' in particular which made his suggestion a generative one for psychologists in the nineteenth century. Yet, when looking at the notion within his scheme, it is clear that *sentiment intérieur* did not fulfil the role of a feeling, this in Lamarck's own view, and belonged much more in his 'faculty discourse'.

The assumption of *sentiment intérieur/sentiment d'existence* first elaborated in the *Philosophie Zoologique* was reworked again in the *Histoire*, but these presentations differed primarily in their emphasis.

The source and cause of movement of lower organisms were outside pressures. They had tropism-like movements, and the reason for it, mentioned in the *Philosophie Zoologique* and elaborated upon in much more detail in Lamarck's *Système*, was the inability to have feelings, and thus to enjoy the concerted bodily movements related to *sentiment intérieur*: "Now, the very imperfect animals under consideration since they do not in anyway possess the *sentiment intérieur* in question, would not know having or have means of eliciting in them the instigating cause of their movements. Obviously, it comes to them from the outside, and certainly it is not at their disposal; also, none of their needs necessitate it being there; that is what I have already shown. All that they need is found within their reach; they are animals only by virtue of being irritable."[78]

Already in the *Philosophie Zoologique*, but much more so in his *Histoire*[79] and in his *Système*, Lamarck tried to stretch the epistemological and cultural ("moral") work that the *sentiment intérieur* could do within his system, and indeed turn it into a full blown social-cultural self, including morals,[80] e.g. to point to a diversity and wide range of needs characteristic of humans which would be amplified by civilization.[81] In addition, Lamarck also tried to relate it to memory, imagination, and intelligence, and its mode of functioning as initiating and bringing about human proclivities – above all that of tending to preserve oneself. In fact, in higher mammals and in humans a wide range of social behaviours could be derived from the latter. Thus, as far as humans were concerned, Lamarck posited within the sensualist framework a feeling of an active "self," which, being a necessary result

---

[78] Lamarck 1801, 251. "Or, les animaux très-imparfaits dont il s'agit, ne possédant nullement le sentiment intérieur en question, ne sauraient avoir ou faire naître en eux la cause excitatrice de leurs mouvemens. Elle leur vient donc évidemment du dehors, et dès lors elle n'est assurément pas à leur disposition; aussi aucun de leurs besoins n'exige qu'elle le soit; ce que j'ai déjà fait voir. Tout ce qu'il leur faut se trouve à leur portée; ce ne sont des animaux que parce qu'ils sont irritables."

[79] As well as in the articles written for the Deterville dictionary.

[80] "Maintenant, il est indispensable de montrer que les penchans des animaux sensibles, que ceux même de l'homme, ainsi que ses passions, sont encore des phénomènes de l'organisation, des produits naturels et nécessaires du sentiment intérieur de ces êtres." Lamarck 1801, 258.

[81] Given that for Lamarck there was an evolutionary fit between possible needs, the specific faculties which would satisfy them and their supporting organs, the absence of *sentiment intérieur* was indicative of the narrow range of the organism's needs, and vice versa, and of the place of its series (a term often used, and much less loaded for him than species). See Daudin 1926.

of organic structures was in that sense, moreover, "innate." However, this feeling of an active "self" was meaningless within human culture without "experiencing" within a specific environment (in that sense a Piaget-like structure) e.g. "No doubt, man is born without ideas, without enlightenment, in possession of nothing but a *sentiment intérieur* and general proclivities which tend to exert themselves mechanically. It is only with time and through education, experience and the circumstances which he encounters that he acquires ideas and knowledge."[82]

In conceiving the *sentiment intérieur* Lamarck was answering questions at the focus of eighteenth century, particularly late eighteenth century, debates, and using eighteenth century means – e.g. materialization of the mental. However, his evolutionary perspective, which assumed both growing complexification and openness to the environment, compelled him to look for a unifying cohering foundation for the individualized entities which populated his theory. This foundation had to cater to the demands of "science'[83] as well as those of 'experience', and thus he produced the first version of an evolutionary 'self', and a new conception of 'internality'.

I shall end with a brief remark: When Lamarck had reached the limits of the then prevalent modes and models of explanation of living nature, he gradually constructed a model which by virtue of combining time, genealogy and causality could provide a better description and analysis of processes of emergence – of the emergence of complexity and of complex self organization – which Lamarck came to see as the principal features of the living entities. This model was at the core of Lamarck's innovations – explicating the 'feelings' and the 'experiencing self' that I have discussed in this paper. It aimed to encompass all the known functioning of the human mind and of the human psyche, and it did so by situating humanity as a component of a natural continuity. The basic analytical units, the mechanisms, the generalizations, the laws, all would have to apply to living organisms at large in order to apply to humans. It "naturalized" human behaviour by putting it on a par with the behaviour of any living organism. The patterns and mechanisms that would apply to living nature would apply to human perception and consciousness, as well as to unconscious, purposeful, planned and spontaneous human activities; to thought, feelings, desires, and in fact to both individual and social behaviour. The problématiques that Lamarck dealt with were certainly those of the natural history and of the physicians-physiologists communities of the time, and were conceived within the broadly defined framework of sensualism–associationism. His questions were defined within this conceptual framework, but the variety of answers given within this framework must have been deeply unsatisfactory to him. His innovative stance, which answered the questions by freeing himself from

[82]"Sans doute, l'homme naît sans idées, sans lumières, ne possédant alors qu'un sentiment intérieur et des penchans généraux qui tendent machinalement à s'exercer. Ce n'est qu'avec le temps et par l'éducation, l'expérience, et les circonstances dans lesquelles il se rencontre, qu'il acquiert des idées et des connaissances" (Lamarck 1817, 279).

[83]The *idéologues* as well as Lamarck certainly worked under the constraint of the evolving conception of science as produced in the 1st class of the *Institut* which Lamarck regularly attended and of which he was an active member until he became totally blind. Without the theological self, the social understanding of man's place in nature also became a matter of scientific knowledge.

that framework embodied a systematic and thoroughgoing application of an evolutionary perspective on the living world. With that he laid a possible foundation for a novel, generative framework.

References

Barsanti, Giulio. 1995. La Naissance de la Biologie: Observations, Théories, Métaphysiques en France 1740–1810. In *Nature, Histoire, Société: Essais en Hommage à J. Roger*, eds. Claude Blanckaert, Jean Louis Fischer, and Roselyne Rey, 197–227. Paris: Klincksieck.

Barsanti, Giulio. 1997a. Lamarck et la Naissance de la Biologie. In *Jean-Baptiste Lamarck 1744–1829*. ed. Goulven Laurent, 349–368. Paris: Editions du CTHS.

Barsanti Giulio. 1997b. Le "Système de la Nature" de Lamarck (1794): Analyse d'un Ambitieux Projet Avorté d'Après un Manuscrit Oublié. In *Le Muséum au Premier Siècle de Son Histoire*, eds. Claude Blanckaert, Claudine Cohen, Pietro Corsi, and Jean-Louis Fischer, 219–228. Paris: Éditions du Muséum national d'histoire naturelle.

Bénichou, Claude and Blanckaert, Claude eds. 1988. Julien-Joseph Virey, Naturaliste et Anthropologue. Paris: J.Vrin.

Blanckaert, Claude. 1992. La Valeur de l'Homme: L'Idée de Nature Humaine Chez Buffon. In *Buffon 88*, ed. Jean Gayon. Paris: Vrin.

Blanckaert, Claude. 2002. Le moment "naturaliste" des sciences des homme. *Revue d'Histoire des Sciences Humaines* 3(2):117–160.

Bourdier, Franck. 1971. L'Homme Selon Lamarck. In *Lamarck Colloque*, ed. Joseph Schiller. Paris:A. Blanchard.

Boury, Dominique. 2008. Irritability and Sensibility: Key Concepts in Assessing the Medical Doctines of Haller and Bordeu. *Science in Context* 21(4):521–535.

Buffon, Georges Louis Leclerc Comte de. 1749–1788. *Histoire Naturelle Générale et Particulière*: avec la description du Cabinet du Roy. Paris, L'Imprimerie royale. http://www.buffon.cnrs.fr.

Burkhardt, Richard. 1977. *The Spirit of System: Lamarck and Evolutionary Biology*. Cambridge, MA: Harvard University Press.

Burkhardt, Richard. 1981. Lamarck's Understanding of Animal Behaviour. In *Lamarck et son temps, Lamarck et Notre Temps*, 11–25. Paris: Vrin.

Burkhardt, Richard. 1995a. Lamarck in 1995, i–xl. In *The Spirit of System: Lamarck and Evolutionary Biology*. Cambridge, MA: Harvard University Press.

Burkhardt, Richard. 1995b (1988). Animal Behaviour and Organic Mutability in the Age of Lamarck. In *Lamarck e il Lamarckismo*, 63–90. Napoli: La città del sole.

Cabanis, Pierre Jean Georges. 1805. *Rapports du Physique et du Moral de l'Homme*, 2ième édition. Paris: Crapart, Caille et Ravier.

Condillac, Etienne Bonnot de. 1746. *Essai Sur l'Origine des Connoissances Humaines*. Amsterdam: Changuion.

Condillac, Etienne Bonnot de. 1755. *Traité des Animaux*. Amsterdam: Ch.Ant.Jombert

Conry, Yvette. 1994. Lamarck, Penseur de frontière. *Nuncius* 9(2):559–592.

Corsi, Pietro. 1982. Models and Analogies for the Reform of Natural History. Features of the French Debate 1790–1800. In *Lazzaro Spallanzani e la Biologia del Settecento: Teorie, Esperimenti, Istituzioni Scientifiche:* atti del convegno, marzo 1981, eds. Walter Bernardi and Antonello La Vergata, 381–396. Firenze: L. S. Olschki.

Corsi, Pietro. 1987. Julien Joseph Virey, le Premier Critique de Lamarck. In *Histoire du Concept d'Espèce dans les Sciences de la Vie*, ed. Scott Atran, et al. 176–187. Paris: Fondation Singer Polignac.

Corsi, Pietro. 1988. *The Age of Lamarck: Evolutionary Theories in France 1790–1830*. Translated by Jonathan Mandelbaum. Berkeley, CA: University of California Press.

Corsi, Pietro. 2005. Before Darwin: Transformist Concepts in European Natural History. *Journal of the History of Biology* 38(1):67–83.

Corsi, Pietro. 2006. Biologie. In *Lamarck Philosophe de la Nature*, eds. Pietro Corsi, Jean Gayon, Gabriel Gohau, and Stéphane Tirard. 37–64. Paris: Presses Universitaires de France.
Corsi, Pietro, general editor of the Lamarck website http://www.lamarck.net.
Corsi, Pietro. 1997. Les élèves de Lamarck: un projet de recherche. In *Jean Baptiste Lamarck 1744-1829*, ed. Goulvent Laurent. 515–526. Paris: Éditions du CTHS.
Corsi, Pietro, Hoquet Thierry, general editors of the Buffon site on the web http://www.buffon.cnrs.fr.
Cuvier, Georges. 1799. *Leçons d'Anatomie Comparée – Recueillies et Publiées sous Ses Yeux.* C. Duméril, ... [Puis] par G. L. Duvernoy, 2 vol. Paris: Baudoouin.
Cuvier, Georges. 1805. *Leçons d'Anatomie Comparée, Recueillies et Publiées sous Ses Yeux.* C. Duméril, an VIII [1800]–an XIV–1805. Paris: Baudouin.
Cuvier, Georges. 1810. *Rapport Historique sur les Progrès des Sciences Naturelles depuis 1789, et sur Leur Etat Actuel,* présenté à Sa Majesté l'empereur et roi, en son Conseil d'état, le 6 février 1808, par la Classe des sciences physiques et mathématiques de l'institut, conformément à l'arrêté du gouvernement du 13 ventôse an x. Paris: l'Imprimerie impériale.
Cuvier, Georges. 1815/1989. *Rapport à l'Empereur sur le Progrès des Science, des Letters et des Arts depuis 1789.* 2. Chimie et Sciences de la Nature, ed. Yves Laissus. Paris: Belin.
Cuvier, Georges. 1860. *Éloges Historiques: Précédés de l'Éloge de l'Auteur,* ed. Pierre Flourens. Paris: E. Ducrocq
Damasio, Antonio R. 1999. *The Feeling of What Happens: Body and Emotion in the Making of Consciousness.* New York: Harcourt Brace.
Damasio Antonio, Damasio, Hanna. 2006. Minding the Body. *Daedalus* 135:3,15–22.
Daston, Lorraine. 2004. Attention and the Values of Nature in the Enlightenment. In *The Moral Authority of Nature,* eds. Lorraine Daston and Fernando Vidal, 100–126. Chicago, IL: University of Chicago Press.
Daston, Lorraine. 2005. Fear and Loathing of Imagination in Science. *Daedalus* 134:4,16–31.
Daston, Lorraine, Gallison Peter. 2007. *Objectivity.* New York, NY: Zone Books.
Daston, Lorraine. 2008. On Scientific Observation. *Isis* 99:97–110.
Daston Lorraine, Pomata Gianna. 2003. The Faces of Nature: Visibility and Authority. In idem eds. *The Faces of Nature in Enlightened Europe*, 1–16. Berlin: Berliner Wissenschaft Verlag.
Daudin, Henri. (1926) 1983. *De Linné à Lamarck: Méthodes de la Classification et Idée de Sèrie en Botanique et en Zoologie (1740–1790).* Paris: Editions des archives contemporaines.
Destutt-De Tracy, Antoine Louis Claude. 1817–1818. *Élémens d'Idéologie,* 3 edn. I[st] Part. Paris: Courcier.
La Décade Philosophique, Littéraire et Politique + La Revue, ou Décade Philosophique, Littéraire et Politique. 1794–1804. Paris: J. B. Say
Dixon, B. Lynne. 1988. *Diderot, Philosopher of Energy.* Oxford: Oxford University Press.
Duchesneau, François. 1982. *La Physiologie des Lumières: Empirisme, Modèles et Théories.* The Hague: Martinus Nijhoff.
Edelman, Gerald M. 2006. The Embodiment of Mind. *Daedalus* 135/3:23–32.
Farber, Paul Lawrence. 1981. Research Traditions in 18th Century Natural History. In *Lazzaro Spallanzani e la Biologia del Settecento*, eds.Walter Bernardi and Antonello La Vergata, 397–403. Firenze: L. S. Olschki.
Fox, Robert. 1990. Laplacian Physics. In *Companion to the History of Modern Science*, eds. Robert Cecil Olby, Geoffrey N. Cantor, John R.R. Christie, and Michael Jonathan Sessions Hodge, 278–294. London, New York: Routledge.
Geoffroy Saint Hilaire, Etienne. 1809/29. *Description des Mammifères Qui Se Trouvent en Égypte* in *Description de l'Egypte* vol ii. Paris: Imprimerie Impériale.
Gillispie, Charles C. 2004. *Science and Polity in France: the Revolutionary and Napoleonic Years.* Princeton, NJ: Princeton University Press.
Gissis, Snait. 2009. Interactions between Social and Biological Thinking: The Case of Lamarck. *Perspectives on Science* 17:3, 237–306.
Goetz Rose. 1986. La "Science de l'Homme" des Idéologues. In *Doctrines, Sciences ou Pratiques Sociales?* Colloque. Nancy: Presses Universitaires de Nancy.

Goetz, Rose. 1993. *Destutt de Tracy*. Genève: Droz.

Gohau, Gabriel. 1997a. L'Hydrogéologie et l'Histoire de la Géologie. In *Jean Baptiste Lamarck 1744–1829*, ed. Goulven Laurent, 137–147. Paris: Editions de CTHS.

Gohau, Gabriel. 1997b. Le Cadre Minéral et l'Évolution Lamarckiennne. In *Colloque International Lamarck*, ed. Joseph Schiller, 105–133. Paris: Blanchard.

Gohau, Gabriel. 2006. Lamarck Philosophe? In *Lamarck Philosophe de la Nature*, eds. P. Corsi, Jean Gayon, Gabriel Gohau, and Stéphane Tirard, 9–36. Paris: Presses Universitaires de France.

Goldstein, Jan. 2002. Mutations of the Self in Old Regime and Post-Revolutionary France. In *Biographies of Scientific Objects*, ed. Lorraine Daston, 86–116. Chicago, IL: University of Chicago Press.

Goldstein, Jan. 2005. *The Post-Revolutionary Self: Politics and Psyche in France, 1750–1850.* Cambridge, MA: Harvard University Press.

Gould, Stephen J. 1999a. A Division of Worms: Jean Baptiste Lamarck's Contribution to Evolutionary Theory, part I. *Natural History* (February).

Gould, Stephen J. 1999b. Branching Through a Wormhole: Reclassifying the Types of 'Worms,' part II. *Natural History* (March).

Gross, Michael. 1979. The Lessened Locus of Feelings: a Transformation in French Physiology in the Early Nineteenth Century. *Journal of the History of Biology* 12/2:231–271.

Hagner, Michael. 1992. The Soul and the Brain Between Anatomy and *Naturphilosophie* in the Early Nineteenth Century. *Medical History* 36:1–33.

Hannaway Caroline, La Berge Ann. 1998. *Constructing Paris Medicine*. Amsterdam: Rodopi.

Hatfield, Gary. 1995. Remaking the Science of Mind: Psychology as a Natural Science. In *Inventing Human Sciences: Eighteenth Century Domains*, eds. Christopher Fox, Roy Porter, and Robert Wokler. Berkeley, CA: University of California Press.

Herbert, Sandra. 2005. The Darwinian Revolution Revisited. *Journal of the History of Biology* 38(1):51–66.

Hodge, Michael Jonathan Sessions 1995. Lamarck: un Grand Changement de Cadre Conceptuel. In *Nature, Histoire, Société: Essais en Hommage à J. Roger*, eds. Claude Blanckaert, Jean Louis Fischer, and Roselyne Rey, 229–239. Paris: Klincksieck.

Israel, Jonathan I. 2001. *Radical Enlightenment: Philosophy and the Making of Modernity, 1650–1750.* Oxford, New York: Oxford University Press.

Israel, Jonathan I. 2006. *Enlightenment Contested: Philosophy, Modernity, and the Emancipation of Man 1670–1752*. Oxford, New York: Oxford University Press.

Jacyna, Leon Stephen 1987. Medical Science and Moral Science: the Cultural Relations of Physiology in Restoration France. *History of Science* 25:111–146.

Jordanova, Ludmila. 1981. La Psychologie Naturaliste et le "Problème des Niveaux": la Notion du Sentiment Interieur Chez Lamarck. In *Lamarck et Son Temps, Lamarck et Notre Temps*, 69–80. Paris: Vrin.

Jordanova, Ludmilla J. 1984. *Lamarck.* Oxford, New York: Oxford University Press.

Jordanova, Ludmilla J. 1989. Nature's Powers: A Reading of Lamarck's Distinction Between Creation and Production. In *History, Humanity and Evolution*, ed. James R. Moore, 71–98. Cambridge: Cambridge University Press.

Kaitaro, Timo. 2008. Can Matter Mark the Hours? Eighteenth-Century Vitalist Materialism and Functional Properties. *Science in Context* 21(4):581–592.

Lamarck, Jean Baptiste Pierre Antoine de Monet de. 1800. *Discours d'Ouverture Prononcé le 12 Floreal a l'An 8*. Site Lamarck www.lamarck.net.

Lamarck, Jean Baptiste Pierre Antoine de Monet de. 1800/1944. Biologie ou Considérations sur la Nature, les Facultés, les Développemens et l'Origine des Corps Vivans, ed. Pierre Paul Grassé. *La Revue Scientifique* 82:267–276.

Lamarck, Jean Baptiste and Pierre Antoine de Monet de. 1801. *Système des Animaux sans Vertèbres*; ou, Tableau Général des Classes, des Ordres, et des Genres de ces Animaux Précédée du Discours d'Ouverture du Cours de Zoologie, Donné dans le Muséum National d'Histoire Naturelle l'An 8 de la République. Paris: Deterville.

Lamarck, Jean Baptiste Pierre Antoine de Monet de. 1802. *Recherches sur l'Organisation des Corps Vivans: et Particulièrement sur Son Origine*. Paris: Chez l'auteur. Maillard.
Lamarck, Jean Baptiste Pierre Antoine de Monet de. 1907. Discours d'Ouverture: an VIII, an X, an XI et 1806. Paris: *Bulletin Scientifique de la France et de la Belgique*.
Lamarck, Jean Baptiste Pierre Antoine de Monet de. 1809. *Philosophie Zoologique*. Paris: Dentu. Site Lamarck. www.lamarck.net.
Lamarck, Jean Baptiste Pierre Antoine de Monet de. 1812, *Extrait du Cours de Zoologie sur les Animaux sans Vertèbres*. Paris: D'Hautel & Gabon. Site Lamarck. www.lamarck.net.
Lamarck, Jean Baptiste Pierre Antoine de Monet de. 1815–1822. *Histoire Naturelle des Animaux sans Vertèbres*. Paris: Deterville. Site Lamarck. www.lamarck.net.
Lamarck, Jean Baptiste Pierre Antoine de Monet de. 1817. Articles from the Various Volumes of the *Nouveau Dictionnaire d'Histoire Naturelle*. Faculté, Fonctions organiques, Habitude, Homme, Idée, Irritabilité, Inteligence, Imagination, Instinct, Judgement, Météorologie, Nature. Paris: Deterville.
Lamarck, Jean Baptiste Pierre Antoine de Monet de. 1820. *Système Analytique des Connaissances Positives de l'Homme*. Paris: Chez l'auteur et Belin.
Lamarck, Jean Baptiste Pierre Antoine de Monet de. 1830. *Catalogue des Livres de la Bibliothèque de Feu M. le Chevalier J. B. de Lamarck*, dont la vente se fera le lundi 19 avril 1830, et jours suivans. Paris: Barrois, Snr.; Benou.
Lamarck, Jean Baptiste Pierre Antoine de Monet de. 1933. *The Lamarck Manuscripts at Harvard*, eds. William Morton Wheeler and Thomas Barbour. Cambridge, MA: Harvard University Press.
Lamarck, Jean Baptiste Pierre Antoine de Monet de. 1972. *Inédits de Lamarck*; d'Après les Manuscrits Conservés à la Bibliothèque Centrale du Muséum National d'Histoire Naturelle de Paris, eds. Max Vachon, Georges Rousseau, and Yves Laissus. Paris: Masson.
Lamarck, Jean Baptiste Pierre Antoine de Monet de. 1984 [1809]. *Zoological Philosophy*, trans. Hugh Elliot. Chicago, IL: University of Chicago Press.
Latreille, Pierre André, Geoffroy Saint Hilaire, Etienne. 1829. *Discours Prononcés sur la Tombe de M. le Chevalier de Lamarck*. Paris: Institut royal de France, Académie royale des sciences.
Lesch, John E. 1984. *Science and Medicine in France: the Emergence of Experimental Physiology 1790–1855*. Cambridge, MA: Harvard University Press.
Moravia, Sergio. 1972. Philosophie et Médicine en France à la Fin du XVIIIe Siècle. *Studies on Voltaire* 472(89):1089–1151.
Moravia, Sergio. 1978. From Homme Machine to Homme Sensible: Changing Eighteenth Century Models of Man's Image. *Journal of the History of Ideas* 39(1):45–60.
Moravia, Sergio. 1980. Enlightenment and the Sciences of Man. *History of Science* 18:247–268.
O'Neale, John C. 1996. *The Authority of Experience: Sensational Theory in the French Enlightenment*. University Park: Pennsylvania State University Press.
Pickstone, John V. 2007. Working Knowledges Before and After Circa 1800. *Isis* 98:489–516.
Reddy, William M. 2001. *The Navigation of Feeling: a Framework for the History of Emotions*. Cambridge: Cambridge University Press.
Reill, Peter Hans. 2005. *Vitalizing Nature in the Enlightenment*. Berkeley, CA: University of California Press.
Rey, Roselyne. 1994 Naissance de la Biologie et Redistribution des Savoirs. *Revue de Synthèse* 1(2):167–197.
Rey, Roselyne. 2000. *Naissance et Développement du Vitalisme en France de la Deuxième Moitié du 18e Siècle à la Fin du Premier Empire*. Oxford: Voltaire Foundation.
Richards, Robert J. 1979. Influence of Sensationalist Tradition on Early Theories of the Evolution of Behavior. *Journal of the History of Ideas* 40(1):85–105.
Richards, Robert J. 1987. *Darwin and the Emergence of Evolutionary Theories of Mind and Behaviour*. Chicago, IL: University of Chicago Press.
Richerand, Anthelme. 1804. *Nouveau Élémens de Physiologie*. 3rd edn. Paris: Crapart, Caille & Ravier.
Riskin, Jessica. 2002. *Science in the Age of Sensibility: The Sentimental Empiricists of the French Enlightenment*. Chicago, IL: University of Chicago Press.

Roger, Jacques. 1979. Chimie et Biologie – des Molécules Organiques de Buffon à la Physiochimie de Lamarck. *History and Philosophy of the Life Sciences* 1:41–64.
Roger, Jacques. (1993) 1997. *The Life Sciences in Eighteenth-Century French Thought*, ed. by Keith R. Benson, trans. Robert Ellrich. Stanford, CA: Stanford University Press.
Sloan, Philip. 1990. Natural History 1680–1802. In *Companion to the History of Modern Science*, eds. Robert Cecil Olby, Geoffrey N. Cantor, John R.R. Christie, and Michael Jonathan Sessions Hodge, 295–313. London: Routledge.
Sloan, Philip. 1995. The Gaze of Natural History. In *Inventing Human Sciences: Eighteenth Century Domains*, eds. Christopher Fox, Roy Portert, and Robert Wokler, 112–151. Berkeley, CA: University of California Press.
Smith, Christopher Upham Murray. 1976. *The Problem of Life*. New York, NY: John Wiley & Sons.
Smith, Roger. 1973. The Background of Physiological Psychology in Natural Philosophy. *History of Science* 11:75–123.
Smith, Roger. 1995. The Language of Human Nature. In *Inventing Human Sciences: Eighteenth Century Domains*, eds. Christopher Fox, Roy Porter, and Robert Wokler. Berkeley, CA: California University Press.
Spary, Emma. 1996. Political, Natural and Bodily Economies. In *Cultures of Natural History*, eds. Nicholas Jardine, James A. Secord, and Emma C. Spary. 178–196. Cambridge: Cambridge University Press.
Spary, Emma. 2000. *Utopia's Garden: French Natural History from Old Regime to Revolution*. Chicago, IL: University of Chicago Press.
Staum, Martin. 1980. *Cabanis: Enlightenment and Medical Philosophy in the French Revolution*. Princeton, NJ: Princeton University Press.
Staum, Martin. 1996. *Minerva's Message: Stabilizing the French Revolution*. Montréal, QC: McGill-Queen's University Press.
Stedman Jones, Gareth. 2002. The New Social History. In *The Age of Cultural Revolutions: Britain and France 1750–1820*, eds. Colin Jones and Dror Wahrman, 94–136. Berkeley, CA: University of California Press.
Steinke, Hubert. 2005. *Irritating Experiments – Haller's Concept and the European Controversy on Irritability and Sensibility 1750–1790*. Amsterdam: Rodopi.
Taylor, Charles. 1989. *Sources of the Self: the Making of the Modern Identity*. Cambridge, MA: Harvard University Press.
Terrall, Mary. 2004. Speculation and Experiment in Enlightenment Life Sciences. In *Cultural History of Heredity in the 18th Century*. Preprint. Berlin: Max Planck Institute for the History of Science.
Thomson, Ann. 2001. Materialistic Theories of Mind and Brain. In *Between Leibnitz, Newton and Kant: Philosophy and Science in the Eighteenth century*, ed. Wolfgang Lefèvre, 149–173. Dordrecht: Kluwer Academic Publishing.
Vidal, Fernando. 2006. *Les Sciences de l'Âme XVIe–XVIIIe Siècles*. Paris: Honoré-Champion.
Virey, Julien Joseph. 1817–1822. Articles from the *Nouveau Dictionnaire d'Histoire Naturelle*: Homme, Nature, Rapports naturels, Règnes de la nature. Paris: Deterville.
Wahrman Dror. 2004. *The Making of the Modern Self: Identity and Culture in 18th Century England*. New Haven, CT: Yale University Press.
Williams, Elizabeth A. 2002. *The Physical and the Moral: Anthropology, Physiology and Philosophical Medicine in France 1750–1850*. Cambridge: Cambridge University Press.
Young, Robert M. 1970. *Mind, Brain and Adaptation in the Nineteenth Century: Cerebral Localization and Its Biological Context from Gall to Ferrier*. Oxford: Clarendon.

Part III
Embodied Minds

Carelessness and Inattention: Mind-Wandering and the Physiology of Fantasy from Locke to Hume

John Sutton

Abstract Associated ideas, complained Locke, follow one another "without any care or attention." In a brilliant inversion of Locke's nervous worries about the perils of misassociation, Hume resolved the sceptical despair brought on by philosophical reasoning only by returning to mindlessness: "carelessness and in-attention alone can afford us any remedy. For this reason I rely entirely on them" (*Treatise*, I.4.2). How did British natural and moral philosophers in the early eighteenth century think about what happens when the mind is elsewhere? How did they theorize the processes by which thoughts, fancies, memories, daydreams, and feelings come to mind without prompting either by reason or reality, by the will or by the world? Examining works by Mead, Harris, Gibbs, and Branch, I detail the role of bodily fluids and nervous spirits in "conveying the mischief" by which imagination tends to ruffle our calm. Minds are often surprised by their own habits, and various forms of regimen were recommended in these works of medical psychology and moral physiology to 'pinion' the imagination and still the roving thoughts. I anchor these local discussions within a broader enquiry into mind-wandering and 'stimulus-independent thought', and sketch a rich neurophilosophical background to Hume's views on the bodily bases of custom and habit.

1 The Restless Mind[1]

Like us, early modern philosophers, both natural and moral, didn't always understand the springs of their own actions. They didn't want to feel everything they felt, and couldn't trace the sources of all their thoughts and imaginings. Events from

[1]The original research on which this paper is based was conducted at the Wellcome Institute way back in February 1999, and I presented initial talks in 1999 to audiences in Sydney and Edinburgh, and at the annual conference of the International Society for the History of the Neurosciences in Lausanne. My work then on early eighteenth-century English medico-psychological writers like

J. Sutton
Macquarie Centre for Cognitive Science, Macquarie University, Sydney

C.T. Wolfe and O. Gal (eds.), *The Body as Object and Instrument of Knowledge: Embodied Empiricism in Early Modern Science*, Studies in History and Philosophy of Science 25, DOI 10.1007/978-90-481-3686-5_12, © Springer Science+Business Media B.V. 2010

past experience come to mind again unwilled: abstract thought is interrupted by fantastical images, like the "winged horses, fiery dragons, and monstrous giants" by which Hume exemplified "the liberty of the imagination."[2] Then, as now, a failure to keep a train of thought on track could be blamed for both personal and social ills, for wasted lives and erratic policies. The ongoing struggle to distinguish the deliverances of reason from what Hume called 'the loose and indolent reveries of a castle-builder"[3] thus required scrutiny of daydream and fancy as much as belief and knowledge.[4] The mind's tendencies to float and to roam were of great interest to early modern philosophers as well as to others concerned with medicine, mental health, morals, education, and taste. This paper sketches one local line of thinking and theorizing about 'mind-wandering' and its bodily causes in British philosophy over the first decades of the eighteenth century, as a small exemplar of a form of cognitive history intended to illuminate independent historical and contemporary concerns about our understanding of mental life. The dual aim is to see problems in our historical material that we might otherwise miss, and to use history to explore phenomena more or less marginalized by modern psychology.[5]

Historians of philosophy often interpret early modern thinkers, in differing traditions and for differing reasons, as tempted by the view that mind requires awareness. Notable exceptions may be acknowledged: this is one reason Hume's claim that experience may produce belief and judgement "by a secret operation, and without once being thought of" was dramatic and puzzling.[6] But awareness and control together are taken to have formed a standing ideal or paradigm for mental life: the ordered mind, at least, would exhibit complete and undivided

Mead, Harris, Gibbs, and Branch was intended to pick up loose threads from my book *Philosophy and Memory Traces* (Sutton 1998), uneasy with my own carping at historical thinkers' resistance to confusion, their struggles with internal division, their desperate attempts to clean out the mind. The Embodied Empiricism project affords a new context for a more constructive line through this material, and I'm most grateful to Charles Wolfe for the encouragement to do so and for his general support. My thanks for help in that earlier phase to Catalin Avramescu, Stephen Gaukroger, L.S. Jacyna, Peter Jones, Jamie Kassler, Doris McIlwain, Gail Kern Paster, Udo Thiel, and Richard Yeo. For thinking through issues about 'mindlessness' and applying intelligence to the reflexes with me in other contexts more recently, I'm grateful to Wayne Christensen, Ed Cooke, Andrew Geeves, Doris McIlwain, Meta Regis, and Evelyn Tribble. Lisa Shapiro's excellent commentary at the Embodied Empiricism meeting in February 2009 was particularly helpful, as were questions from Dominic Murphy and Richard Yeo. This is also an opportunity to acknowledge the enormous influence of the boundary-spanning work of G.S. Rousseau on imagination and 'discourses of the nerve'. If I engage explicitly with his writings less here than on some previous occasions, it's only because this is, in a sense, so thoroughly his topic, and I can only point readers to the essays now helpfully collected in Rousseau 2004, and especially to his remarkable recent essay on eighteenth-century 'brainomania' (Rousseau 2008).

[2] Hume 1739/1978, I.1.3, at p. 8.

[3] Hume 1739/1978, 624.

[4] See also Tierney-Hynes 2007 on the 'castle-builder'.

[5] On cognitive history compare Richardson 2001; Lloyd 2007; Smail 2008; Sutton 2000, 2002, 2007a; Tribble 2005. We hope that the risks taken in work like this of catching 'the virus of the precursor' are outweighed by the benefits.

[6] 1739/1978, 104; Traiger 1994.

mindfulness, and be regulated by the agent's will in harmony with reason. The prevalence of these default assumptions created trouble, firstly, for philosophers and moralists who wanted to construe the old conflict between reason and the passions as occurring *within* the mind, rather than between the self and entirely alien forces.[7] Further, it was hard to find theoretical room for the occurrence of any ideas, memories, decisions, or feelings with no or diminished awareness ("without once being thought of"), or in the absence of voluntary regulation and direction. Ordinary mind-wandering, daydreaming, and fancy remained mysterious and poorly-theorized: it was difficult to identify the distinct dimensions on which these nebulous psychological phenomena need to be studied. Yet we can still identify many distinctive historical attitudes, in both theory and practice, to the relations – conflictual or interactive, in competition or in coordination – between what we might think of as the attentive mind and the floating mind. Perhaps the dominant strains of Western moral psychology have privileged reflection and control, encouraging us to be suspicious of and to minimise the influence of unguided thinking: but there have also often been alternative views, both mystical and naturalistic, which value both mind-wandering and habitual flow.

In the twentieth century's quite different intellectual context, brave alliances between psychodynamic and cognitive approaches to the unconscious were needed gradually to initiate the scientific study of unguided thought flow, zoning-out, and mind-wandering.[8] Anti-dualist consensus notwithstanding, executive control (over thought and action alike) is still often seen as requiring both awareness and intent: this has rightly been blamed for the prevalent psychological neglect of daydreaming and fantasy.[9] Such processes are often precisely driven by the agent's current concerns, by ongoing or unfinished goals[10], yet are initiated and maintained without explicit intention and (sometimes) without ongoing awareness. So official theories which yoked agency to intention or awareness rendered such phenomena barely visible.

Again, of course, there are strong counter-movements, reaching well beyond psychoanalytic theory, which do encourage the incorporation of the tacit realm *within* our psychology. But, despite helpfully attending to inattention, in some cases these alternative lines of thought reinforce key dichotomies from their rationalist targets. Philosophers of various persuasions argue against over-reliance (in theory and in practice) on attention and top-down control, suggesting that "mindedness is the enemy of embodied coping,"[11] or that a wandering mind "is conducive to effective action because of its responsiveness to the objective demands of one's materials and circumstances;"[12] while both philosophers and cognitive psychologists underline the pervasiveness of automaticity in everyday life,[13] and the ironic or self-refuting

[7] James 1999; Schmitter 2006.

[8] Singer 1966; Antrobus et al 1970; Berntsen 2009.

[9] Smallwood and Schooler 2006; Schupak and Rosenthal 2009.

[10] Klinger 2008.

[11] Dreyfus 2007, 353.

[12] Velleman 2007, 184.

[13] Gendler 2008; Bargh 1997.

tendencies of attempts at mental control.[14] But even when nonconscious thoughts and feelings are no longer seen as entirely outside the cognitive realm, they are too often still construed, as in early modern discussions, as lacking in both knowledge and control. These twin pillars of the mind were and are often yoked together: then described as reason and will, now as (say) declarative knowledge and executive control. Habitual or grooved thoughts and actions operate in the main, the idea goes, without access to explicit background beliefs or factual memory, and often without in turn leaving any explicit trace in memory; and they characteristically operate 'automatically', without the need for deliberate initiation or conscious online guidance. These views are buttressed by some neuroscientific work, which sees sequence memory as crystallized and inflexible once learned, with the components of kinaesthetic sequences chunked as single entities in memory, hard to uncouple and selective redeploy, and habit memory entirely "controlled by antecedent stimuli," evacuated of awareness so that we act on its basis "without anticipating the consequences."[15]

Studying the diverse phenomena of mind-wandering – of carelessness and inattention – can, I suggest, help us undermine these dichotomies between goal-directed and automatic action, and between controlled and habitual thinking. Between the basic reflex and fully reflective, deliberate, self-aware action lie extraordinarily diverse arrays of distinct psychological phenomena, which vary on many different dimensions.[16] Neither awareness nor control, neither knowledge nor intention, neither reason nor will, need be seen in an all-or-nothing manner.

This historical study is thus intended to complement a more promising recent wave of empirical research on "the restless mind" and on daydreaming, unguided thought flow, zoning out, fantasy and mind-wandering, phenomena which are "ubiquitous in mental life"[17] and of considerable theoretical, personal, and moral interest. How often do we fail to notice our minds wandering? When and how do we sometimes maintain performance on mundane tasks, even quite tricky ones like driving, when our minds are off and away? How do we catch ourselves in the act of fantasy, and what changes when we do? Can unguided imaginative wandering help in solving problems? What happens when we entrust key actions or decisions to such mere habits of mind? Two questions about mind-wandering above all interest us, as they did the early modern writers I discuss here, with their different frameworks and terminologies and explanatory options. What causes such unguided shifts in the flow or thread or sequence of thoughts and feelings, which seem to be internally-generated yet apparently involuntary? And, secondly, how (both in general and in specific instances) can we better direct, fully inhabit, or align ourselves with our roaming minds?

---

[14] Wegner 1997.

[15] Graybiel 1998; Ennen 2003; Yin and Knowlton 2006.

[16] Lambie and Marcel 2002; Sutton 2007b.

[17] Smallwood and Schooler 2006, 946.

At least four distinct dimensions are at issue in this recent literature:[18] feelings and thoughts may be more or less fanciful and wishful (as opposed to realistic), more or less cut off from the current environment in forms of 'task-unrelated' and 'stimulus-independent' thought,[19] unintended or spontaneous (rather than deliberate) in their initiation and direction, and either accessible in awareness or not (I can be surprised to find that I've been thinking about something else for some time, while on other occasions I'm perfectly well aware of my ongoing meandering stream of thought).

These topics might seem remote from the official concerns of a historical investigation into a tradition of 'embodied empiricism', concerned as it is with the life sciences and medicine, with anatomy, fevers, and hysteria. But the kind of mind in question here, in early eighteenth-century phenomenological and psychosomatic inquiries into wandering thoughts and stray feelings, is of course entirely different from the a-historical, disembodied, isolated mind sometimes said to have been set at the heart of an official theory called 'empiricism'. Most generally, the writers I'll discuss, between Locke and Hume, are always treating the dynamics of body and mind together: even if they do not assume the psychophysical identities of earlier humoral materialisms, they still illustrate, worry at, and offer prescriptions to work with the intimately interactive relations of nerves and thoughts, passions and pores. More specifically, although these early eighteenth-century texts are mostly under the influence of what historians label 'iatromechanism', they do not exhibit some of its textbook characteristics. The body-machine is no more a rigid, inflexible clock, always responding in the same way to the same stimulus, than it was for Descartes.[20] Whether the activity was attributed more to the body's liquors and fluids and juices, or (with the advance of solidism) to the elastic and restorative powers of the fibres and tubes and pipes through which such spirits flowed, physiological processes were seen as exhibiting their own dynamics, both intrinsic and involuntary.[21] These inner elasticities and vibrations could be precisely the source of undirected psychological activity: if either medicine or philosophy was ever successfully to calm the mind or society, then the task of psychosomatic regimen was to improve the bodily conditions for their optimal exercise.[22]

---

[18] Smallwood and Schooler 2006; Mason et al 2007; Klinger 2008; Berntsen 2009; Schupak and Rosenthal 2009.

[19] As we'll see, early modern thinkers treated stimulus-independent forms of mind-wandering alongside the different cases in which attention is easily captured by current stimuli.

[20] English versions of iatromechanism are now often seen by revisionary historians as more flexible, biologically-oriented, and contextually anchored than on earlier interpretations (Ishizuka 2006). Yet the contrast is still often drawn with a rigid Cartesian model in which dynamics, sentience, and life had been evacuated from the body. For a different account of Descartes' ideas about the complexity and flexibility of 'automatic' processes, and about our open organic interactions with the environment, see Sutton 1998, chapter 3; 2000.

[21] Ishizuka, 2006.

[22] Cunningham 1990; Sutton 1998, chapters 2, 5, 7, 9.

2 Carelessness and In-Attention

We start with an under-noticed aspect of the chapter on association which Locke added to the 4th edition of his *Essay Concerning Human Understanding* in 1700. The mind makes strong combinations of ideas in itself, says Locke, "either voluntarily or by chance, and hence it comes in different Men to be very different."[23] Such individual differences arise, in other words, because while some couplings of ideas are due to reflection, others are "wholly owing to Chance or Custom."[24] This explains both particular errors, and the more general "degree of madness" in most of us.[25] Ideas which ought to be "loose and independent one of another" connect wrongly, under the rowdy influences of "Education, Custom, and the constant din of their Party," and so "set us Awry in our Actions, as well Moral as Natural."[26] These unfortunate outcomes occur even "in very sober and rational Minds," says Locke in striking terms, just because ideas once associated will "follow one another … without any care or attention."[27]

In this scheme, then, Locke links volition and reflection with care and attention as labels for the *appropriate* forms of internal guidance of our sequences of ideas, as the opposite of misassociation. This concept of 'attention' was relatively novel. In seventeenth-century English, you are attending when you notice, take heed of, or direct the mind towards objects or events in the external world. But in Locke, attention operates (or fails to) on the *inner* world: "when the *Ideas* that offer themselves … are taken notice of, and, as it were, registred in the Memory, it is *Attention*."[28]

Note also the physiological grounding of Locke's detailed picture of the processes of remembering and associating ideas, a far cry from his official neutrality about the physical operation of the mind.[29] In the absence of 'care and attention', the grooved sequences of associated ideas are based in (or even just are) grooved sequences of patterned motions of nervous fluids:

> Custom settles habits of Thinking in the Understanding, as well as of Determining in the Will, and of Motions in the Body; all which seems to be but Trains of Motions in the Animal Spirits, which once set a-going continue on in the same steps they have been used to, which by often treading are worn into a smooth path, and the Motion in it becomes easy and as it were Natural.[30]

---

[23]Locke 1690/1975, *Essay* 2.33.6.

[24]Locke 1690/1975, 2.33.5.

[25]Locke 1690/1975, 2.33.4.

[26]Locke 1690/1975, 2.33.9, 18.

[27]Locke 1690/1975, 2.33.3, 6.

[28]Locke 1690/1975, 2.19.1. The OED ambitiously characterizes this passage as the first instance of a distinct and metaphorical use of 'attention'.

[29]Locke 1690/1975, 1.1.2; Sutton 1998, chapters 7 and 9.

[30]Locke 1690/1975 2.33.6. This is the passage brilliantly echoed by Sterne in the first chapter of *The Life and Opinions of Tristram Shandy* 1759/1983, 5; see Myer 1984, Sutton 1998, 207–213.

Such psychophysical pairings – in remembering, imagining, and thinking alike – are symptoms of our deepest cognitive failings, eliciting Locke's sad realism about the cognitive effects of the Fall. As angels need no memory, they are free of the need to encode and retrieve. For us, in contrast, memory often goes astray either through "the temper of the Body" as "the Imagery moulders away"[31], or because of failures of attention: we don't register things, notes Locke, which "have been little taken notice of; the Mind, either heedless, as in Children, or otherwise employ'd, as in Men."[32]

The concern here is with what happens when the mind's away, when – through incapacity or overload – ideas turn over without control or even care. Descartes had argued that although animals do see and feel, they "do so not as we do when we are aware that we see, but only as we do when our minds are elsewhere."[33] This is just what's happening, for Locke, in misassociation, and why it is dangerous: without heed or care, when 'otherwise employ'd' or else driven by custom, history, and ingrained embodied tendencies, the mind is barely present at all, its influence on our actions severely eroded. It's just this nervous worry that Hume echoes and brilliantly inverts in his *Treatise of Human Nature.*

Through book 1, Hume elicits our sceptical doubts about both reason and the senses. He suggests that such doubt "can never be radically cur'd, but must return upon us every moment, however we may chace it away, and sometimes may seem entirely free from it."[34] It is not only in the notorious conclusion to Book 1 that Hume evokes the 'forelorn' and 'disconsolate' mood brought on by intense reflections.[35] He was not many years off an unsuccessful course of "Anti-hysteric Pills," described in a 1734 letter which compared Hume's history of nervous disorders, due to his "profound reflections" with their "warmth or Enthusiasm," to religious fanatics and "French Mysticks" whose "rapturous Admirations might discompose the Fabric of the Nerves & Brain."[36] But although neither drugs nor backgammon can enhance the overzealous philosopher's mood, there is an unexpected solution to the perils of thinking: "Carelessness and in-attention alone can afford us any remedy. For this reason I rely entirely on them."[37] By allowing heedlessness, encouraging the mind to be elsewhere or otherwise employed, Hume finds temporary respite from reflection. In thus embracing carelessness and

---

[31] Locke 1690/1975, 2.10.4–5.

[32] Locke 1690/1975, 2.10.4.

[33] Descartes to Fromondus, in Descartes 1991, 61–2; and see Gaukroger 1995, 287–8.

[34] Hume 1739/1978, 1.4.2, 218.

[35] Hume 1739/1978, 1.4.7, 264.

[36] Hume 1734/1993, 349. This language is echoed in the *Treatise:* Hume complains that metaphysical reasonings have "heated my brain" (1.4.7, 266). I agree with Marina Frasca-Spada (2003) that much of the terminology which Hume applies to ideas in the *Treatise* has been transformed with little alteration from a standard physiological idiom: ideas "flow in upon the mind" in a forcible or lively manner, for example, while others are "faint and languid" (Hume 1739/1978, 9), while the vividness of certain conceptions "diffuses itself ... and is convey'd, as by so many pipes or canals" (*ibid.,* 122; Frasca-Spada 2003).

[37] Hume 1739/1978, 1.4.2, 218.

inattention, Hume recommends trust in both instinct and experience, accepting the effects of education and custom, everything which is "independent of all the laboured deductions of the understanding."[38]

The verbal echo is a neat new way to catch the difference between Locke's restricted use of association to explain error, and Hume's radical extension of the principles of association. Hume exhibits and recommends trust in the tacit realm, in the deliverances of custom and habit, in light of his retreat from care and attention. But there's also a richer local history to spy into here, a history of mind-wandering, medicine, and moral physiology, of habit and body and brain, of embodied empiricism between Locke and Hume. Through the early years of the eighteenth century, how did other British natural and moral philosophers think of these processes by which thoughts, fancies, memories, daydreams, and feelings come to mind without being prompted either by the world or by the will, by reason or by reality? Without denying the novelty of Hume's case that reason itself is, or is the product of, natural habitual and affective processes, we can identify a discourse that was more broadly shared by natural and moral philosophers in the years before the *Treatise.* The consensus lay in ways of talking about, and in many cases theorizing, the many ways in which the causes, contents, and course of mental life are out of our control: many writers sought a way to think about a multiplicity of causes of thought and feeling, among which reason and will would have to struggle for influence. For the odd historical reason that both Locke and Hume sought to distance their epistemological work from the contemporary natural-philosophical and medical frameworks with which they were familiar, it's too easy for us to lose the sense that they were fully aware of the embodied roots of mind-wandering.[39]

3 Pinnioning the Imagination

The late twentieth-century study of mind-wandering and daydreaming arose in part from work on (night-time) dreaming, with similar ambitions to unify depth-psychological and cognitive perspectives,[40] and it continues alongside the

---

[38] Hume 1975, 55. For Bertrand Russell, Hume's recommendation of carelessness and inattention was not only the ultimate in self-refutation for a philosopher in particular, but also quite generally "the complete bankruptcy of reason" (Russell 1997, 239). Our default interpretive stance now is more naturalistic and affective, and thus more sympathetic.

[39] So I agree with James A. Harris's case, in an excellent recent paper, that the first two books of Hume's *Treatise* in particular should be read against the background of many early eighteenth-century books "devoted to showing how philosophy could help with living a happier and more virtuous life, by showing the way to better regulation of the passions," most of which books are "completely unread today" (Harris 2009). But Harris's focus is on straightforward moral philosophy as the context, rather than on the medical-psychological and moral-physiological literature from which I sketch just a few themes in this paper. In these latter fields, there were signs prior to Hume of his idea, nicely described by Harris, that the old contest between reason and desire would be better seen as "the interaction of a panoply of feelings," to be registered and explored in their mysterious workings by the analyst of human nature. Frasca-Spada (2003) rightly notes that Hume's few references to common physiological theory are "impeccably well informed."

[40] Antrobus et al, 1970.

equally challenging and speculative multidisciplinary sciences of dreams.[41] One highly contested ongoing debate concerns just how bizarre and fantastical dreams are. The dominant view is that dream narratives are intrinsically implausible, utterly unrealistic delusions or psychoses resulting from "a mental readout of the chaotic brainstem activity of REM sleep."[42] But among a number of challenges to this mainstream theory are results from systematic content analyses of dreams, which compare them not to objective real-life events or actions but to waking *mental* life: G. William Domhoff suggests, for example, that "there is far more discontinuity, drift, and inattention in waking thought than is implied by the claim that changes in dream scenes or settings are inherently bizarre."[43] Again, we can use this contemporary debate as a historical clue: in asking what pictures of the inattentive waking mind were available to our early eighteenth-century thinkers, we can use their views about the similarities and differences between waking and dreaming thoughts and feelings.

This quest takes us to an appropriately obscure exemplar, after our two canonical texts on carelessness and inattention. Thomas Branch, a writer of whom we know next to nothing, responded eloquently in his *Thoughts on Dreaming* (1738) to views defended in Andrew Baxter's 1733 *Enquiry into the Nature of the Human Soul*.[44] Baxter had argued that dreams derive from supernatural agents: Branch responds that what might appear to be supernatural is in fact inside us.[45] In doing so, he expresses forcefully the view that ordinary waking mental life is more confused than regular, anchored more in a fantastical than an objective realm.

---

[41] Sutton 2009.

[42] Hobson & Stickgold 1994, 10–11.

[43] Domhoff 2003, 153; see also Flanagan 2000, 58–61.

[44] Branch also published a compendium of legal sayings in 1753, and may have been alive still in 1769. A second edition of Baxter's *Enquiry* had appeared in 1737, the year before Branch's book. For background on the Baxter-Branch debate see Dacome 2004. Dacome's overarching case is that dreams were gradually medicalized and pathologized through the eighteenth century, as moral physiologists sought to establish "a new model of the credible mind, one in which the elimination of the vagaries of the mind was to be carried out by means of body policing," Dacome 2004, 397. See Daston 1998 for a related broader narrative of the pathologizing of imagination in the Enlightenment. These early eighteenth-century texts also exemplify the spread of discussion about these further reaches of mental life well beyond philosophy, then as now. But further work is needed to piece together the impact and reception of works like these, and to understand how they related to moral, imaginative, cognitive, and social practices of the time.

[45] In responding to Baxter on dreaming, Branch also offers full-scale theories of perception and memory, in seeking to demonstrate just how much the soul can do without external guidance, to prove that "our Dreams may be our own" rather than implants from spiritual beings. He also offers a rich phenomenology of dreaming, raising and effectively answering sixteen objections to his core idea that dreams are just thoughts during sleep: just as he denies their supernatural origin, so he denies that they are brute biological givens, for they take considerable psychological sophistication. For this reason Branch at least does not neatly fit Dacome's account of the Enlightenment pathologizing of dreams, which I would argue also neglects the developmental-cognitive-affective accounts of memory and dreams in David Hartley's *Observations on Man* 1749: compare Sutton 1998, chapter 13. Theories of dreams were no more homogeneous and unified (from either conceptual or applied points of view) than they are now.

Baxter had argued that the bizarre content of dreams means that they cannot be accounted for by natural causes, there being insufficient material in "the Business and Thoughts of the Day" to furnish our dreams. Branch's strategy in reply is to challenge the distance between mental life in waking and dreaming. Like Domhoff in the modern debate, he asks us to consider that many daytime thoughts of internal origin, driven neither by perception of the world, nor by reason, are just as wayward.

"Consider," Branch requests, "with what great Difficulty it is that we fix it [the Mind] long, whilst awake, on one Subject; and that in Opposition to our best Endeavours." When awake, we can fixate ourselves by using external props – objects, activities, or other people – as scaffolding for our thought: in reading books, conversing, or putting our views on paper we use prostheses to support our attention. But, Branch laments, the mind is "ever and anon flying off, and will hardly be held in." So when such external supplements to thought are absent, as in sleep, "it is far from being strange, that the Mind, naturally a Wanderer, should rove at large."[46]

The vast and complex landscapes of our dreams are parallel to the fiery productions of imagination. Our thoughts can indeed *seem* to be of alien origin: every man each day has "Imaginary Forms brought before him, which he knows not of going in search after, and even wonders how they were introduced".[47] But their origin is in fact internal, produced by the compounding and mixing of ideas, the continuing business of imagination.

Branch draws a sharp distinction between voluntary invention and involuntary imagining. The soul *can* deliberately 'confine' and 'rectify' imagination for a particular purpose, or select the Forms it brings, by judgement, in an act of creation. But this is not easy, and "is certainly," admits Branch, "a work of fatigue." When, on the contrary, "we control not the imagination, but let it fly at all, and pursue its own Game, this costs us no Pains; many Persons find much more in pinnioning it."[48]

So in the course of arguing against the attribution of dreams to "foreign agents"[49] or other alien sources, Branch underlines the complexity and heterogeneity of the internal origins of our mental life. He depicts mind-wandering as our default psychological

---

[46] Branch 1738, 45–46. The idea that external artifacts play key roles in distributed cognitive systems, transforming the demands on individual psychological resources, has been widely revived recently (Hutchins 1995; Clark 1997; Sutton 2002), but of course has itself a long history (Donald 1991; Tribble 2005; Sutton 2007a). Branch links his sense that the mind is fluid, and prone to rove, to the fact that we rely on more stable external cognitive artifacts (compare Sutton 2008). His more original point is that the residual differences between waking and dreaming mental life are due not to intrinsic physiological differences, but to the absence of social and material supports in sleep. The sociologist Maurice Halbwachs (1925/1992) likewise ran an extended analogy between dreaming, with its fragmentary, torn, confused raw materials, and the mental life of a non-social individual, to demonstrate that our waking mental life is permeated by and thoroughly sculpted by our social frameworks. Branch perhaps has less faith than Halbwachs in the coherence and stability provided in waking thought by social networks.

[47] Branch 1738, 65.

[48] Branch 1738, 66. Dreaming is thus, for Branch, closer in character to imagining than to hallucinating. Compare Foulkes 1999, against Hobson's account of dreams as delusions.

[49] Branch 1738, 26.

mode, vigilance against which comes at some cognitive cost.[50] It is natural for the mind to be off on a frolic of its own. In dreaming, lacking direction from both reality and rationality, we are entirely unable to pinion the imagination. On the picture of waking life which thus emerges in parallel, executive control – the exercise of due care and attention, or effort and inhibition – is not impossible, but it is rare and costly.

4 Conveying the Mischief: Body Fluids and Openness to Influence

The idea that many of the sources of disorder are within was also backed by prevailing psychophysiological theory, which I sketch here using works written early in the eighteenth century by the Newtonian Richard Mead, the encyclopedist John Harris, and the Cornish physician James Gibbs.[51] Although Mead, in his 1702 work *A Mechanical Account of Poisons,* officially characterises 'mathematical learning' as the distinguishing mark of a genuine physician, he offers in fact only richly verbal and irredeemably qualitative accounts of the paths of transmission within body and nervous system. Mental life is not protected or insulated from any trouble and taint in the "small Tubes all over the Body," Mead notes, for the fluid of the nerves, "Undulating continually towards the Brain, and being the chief Instrument of Motion and Action, may sometimes more immediately convey the Mischief."[52] In this section, I briefly rehearse the widely shared picture of the array of interconnected body fluids and vessels along which mischief of various kinds is conveyed. If pathologies could be physical and psychological at once, then philosophers, moralists, and physicians needed to map and inhabit all these richly interconnected psychosomatic phenomena.

There were increasing doubts about the ontology of certain physiological fluids, notably nervous or animal spirits, invisible and "immechanical" agents that "elude all art," as the corpuscularian physician Thomas Morgan complained.[53] But the

[50] Compare Mason et al 2007.

[51] On Mead and Gibbs see also Roos 2000. For Harris's *Lexicon Technicum: or, an universal English dictionary of arts and sciences* (1704) I've used the 2nd edition (2 volumes, 1708 & 1710). Harris, who had been Boyle Lecturer in 1698, and was Secretary of the Royal Society for a year in 1709–1710, wrote an array of hack works: the DNB (IX, 13–14) says that "Harris was culpably improvident, and was generally in distress," noting sadly that his 1719 history of Kent is "extremely inaccurate." Thanks to Richard Yeo for advice on Harris.

[52] Mead 1702, 20–21.

[53] Morgan 1735, 152–4. For the earlier history of debates about animal spirits, and more detailed accounts of eighteenth-century controversies about their existence, see Jacyna 1995; Clower 1998; Sutton 1998, chapters 2, 8, 10; Rousseau 2004. Among our other current writers, Gibbs nicely compares the deniers of animal spirits to atheists. Observability is entirely irrelevant: although we can't see God, we know he exists, so the fact that no cavity can be discovered in tubes of nervous fibre doesn't matter, because "if the Hole was discernable, by which the Spirits pass thro' a Fibre, it might be unfit for the Passage of so fine and rarify'd a Fluid, as the Spirits are." Gibbs 1712, 27.

solid parts could be the subject of just as many and as complex psychologized properties and variables, as the strength and vigour and harmony of the composition of the nerve had to be maintained.[54] And even for solidists, the condition of interconnected body fluids – blood, bile or gall, chyle, lymph, spittle, pancreatic juice, semen, as well as any "peculiar Juice in the nerves" – remained vital in distending or altering the body's elastic fibres, so that "flow and obstruction" remained at the heart of the "economy of circular physiology."[55] As Mead put it, "the Vessels are rarely obstructed, unless it be from the fault of the Liquid they carry."[56] So despite differences across physiological schools, which in other contexts we'd want to investigate closely, here we can focus on the existence of unified schemes and language for thinking about mood and emotion, involuntary thoughts and memories, imaginings and fantasy, alongside disease and health. Linking the innards both to practices and exercise and regimen, and to mental life, this language of quick and nimble, fleeting spirits and fluids, which could be low, sunk, broken, oppressed, dejected, petulant, harassed, "ruffled beyond description," hurried, or roused was used to think through psychological confusion and distress, and in explaining every disorder of the animal machine.

We can trace the possible paths of influence which transmitted mischief or disorder in its various forms. Mead's mechanical account of poisons exemplifies the operation of external sources. After reading new Italian theories of vipers, and Tyson on the rattle-snake, Mead wanted "to hint something concerning the Nature of Fluids in General."[57] The salts of venom irritate and fret the sensile membranes, creating an excess of animal juices. This 'disjoins' parts of the blood, altering its mixture. Poison changes mainly the arterial blood, but the fluid of the nerves may be considerably changed as well. Most generally, we can expect to explain all disorder in the body through "the doctrine of the Mixture of Heterogene Fluids, and their Separation."[58] We are working with a diverse array of continuous variables. For Mead, there can be "a vast *variety* … in the Fermentations even of one and the same Fluid," because these are simply "*Changes* made in the *Cohaesion* of the compounding Particles," and are thus "capable of as many Alterations as *Motion* in its *Degrees* and *Directions* can admit of, which are really infinite."[59]

[54] Arguing in favour of the solids, David Baynes/Kinneir explicitly recommends metaphorising the spirits, so to talk of someone being in good spirits would mean they are in health (1738, 11–12). In fact the incorporation of the language of animal spirits into economics had already begun, foreshadowing their post-Keynesian career as markers of consumer confidence (Winslow 1986, Akerlof and Shiller 2009).

[55] Ishizuka 2006, 438–440. For more general interpretations of the phenomenology of humoral materialism, see for example Duden 1991 on the sensed "kinesthetic system of oriented flows"; Paster 1993, 1997; Rublack 2002; Seuntjens 2006; Sutton 2007a.

[56] Mead 1702, 19.

[57] Mead 1702, 13.

[58] Mead 1702, 19.

[59] Mead 1702, 17.

Likewise, there were puzzlingly interconnected effects of purely internal processes of fermentation or ebullition. Across the many entries on the interconnections of body fluids in his *Lexicon Technicum*, Harris draws on diverse recent writers to update and mechanize earlier accounts of the stages of purification of bodily spirits. Food affects the blood, for example, in many ways such as in "Chylification," which depends first the existing state of the stomach and the guts, and then the "various Mixtures and Preparations of Chyle" as it is dissolved and fermented from food: in "Sanguification," then, as blood and chyle mix, it is easy for particles of blood to be "intangled and detained from flight," or for the heat of ebullition to become so great that "it often endangers the Vessels they are contained in." Harris too is attracted in principle by the ideal of geometrizing the influence of airs, waters, and places on body fluids: the nature of secretions in general depends on the diameter of the orifice of the secreting duct, on the angle of incidence of the duct with the vessel, and on the different velocities at which fluids arrive at the orifices.

Advertising his mixtures for the cure of scrofulous distempers, James Gibbs tells of a girl from Truro who in 1706 when 16 years old had "an hysteric disorder of her spirits at 8 p.m., plus loss of appetite."[60] Gibbs identifies two possible causes. Sometimes "the Passages of the Spirits are so obstructed in the Nerves, as to produce Paralytic Impediments," while "at other times the Spirits are irritated into Convulsive Ferments." Fortunately his preparation attacks the common causes of both. All nervous diseases are caused by "the Depravations of the *Nervous Juice*" – humours are often "frothed up" as they leave the glands which secrete them, and animal spirits are stagnated or paralysed, preventing the natural office of the fluids, which is "chiefly to *lubricate* and fill the Interstices of the Fibres of the Nerves."[61] The spirits can be affected or 'diminished' equally by acids and by sadness.

In these writers, we see mechanized versions of older cosmobiologies. The Newtonians identify the mechanics of cosmic and of bodily fluids. For Mead, the same principles of action operate in the Universe and "in the most minute and finest Corpuscles" of any internal vessel with its "very subtile and elastic Fluid."[62] Whatever the precise ontological commitment (to fluids or vibrating ethers, for example), there are not just analogies but identities across the whole realm of subtle substances. As well as advice on musical and other exercises, this drives ideas about cosmic influence in the 'lunar medicine' of this early eighteenth-century period.[63] Gibbs explains how 'the moon has a considerable influence on the constitutions of some persons': disorders of the eye, for instance, increase after every full moon because the spirits of the optic nerve are 'dispos'd directly to receive' particles of the aetherial fluid which may compress and restrain their turgescence.[64] In discussing effluvia and influences, these writers cite Boyle who,

[60] Gibbs 1712, 10.

[61] Gibbs 1712, 8–12, 38–39.

[62] Mead 1702, 14–15.

[63] See also Roos 2000.

[64] Gibbs 1712, 54–64.

according to Harris, 'is inclined to believe that the Planets may have some Physical Influence or Operation on Bodies of our Globe," so that (for example) thin air, when "altered by these planetary virtues, must needs variously impress, move, agitate and infect the Spirits or Subtiler parts of all Bodies within its Reach," giving rise to sudden cramps, convulsions, blights, colds, or pestilential invasions which "often, as it were in an instant," seize on our Bodies."[65]

The forms of cosmobiological holism in play in these iatrophysical works, with all their claims to novelty, are just as extensive and as concrete as in earlier eclectic medico-physiological syntheses, and just as easily turned towards a range of practical interventions. Because of the multiplicity of relevant parameters behind bodily changes, all changing at different rates, there were many ways to try to change or influence these interconnected processes between world and body, and within the body. We can now underline the point that these explanatory schemes also and inevitably included psychological disorders and diversions. No matter how much mechanical or Newtonian physiologists wished to discuss the operations of body, brain, and nerves in purely quantitative terms, those who strayed into the morally and commercially intriguing domains of medical psychology – the chancy operations of feeling, remembering, imagining, reasoning, and even perceiving – could not avoid richer boundary-spanning language and theorizing.

5 Surpriz'd by Habit: Control and Error in Moral Physiology

In turn, composed or depraved fluids affect the mind. This returns us to mind-wandering. When unguided and undirected, thoughts and feelings are driven by our embodied habits, by the grooved tendencies embodied in our internal vessels and the fluids they conceal. What's appropriate and objective differs from what's improper or corrupt only in its distal causes: the immediate neural precursors of any thoughts are the same in both cases.

In normal operation, mental processes and the actions which they cause are guided by the twin supports of perception and reason. These offer external and internal sources of direction for thought and action. Objectivity is provided on the one hand by the external world, as our senses give us fallible but mostly trustworthy knowledge of reality. And on the other hand, the inner foundation offered by reason delivers clear judgements which, which combined with the impetus of the will, can guide us in practical action.

But these twin sources of direction provided by reality and reason, by the world and the will, do not exhaust the possibilities. After the Fall at least, our own cognitive capacities include a range of mechanisms of distortion. Remembering, imagining, and dreaming, as well as psychological processes directly caused by specific bodily disturbances, all open up the possibility that the ideal transparency of our attunement to the world can be subverted from within. Again, this theme in

[65] Harris 1710, vol II, s.v. 'spirits'.

our 'empiricists' is clearly present in Descartes too: in *L'homme*, the ordinary mechanisms of corporeal memory are depicted as intrinsically giving rise to fantasy. It "usually happens," notes Descartes, that in the flow of animal spirits over time through the pores of the brain, "several different figures are traced in this same region of the brain almost equally perfectly." This means that

> the spirits will acquire a combined impression of them all ... It is thus that chimeras and hypogryphs are formed in the imaginations of those who daydream, that is to say who let their fancy wander listlessly here and there without external objects diverting it and without the fancy's being directed by reason.[66]

In addition to these importunities of imagination, the same processes explain the intrusions of old unwanted memories into present mental life: "it is thus that past things sometimes return to thought as if by chance [*comme par hazard*] and without the memory of them being excited by any object impinging on the senses."[67]

Noting again, in passing, that this doesn't sound at all like the kind of storage system which might be used by the kind of 'Cartesian automaton' described in textbook accounts of mechanism (see also Sutton 2000), we can move on to examine a standard account, shared by so-called Cartesian and Newtonian physiological psychologists, of the relations between perception, inner processes, volition, and action. Mead describes the normal operation of the process:

> upon this Representation [of outward Objects to the Common Sensory], at the Command and Pleasure of the Soul, part of the same [nervous] Fluid is determin'd into the Muscles, and mixing with the Arterial Blood there, performs all the Variety of voluntary Motions and Actions.[68]

When developed appropriately over time, this process *can* operate successfully even without the active online involvement of reason. Because of "the Constancy of this Order in us," without reason the representations made to the mind can still "immediately and necessarily produce suitable Motions in the Bodily Organs." Mead is envisaging something very like Locke's association, though in a distinctive iatrophysical language. Like Locke he realizes that we are opened up to the possibility of error by this otherwise useful tendency of habitual processes to continue on without conscious intervention. Patterns of both action and thought which have come to be linked together, if prompted or triggered by causes other than perception or reason, bring disorder and, in the extreme, delirium, which he describes as

> the Representation and Various Composition of several Species to the Mind, without any Order or Coherence; together, at least most commonly, with irregular, or, as it were, undesigned Motion of the Body; that is, such a wandering and irregular Motion of the Nervous Fluid, whereby several Objects are represented to the Mind, and upon this Representation divers Operations performed by the Body, though those Objects are not Impressed upon the Organs, nor those Operations or Motions deliberately Commanded by the Soul.[69]

---

[66] Descartes 1972, 96. For commentary on this passage see Landormy 1902, 280–1; Krell 1990, 72–3 (on these impressions absorbed "higgledy-piggledy" as "prone to moral turpitude, lassitude, lethargy, and benumbment"); Sutton 1998, 61–2.

[67] Descartes 1972, 96.

[68] Mead 1702, 61.

[69] Mead 1702, 61–62.

This ‘wandering’ is internally generated. In theory, “the Mind is the first Principle of all Muscular Motion,” but here it appears that much of what goes on in us, in driving thought and action, is foreign to it:

> in such Cases as these, its Promptitude to Action or Habit being so great, it is in a manner surpriz’d, and cannot recover itself after the Spirits are with violent force determined pursuant to the Representations of the Species.[70]

Surprised by its own habits, the mind is the victim of its own idiosyncratic history, roaming along with its delicate or delirious spirits. Mead offers detailed diagnoses of the distinctive ways in which “the Hurry and Confusion of the Spirits” can render the mind overly vulnerable to certain stimuli – colours, particular emotions, trivial entertainments, or obscene talk and actions.[71] Error takes many forms: insensitivity and oversensitivity to the world are equal risks which were increasingly theorized as part of the physiology of consumerism. Those with weak or slender nervous fibres are too easily acted on by external objects, George Cheyne for example being too “easily ruffled on a surprise.”[72]

6 Remedies for Reveries

But just as an extraordinary variety of contextual factors could distract or capture the mind, opening it to the influences of habit and the body, so the plasticity of psychosomatic interplay still allowed for a wide array of remedies. These ranged from chemical preparations and anti-hypochondriack pills through musical cures and physical activity and baths and spas to the various forms of exercise recommended by these iatromechanist physicians. As Ishizuka argues, ‘exercise’ in this period could include anything which imitated or encouraged the internal motions of the fibres, taking drugs as much as riding, because the non-voluntary internal motions which ground both motor and cognitive habits could be exercised and altered in many different ways.[73]

So there could be no distinction between physical and psychological cures: our writers focus on the general idea of gradually coming to know and indirectly influence your own habits by any means possible. They often employed stories about ingrained links between specific thoughts or actions and particular contexts. Apologizing for the “pleasant oddness” and “comical Circumstances” of the tale, Locke tells us of a man who learned to dance in a room where stood a remarkable

---

[70] Mead 1702, 62.

[71] Mead 1702, 65–67. Mead also offers a geo-sexual climatology: sometimes spirits will “without any manifest Cause at all, be hurried towards those Organs, to which at other Times they have been most frequently determined; and every one knows which they are in hot Countries and Constitutions”, 67. See Floyd-Wilson 2003 on related earlier geohumoralist assumptions.

[72] Quoted in Barker-Benfield 1996, 9–11.

[73] Ishizuka 2006, 452–3.

old trunk, and could not perform in any other place.[74] Descartes had offered his own recipes for training the brain. Until a man realizes that the reason he wants to cry at music which makes others want to dance is that he "has never heard a galliard without some affliction befalling him," and that this is because "it evokes ideas in the memory," he has no chance of altering this response. But with *industrie* – effort, or psychological work – to identify and alter our idiosyncratic *habitudes,* we can help ourselves deal with "all the contingencies of life."[75] Likewise, Mead introduces his defence of musical cures, in which a "strong pulse" brings an "increased Influx of the Liquor of the Nerves into the Muscles," by repeating a story which Boyle repeated from Scaliger about the knight of Gascony who had to piss whenever he heard the sound of bagpipes.[76] Modulating these psychophysical connections can be done just as effectively, argues Mead, by indirect means as by the power of the will, since both means have to operate through the same "shaking of the nerves."

Although Locke saw association as primarily the root of trouble – aversion to foods, fear of darkness, unwarranted hatreds, dislike of books, ingrained political prejudices – he also saw a different side to custom and habit. Locke distinguishes motor-based associations from cognitive habits, though he then attributes them to the same cause in the trains of motion of the animal spirits. A musician finds a melody playing itself out in his understanding just "as regularly as his Fingers move orderly over the Keys of the Organ to play out the Tune he has begun, though his inattentive thoughts be elsewhere a wandering." This case helps us to conceive, says Locke, of what he calls "Intellectual Habits, and of the tying together of Ideas."[77]

In turn, custom and habit then for Hume are labels for characteristics of the imaginative processes which produce belief. These are non-rational and non-reflective propensities or tendencies. Some, like our beliefs in causation and in the continuing existence of the external world, are probably permanent and irresistible natural fictions. Others are changeable, more or less weak and irregular, offering opportunities for cognitive and practical reform by way of a change of habits, the implanting of different inclinations by changing our habits of belief. The value of philosophy is that, properly undertaken, "it insensibly refines the temper, and it points out to us those dispositions that we should endeavour to attain, by a constant bent of mind, and by repeated habit." On naturalizing interpretations of Hume, at least, the authority of custom and habit is proper. Beliefs are not formed by reasoning, but in the main by the history-dependent and body-dependent mechanisms of the restless mind.

Further rereading of Hume in light of our consideration of mind-wandering is just one of the threads left open for further research: having identified residues in his work of these discourses of moral physiology and medical psychology, it is tempting to think, with Frasca-Spada (2003), that "the avoidance of physiological

---

[74] Locke 1690/1975, 2.33.16.

[75] Sutton 2000.

[76] Mead 1702, 70.

[77] Locke 1690/1975, 2.33.6.

accounts in his pages is an oddity calling for an explanation." Likewise, we need to incorporate the many practical and commercial ways in which the multicausal psychosomatic frameworks discussed here still influenced what and how much early eighteenth-century people ate and drank, how and when they slept, took holidays, conversed, what recipes and medicines they took. What I have done here barely scratches the wonderful material in these medico-moral 'mixed discourses' of spirits, body, and self, or of brain, mind, and soul.[78] But by identifying mind-wandering, fantasy, and inattention as a specific domain of enquiry and debate for these historical actors, I hope to have brought to clearer visibility a wide and intriguing range of phenomena which are neither wholly conscious and controlled, nor entirely brute and automatic.

References

Akerlof, George A. & Robert J. Shiller. 2009. *Animal Spirits: How Human Psychology Drives the Economy.* Princeton, NJ: Princeton University Press.

Antrobus, John S, Jerome L. Singer, S. Goldstein, and M. Fortgang. 1970. Mindwandering and Cognitive Structure. *Transactions of the New York Academy of Sciences* 32, 242–252.

Bargh, John A. 1997. The Automaticity of Everyday Life. In *The Automaticity of Everyday Life: Advances in Social Cognition*, ed. R.S. Wyer, vol.10, 1–61. Mahwah, NJ: Lawrence Erlbaum.

Barker-Benfield, G.J. 1996. *The Culture of Sensibility: Sex and Society in Eighteenth-Century Britain.* Chicago, IL: University of Chicago Press.

Baynes (Kinneir), David. 1738. *A New Essay on the Nerves, and the Doctrine of the Animal Spirits Rationally Considered; Shewing the Great Benefit and True Use of Bathing ....* Bath: W. Innys and R. Manby

Berntsen, Dorthe. 2009. *Involuntary Autobiographical Memories: an Introduction to the Unbidden Past.* Cambridge: Cambridge University Press.

Branch, Thomas, 1738. *Thoughts on Dreaming.* London: Dodsley and Jolliffe.

Clark, Andy. 1997. *Being There: Putting Brain, Body, and World Together Again.* Cambridge, MA: MIT Press.

Clower, William T. 1998. The Transition from Animal Spirits to Animal Electricity: a Neuroscience Paradigm Shift. *Journal of the History of the Neurosciences* 7, 201–218.

Cunningham, Andrew. 1990. Medicine to Calm the Mind: Boerhaave's Medical System, and Why It Was Adopted in Edinburgh. In *The Medical Enlightenment of the Eighteenth Century*, eds. A. Cunningham & R. French, 40–66. Cambridge: Cambridge University Press.

Dacome, Lucia. 2004. 'To What Purpose Does it Think?': Dreams, Sick Bodies and Confused Minds in the Age of Reason, *History of Psychiatry* 15, 395–416.

Daston, Lorraine. 1998. Fear and Loathing of the Imagination in Science. *Daedalus* 127, 73–85.

Descartes, René. 1972. *Treatise of Man*, ed. & trans. T.S. Hall. Cambridge, MA: Harvard University Press.

Descartes, René. 1991. *The Philosophical Writings of Descartes, vol. III: Correspondence*, trans. J. Cottingham, R. Stoothoff, D. Murdoch, & A. Kenny. Cambridge: Cambridge University Press.

Domhoff, G. William. 2003. *The Scientific Study of Dreams: Neural Networks, Cognitive Development, and Content Analysis.* Washington, DC: American Psychological Association.

[78] Rousseau 2008.

Donald, Merlin. 1991. *Origins of the Modern Mind.* Cambridge, MA: Harvard University Press.
Dreyfus, Hubert L. 2007. The Return of the Myth of the Mental. *Inquiry* 50, 352–365.
Duden, Barbara. 1991. *The Woman Beneath the Skin: a Doctor's Patients in Eighteenth-Century Germany.* Cambridge, MA: Harvard University Press.
Ennen, Elizabeth. 2003. Phenomenological Coping Skills and the Striatal Memory System. *Phenomenology and the Cognitive Sciences* 2, 299–325.
Flanagan, Owen. 2000. *Dreaming Souls: Sleep, Dreams, and the Evolution of the Conscious Mind.* Oxford: Oxford University Press.
Foulkes, David. 1999. *Children's Dreaming and the Development of Consciousness.* Cambridge, MA: Harvard University Press.
Floyd-Wilson, Mary. 2003. *English Ethnicity and Race in Early Modern Drama.* Cambridge: Cambridge University Press.
Frasca-Spada, Marina. 2003. Belief and Animal Spirits in Hume's *Treatise. Eighteenth-Century Thought* 1, 151–169.
Gaukroger, Stephen. 1995. *Descartes: an Intellectual Biography.* Oxford: Clarendon.
Gendler, Tamar Szabo. 2008. Alief in Action (and Reaction). *Mind & Language* 23, 552–585.
Gibbs, James. 1712. *Observations of Various Eminent Cures of the King's Evil ... to Which Is Added an Essay, Concerning the Animal Spirits, and the Cure of Convulsions.* London: Ralph Simpson and W. Innys, and Exeter: Philip Yeo.
Graybiel, Ann M. 1998. The Basal Ganglia and the Chunking of Action Repertoires. *Neurobiology of Learning and Memory* 70, 119–136.
Halbwachs, Maurice. 1925/1992. The Social Frameworks of Memory, in Halbwachs. In *On Collective Memory,* ed. L.A. Coser. Chicago, IL: Chicago University Press.
Harris, James A. 2009. 'A Compleat Chain of Reasoning': Hume's Project in *A Treatise of Human Nature,* Books I and II. *Proceedings of the Aristotelian Society* 109, 129–148.
Harris, John. 1708–1710. *Lexicon Technicum: or, an Universal English Dictionary of Arts and Sciences*, 2nd edition, 2 volumes, London: Brown et al.
Hartley, David. 1749/1971. *Observations on Man, His Frame, His Duty, and His Expectations.* London: repr. New York: Garland.
Hobson, J. Allan & Robert Stickgold. 1994. Dreaming: a Neurocognitive Approach. *Consciousness and Cognition* 3, 1–15.
Hume, David. 1734/1993. A kind of history of my life (letter to Arbuthnot or Cheyne). In *The Cambridge Companion to Hume*, ed. D.F. Norton, 345–350. Cambridge: Cambridge University Press.
Hume, David. 1739/1978. *A Treatise of Human Nature*, ed. P.H. Nidditch. Oxford: Clarendon.
Hume, David. 1975. *Enquiries Concerning Human Understanding and Concerning the Principle of Morals*, ed. P.H. Nidditch. Oxford: Clarendon.
Hutchins, Edwin. 1995. *Cognition in the Wild.* Cambridge, MA: MIT Press.
Ishizuka, Hisao. 2006. The Elasticity of the Animal Fibre: Movement and Life in Enlightenment Medicine. *History of Science* 44, 435–468.
Jacyna, L.S. 1995. Animal Spirits and Eighteenth-Century British Medicine. In *The Comparison Between Concepts of Life-Breath in East and West,* eds. Y. Kawakita, S. Sakai, & Y. Otsuka, 139–161. Tokyo, Japan & St Louis, MO: Ishiyaku Euro-America.
James, Susan. 1999. *Passion and Action: the Emotions in Seventeenth Century Philosophy.* Oxford: Oxford University Press.
Klinger, Eric. 2008. Daydreaming and Fantasizing: Thought Flow and Motivation. In *Handbook of Imagination and Mental Simulation,* eds. K.D. Markman, W.M.P. Klein, & J.A. Suhr, 225–239. New York: Psychology Press.
Krell, David Farrell. 1990. *Of Memory, Reminiscence, and Writing: on the Verge.* Bloomington: Indiana University Press.
Lambie, John A. & Anthony J. Marcel. 2002. Consciousness and the Varieties of Emotion Experience: a Theoretical Framework. *Psychological Review* 109, 219–259.
Landormy, Paul. 1902. La Mémoire Corporelle et la Mémoire Intellectuelle dans la Philosophie de Descartes. *Bibliothèque du Congrès International de Philosophie* 4, 259–298.

Lloyd, G.E.R. 2007. *Cognitive Variations: Reflections on the Unity and Diversity of the Human Mind.* Oxford: Oxford University Press.

Locke, John. 1690/1975. *An Essay Concerning Human Understanding,* ed. P.H. Nidditch. Oxford: Clarendon.

Mason, Malia F., Michael I. Norton, John D. Van Horn, Daniel M. Wegner, Scott T. Grafton, and C. Neil Macrae. 2007. Wandering Minds: the Default Network and Stimulus-Independent Thought. *Science* 315 (19 January), 393–395.

Mead, Richard. 1702. *A Mechanical Account of Poisons in Several Essays.* London: Ralph South.

Morgan, Thomas. 1735. *The Mechanical Practice of Physick ....* London: T. Woodward.

Myer, Valerie Grosvenor. 1984. Tristram and the Animal Spirits. In *Laurence Sterne: Riddles and Mysteries*, ed. V.G. Myer, 99–112. London: Barnes and Noble.

Paster, Gail Kern. 1993. *The Body Embarrassed: Drama and the Disciplines of Shame in Early Modern England.* Chicago: Chicago University Press.

Paster, Gail Kern. 1997. Nervous Tension: Networks of Blood and Spirit in the Early Modern Body. In *The Body in Parts*, eds. D. Hillman & C. Mazzio, 107–125. London: Routledge.

Richardson, Alan. 2001. *British Romanticism and the Science of the Mind.* Cambridge: Cambridge University Press.

Roos, Anna Maria. 2000. Luminaries in Medicine: Richard Mead, James Gibbs, and Solar and Lunar Effects on the Human Body in Early Modern England. *Bulletin of the History of Medicine* 74, 433–457.

Rousseau, G.S. 2004. *Nervous Acts: Essays on Literature, Culture and Sensibility.* London: Palgrave.

Rousseau, G.S. 2008. 'Brainomania': Brain, Mind and Soul in the Long Eighteenth Century. *British Journal for Eighteenth-Century Studies* 30, 161–191.

Rublack, Ulinka. 2002. Fluxes: the Early Modern Body and the Emotions. *History Workshop Journal* 53, 1–16.

Russell, Bertrand. 1997. *The Collected Papers of Bertrand Russell*, vol.11. London: Routledge.

Schmitter, Amy M. 2006. 17th and 18th Century Theories of Emotions. *The Stanford Encyclopedia of Philosophy.* (Summer 2006 Edition), Edward N. Zalta (ed.), http://plato.stanford.edu/archives/sum2006/entries/emotions-17th18th/.

Schupak, Cynthia and Jesse Rosenthal. 2009. Excessive Daydreaming: a Case History and Discussion of Mind Wandering and High Fantasy Proneness. *Consciousness and Cognition* 18, 290–293.

Seuntjens, Wolter. 2006. Vapours and Fumes, Damps and Qualms: Windy Passions in the Early-Modern Age (1600–1800). *English Studies* 87, 35–52.

Singer, Jerome L. 1966. *Daydreaming: an Introduction to the Experimental Study of Inner Experience.* New York, NY: Random House.

Smail, Daniel Lord. 2008. *On Deep History and the Brain.* Cambridge, MA: Harvard University Press.

Smallwood, Jonathan and Jonathan W. Schooler. 2006. The Restless Mind. *Psychological Bulletin* 132, 946–958.

Sterne, Laurence. 1759/1983. *The Life and Opinions of Tristram Shandy, Gentleman*, ed. I.C. Ross. Oxford: Oxford University Press.

Sutton, John. 1998. *Philosophy and Memory Traces: Descartes to Connectionism.* Cambridge: Cambridge University Press.

Sutton, John. 2000. The Body and the Brain. In *Descartes' Natural Philosophy*, eds. S. Gaukroger, J.A. Schuster, & J. Sutton, 697–722. London: Routledge.

Sutton, John. 2002. Porous Memory and the Cognitive Life of Things. In *Prefiguring Cyberculture*, eds. D. Tofts, A.M. Jonson, & A. Cavallero, 130–141. Cambridge, MA: MIT Press.

Sutton, John. 2007a. Spongy Brains and Material Memories. In *Embodiment and Environment in Early Modern England,* eds. M. Floyd-Wilson & G. Sullivan, 14–34. London: Palgrave.

Sutton, John. 2007b. Batting, Habit, and Memory: the Embodied Mind and the Nature of Skill. *Sport in Society* 10, 763–786.

Sutton, John. 2008. Remembering. In *The Cambridge Handbook of Situated Cognition*, eds. P. Robbins & M. Aydede, 217–235. Cambridge: Cambridge University Press.
Sutton, John. 2009. Dreaming. In *The Routledge Companion to the Philosophy of Psychology*, eds. P. Calvo & J. Symons, 524–544. London: Routledge.
Tierney-Hynes, Rebecca. 2007. Hume, Romance, and the Unruly Imagination. *Studies in English Literature* 47, 641–658.
Traiger, Saul. 1994. The Secret Operations of the Mind. *Minds and Machines* 4, 303–316.
Tribble, Evelyn. 2005. Distributing Cognition in the Globe. *Shakespeare Quarterly* 56, 135–155.
Velleman, J. David. 2007. The Way of the Wanton. In *Practical Identity and Narrative Agency*, eds. C. Mackenzie & K. Atkins, 169–192. London: Routledge.
Wegner, Daniel M. 1997. Why the Mind Wanders. In *Scientific Approaches to Consciousness*, eds. J. Cohen & J.W. Schooler, 295–315. Hillsdale, NJ: Lawrence Erlbaum.
Winslow, E.G. 1986. Keynes and Freud: Psychoanalysis and Keynes's Account of the Animal Spirits of Capitalism. *Social Research* 53, 549–578.
Yin, Henry H. & Knowlton, Barbara J. 2006. The Role of the Basal Ganglia in Habit Formation. *Nature Reviews Neuroscience* 7, 464–476.

Instrumental or Immersed Experience: Pleasure, Pain and Object Perception in Locke

Lisa Shapiro*

Abstract This paper aims to draw out two distinct strands in Locke's account of our simple ideas of experience: an instrumental and an immersed model of experience. The place of pleasure and pain in sensation is key to the distinction between these two models. After showing this equivocation in Locke's account, I consider its implications for his account of object perception, or our ideas of particular substances, and suggest that considering these issues in Locke might afford insight into contemporary discussions of the Binding Problem. I conclude by showing how Berkeley and Condillac resolve this equivocation in Locke and considering why Locke himself might have failed to do so.

1 Introduction

In this paper I aim to problematize a pervasive yet unstated assumption about Locke's account of, at least, our simple experience. Most readers take Locke to conceive of simple experience as simply conveying information about the properties objects in the world are meant to have independently of human efforts to engage with that world. It is as if our sense organs are like instruments designed to detect the properties of objects, and our experience of the world might as well be disembodied. While this line of thought is to be found in Locke's *Essay Concerning Human Understanding,*[1] I aim to show that Locke's view wavers between this disembodied or instrumental conception of our experience and one that takes us to

---

* This paper has benefited greatly from the discussion at the Embodied Empiricism workshop at the University of Sydney and at a conference on early modern accounts of the passions at University of San Francisco. I owe particular thanks to Gideon Manning for his extensive commentary, and to Donald Ainslie, Peter Anstey, Hal Cook, Ofer Gal and John Sutton for their remarks. Research for the paper was supported by a grant from Social Sciences and Research Council of Canada.

[1] Cited internally as ECHU with reference to book, chapter and paragraph.

L. Shapiro
Department of Philosophy, Simon Fraser University

C.T. Wolfe and O. Gal (eds.), *The Body as Object and Instrument of Knowledge: Embodied Empiricism in Early Modern Science*, Studies in History and Philosophy of Science 25,
DOI 10.1007/978-90-481-3686-5_13,

be essentially immersed in the world and our experience to be essentially affective. Once I establish the points of tension in Locke's account of our simple experience, I explore one implication of this tension: Locke's account of how we come to perceive objects, or what he terms particular substances, from our simple ideas. Contemporary cognitive psychologists refer to this issue as the Binding Problem – that of how information from different sense modalities gets bound together – and understanding the challenges of Locke's account might afford insight into contemporary discussion of similar issues. Just as we can ask what contemporary import attention to this tension in Locke has, so too can we ask why Locke himself did not resolve this tension, especially since several Lockeans adopt a model of immersed experience. I present some thoughts on this matter as well. However, my focus is to elucidate the tension in Locke's account of experience between an instrumental and an immersed conception of embodiment.

In Section 2, I clarify the sense in which Locke's conception of simple experience might as well be disembodied by considering what Locke might mean by asserting that our senses are 'fitted' to the world around us. On one reading, while our simple ideas might well depend on the body we have, our embodied condition does not affect the information those simple ideas provide us with – their content – and in it is in this sense that our experience might as well be disembodied. There is a second possible reading of Locke's notion of 'fittedness', one which is aligned with an alternative immersed conception of simple experience that was prevalent in the early modern period in thinkers such as Descartes, Spinoza, as well as in canonical empiricists such as Berkeley and Condillac. On this alternative account, sensory experience also conveys information about the world, but that information incorporates the way things affect our well-being; pleasure and pain are integral to our sensory ideas. On this line, our embodied condition does affect the content of our sensory ideas. With this distinction in mind, I turn in Section 3 to consider Locke's account of our ideas of pleasure and pain, and Identify some tensions and equivocations in that account. In Section 4, I argue further that these issues infect Locke's account of our perceptions of objects. In this section I also consider how contemporary cognitive scientists might gain insight from this point. In Section 5, I briefly note how some empiricists who self-consciously follow Locke resolve his equivocation regarding our ideas of pleasure and pain. They seem to adopt the model of immersed experience. Finally, in Section 6 I speculate about the reasons Locke himself might not have resolved the tension between instrumental and immersed experience in his work.

2 Locke's Account of Sensation as Fitted to our Surroundings

I have claimed that insofar as Locke models sensation as importing information about properties of objects instrumentally, he might as well take sense perception to be disembodied. This claim demands an obvious qualification. It has been widely remarked that Locke is deeply engaged with the scientific developments of the latter half of the seventeenth century.[2] This engagement with the contemporary scientific landscape also impacts his discussion of sensation. Not only does Locke admit that we might have had different sorts of simple ideas had we different sensory organs than we do have,[3] he also considers the impact of the constitution of the sensory organs we do have on the content of the sorts of simple ideas humans have. He writes:

> Had we senses acute enough to discern the minute particles of bodies and the real constitution on which their sensible qualities depend, I doubt not but they would produce quite different ideas in us; and that which is now the yellow colour of Gold would then disappear, and instead of it we should see an admirable texture of parts of a certain size and figure... (ECHU 2.23.8)[4]

Locke, here, is explicitly concerned to make sense of the microscope and the ways in which that piece of technology changes our perceptions, but the general point is nonetheless instructive. Locke seems to acknowledge that we perceive the qualities of things we do because we have the sense organs we have. In this way, Locke does suggest that the content of our sensations is importantly tied to the bodies we have.

This simple sense in which we can understand our bodies as figuring in sensation is, not, however, opposed to an instrumental conception of embodiment. To clarify the distinction between what I am calling instrumental and immersed embodiment and experience, let me consider in more detail the context of the passage quoted above. There Locke asserts that our senses "are fitted … for the neighborhood of the bodies that surround us, and we have do with" (ECHU 2.23.13). We can elucidate the distinction at issue here by fleshing out this notion of 'fittedness'.

It is clear that, for Locke, our senses are 'fitted' to our surroundings insofar as they serve our purposes. As Locke puts it: "We are able, by our senses, to know, and distinguish things; and to examine them so far, as to apply them to our uses, and several ways to accommodate the exigencies of this life" (ECHU 2.23.12). But we can distinguish two distinct ways in which our sensory experiences can

---

[2] This engagement is announced in the *Epistle to the Reader* of the *Essay,* with Locke's self-description as "an under-labourer in clearing ground a little, and removing some of the rubbish that lies in the way to knowledge" of the sort that Boyle, Sydenham, Huygens, and 'the incomparable Mr. Newton' were pursuing, and is evident in his appropriation of Boyle's brand of the mechanical philosophy in detailing the inner workings of bodies. In recent years, Locke's scientific interests have been well-discussed. Relatively early discussions include: Sanchez-Gonzalez 1990. Wilson 1995, Stein 1990, and Downing 1997, 1998. For more recent discussion see: Anstey and Harris 2006, Jones 2007, Keating 2002, Meynell 2003, Milton 2001 and Walmsley 2009.

[3] See for instance ECHU 2.2.3: "This is the reason why, though we cannot believe it impossible to God to make a creature with other organs, and more ways to convey into the understanding the notice of Corporeal things, than those five, as they are usually counted, which he has given to man. Yet I think it is not possible for any one to imagine any other qualities in bodies, however constituted, whereby they can be taken notice of, besides, sounds, tastes, smells, visible and tangible qualities."

[4] See also ECHU 2.1.23–24 and 2.23.13.

serve our purposes. On the one hand, our senses can be said to serve our purposes insofar as they provide us with information that we can go on to use to achieve our ends. On this view, there is no essential connection between the sensory information we receive and the purposes we have. It just happily turns out that the two fit together. Our sense organs are instruments that are capable of detecting certain properties, or qualities, in the world. While these instruments might be designed so as to detect the properties they do, and so 'be fitted to our surroundings', nothing in that design requires us to *use* that information. My eyes might be fitted to convey information about the numbers of hairs on each person's head, but that information need not serve any purpose. If I decide, quite independently, that I want to be a hairdresser, then maybe I can put that information to good use. It is central to this *instrumental* way of understanding the way our bodies figure in sensation that the sensory content, the information we receive, is not in principle tied to the purposes to which we might put that information. In the first instance, the information we get is not dependent on any purposes we might have, nor need our purposes derive from the information we gain through sensory experience.

This model of fittedness does seem to be what Locke intends in the passages I have pointed to. Indeed, the most familiar parts of Locke's account of sensation accord well with this conception of our sensory experience. Locke characterizes the origin of our simple ideas as a matter of the senses 'conveying into the Mind' those perceptions proper to the way sensible objects affect our sense modalities (ECHU 2.1.3), thereby suggesting that sense perception amounts to the input of information about objects, information which is meant to be true of them irrespective of the receiver of that information, and his, her or its purposes. This same line of thought would seem to permeate Locke's discussion of the distinction between primary and secondary qualities and our ideas of them. Qualities, for him, are simply powers to produce simple ideas in us; we get information about the properties of bodies straightforwardly in our ideas of primary qualities in virtue of the resemblance between those ideas and the qualities themselves. This resemblance relation has nothing to do with the uses to which we might put the information about the primary qualities of bodies. Moreover, it seems that Locke takes our ideas of secondary qualities to provide us with some information about properties of bodies even though those ideas do not resemble their causes. After all, our ideas of secondary qualities are still produced in us by some complicated array of primary qualities, and so they would seem to encode some information about those primary qualities. We might have to crack that code, but the information that is conveyed, again, makes no reference to the uses to which that information might be put.[5]

---

[5]There is no doubt much more to say here. Understanding the relation between our ideas of primary and secondary qualities for Locke is a vexed topic. As I note below, sometimes Locke does suggest that secondary qualities are analogous to pleasures and pains, and so do potentially give us some information bearing on our welfare. I say 'potentially' here because not only will this depend on the dimensions of analogy but also on the conception of pleasure and pain in play, as will become clear in the next section.

There is, however, a second way in which we might take our sensory experiences to be fitted to our surroundings. On this view, the very content of our simple sense perceptions *is* tied to our purposes, and in particular, to our end of self-preservation. Immediately prior to Locke, this sort of account of sensory experience is seen most clearly in Descartes, though it is also found in Spinoza, and arguably Hobbes, as well.[6] In the Sixth Meditation, Descartes maintains that

> the proper purpose of the sensory perceptions given me by nature is simply to inform the mind of what is beneficial or harmful for the composite of which the mind is a part; and to this extent they are sufficiently clear and distinct. But I misuse them by treating them as reliable touchstones for immediate judgements about the essential nature of the bodies located outside us. (7:83; 2:57–8)[7]

Rather than informing us about properties of things in the world, for Descartes, our sensations inform us about how things are *in relation to us*, how they benefit and harm us, and in general how they affect our well-being.[8] Though Descartes recognizes we can be mistaken about the benefits and harms things offer us – for instance, we can feel thirsty when we ought not to take in more fluids – for him, we still experience the sensations we do in accord with the system which "is most especially and most frequently conducive to the preservation of the healthy man" (7:88; 2:60).

One might well ask how, on this view, our sensations give us this kind of information. Does the input into one sense modality or another contain information about the way things in the world impact us? Some medieval thinkers seem to have espoused a model somewhat like this. According to Aquinas, for instance, all animals were able to register the way things in the world stood to affect their very existence through a separate sensory faculty, the *vis estimativa*. Famously, through this faculty, a sheep is able to perceive a wolf as dangerous – that is, as capable of causing that sheep harm.[9] Similarly, through this faculty we able to perceive various threats in our environment, as well as those things that stand to benefit us.

Descartes, however, does not want to go the route of this simple sort of sensory model. Rather, for him, two aspects of our sensory experience serve to provide us with information. First, the *variation* in sensory input conveys information about the ways things benefit and harm us as unions of mind and body. He writes:

---

[6] It is hard to know just what Spinoza's account of sensation is – he himself does not use the term – but Spinoza is clear that "the ideas which we have of external bodies indicate the condition of our own body more than the nature of the external bodies" (*Ethics* IIP16Cor2 in Spinoza 1985). At the basis of this claim is Spinoza's claim that our conatus is our essence (EIIIP7). Like it is for Spinoza, *conatus* is at the centre of Hobbes's philosophy, and one might well argue that our drive to persevere shapes the content of our sensory experiences had through the working of the machine of our body.

[7] Citations of Descartes's works follow this format: (Volume: page of AT; Volume: page of CSM/K). 'AT' refers to Descartes (1996/1908). 'CSM' and 'CSMK' refer respectively to Vols 1–2 and to Vol 3 of Descartes (1985–1991).

[8] Simmons 1999 makes a similar point.

[9] See, for instance, *Summa Theologiae* I, q.78, art.4. (Aquinas 2002).

> And from the fact that I perceive by my senses a great *variety* of colours, sounds, smells and tastes, as well as differences in heat, hardness and the like, I am correct in inferring that the bodies which are the source of these various sensory perceptions possess differences corresponding to them, though perhaps not resembling them. Also, the fact that *some of the perceptions are agreeable to me while others are disagreeable* make it quite certain that my body or rather my whole self, in so far as I am a combination of body and mind, can be affected by the various beneficial or harmful bodies which surround it. (7:81; 2:56; emphasis added)

Thus, our sensations of colour all on its own does not tell us anything about the world, let alone about any real colours in the world. Rather, variations in our sensations of colour convey information about real variations in the world. Second, in this passage Descartes suggests that these sensations themselves are either agreeable or disagreeable, and through this aspect of our sensations we are steered towards what is beneficial and away from what is harmful to us. For Descartes, sensations seem *intrinsically* to involve pleasure and pain. It is thus no accident that in the Sixth Meditation, the paradigm sensations are hunger and thirst, two sensations that do seem to involve some kind of pain. Insofar as Descartes does take our sensory experiences to have this intrinsic affective aspect he builds a certain sort of teleology into our sense perception. For we feel pleasure and pain as is most conducive to our preservation. That is, for Descartes, the content of our sensations *contains* an end of self-preservation. In this important way our experiences contain our purposes, or at least one central purpose. On this model, our cognitive contact with the world comes as we are immersed in a world, situated with respect to things that impact on our well-being; our senses are fitted to our surroundings insofar as they are geared to enabling us to achieve our end of self-preservation.

From the perspective of this second way of understanding our senses as fitted to our surroundings, one thing is quite notable in Locke's account of our simple ideas of sensation, at least as I have presented it so far: Locke makes no mention of pleasure and pain in his discussion of the ideas we receive through the five canonical sense modalities. Insofar as Locke would seem to think that our sensory ideas, in and of themselves, do not involve pleasure and pain, he would also take it that our sensory ideas contain no information about how things in the world impact on our welfare, and in general our ability to preserve ourselves.

3 Pleasure and Pain in Locke's Account of Our Simple Ideas of Sensation

Of course, Locke does not ignore the fact that we have sensations of pleasure and pain. Indeed, Locke counts pleasure and pain as those simple ideas that we can have either through sensation and reflection. Locke's discussion, however, is far from systematic. In this section, I identify two equivocations in Locke's account.

First, it is unclear whether Locke takes pain and pleasure to contain intrinsically information about our well being. On the one hand, when Locke first introduces the simple ideas of pain and pleasure, he seems to tie those perceptions to our welfare. The fact that we often feel both pleasure and pain towards the same object

> gives us new occasion of admiring the wisdom and goodness of our Maker: Who, designing the preservation of our being, has annexed pain to the application of many things to our bodies, *to warn us of the harm that they will do*, and as advices to withdraw from them. (ECHU 2.7.4, emphasis added)

Pleasure and pain thus seem to carry with them information about the ways things can benefit and harm us. Yet Locke defines good and evil with reference to pain and pleasure.

> That we call *good*, which *is apt to cause or increase pleasure, or diminish pain in us; or else to procure or preserve us the possession of any other good, or absence of any evil.* … [W]e name that *evil*, which *is apt to produce or increase any pain, or diminish any pleasure in us; or us to procure us any evil, or deprive us of any good.* (ECHU 2.2.2)

It seems here that Locke is taking pleasure and pain as primitive feelings which themselves cannot be explained. Pleasure and pain, on this line, are not indicators of any objective benefit or harm that might visit us. Rather benefits and harms, or good and evil, are defined in virtue of what we happen to feel. If putting a hand in a roaring fire brings pleasure, then it is by definition good. Yet in the very next paragraph Locke seems to switch tacks to assert that good and evil *cause* pleasure and pain, as if there were facts of the matter about what was good and bad, and our feelings of pleasure and pain indicate those matters of fact to us (ECHU 2.20.3). This claim, however, does not sit well with the definitions of the various passions that follow. In those definitions it seems that these species of pleasure and pain are a matter of mere subjective feeling, which then by accident shape our ascriptions of good and evil. For instance, we love grapes simply because their taste brings delight. Without that feeling of delight, the grapes cease to be good, and we no longer love them (ECHU 2.20.4). It seems then that we are supposed to read the first half of the first sentence of ECHU 2.20.3 as concerning what we *call* good and evil rather than any real relation of benefit or harm.[10]

A similar tension appears in Locke's remarks concerning skepticism about the existence of the external world. In Book IV, when Locke firsts considers our knowledge of 'real existence', he challenges anyone who experiences the pain of being in a fire to assert that he is dreaming. His point here is that some ideas, simply in their phenomenology, give us sufficient information to conclude that a world exists beyond our ideas. However, he does not maintain that in learning that something exists which also causes us (in this case) pain we also learn something about the way that something might benefit or harm us.[11] Yet when Locke turns to consider our 'knowledge of the existence of other things' in more detail in ECHU 4.11, he seems to admit that pleasures and pains *do* tell us about various benefits and harms in the world. He writes:

[10] ECHU 2.21.42 supports this reading. There Locke asserts that "whatever has an aptness to produce pleasure in us, is that we call *Good*, and what is apt to produce Pain in us, we cal *Evil*, for no other reason, but for its aptness to produce pleasure and pain in us."

[11] See for instance ECHU 4.2.14.

> That *the certainty of* things existing *in rerum Natura*, when we have the testimony of our senses for it, is not only *as great* as our frame can attain to, but *as our condition needs*. (ECHU 4.11.8; original emphasis)

And as he fleshes this out, it is clear that the 'condition' he refers to is that of our general well-being.

> For he that sees a candle burning, and hath experimented the force of its flame by putting his finger in it, will little doubt that this something existing without him, *which does him great harm, and puts him to great pain*: which is assurance enough, when no man requires greater certainty to govern his actions by that what is as certain as his actions themselves. (ECHU 4.11.8, emphasis added)

Locke thus seems in the end to allow that pleasures and pains do, for the purposes of action at least, inform us about the ways things in the world can benefit or harm us. Insofar as he does, Locke might well conceive of our bodies not as mere instruments with which to detect properties in the world but rather as immersed in it. At least our sensations of pleasure and pain inform us of how other bodies stand in relations to us and our purpose of self-preservation.

Even if we do allow that Locke takes our sensations of pleasure and pain to contain information about the way things benefit and harm us,[12] there is a second

[12] There might well be a third alternative to consider here. Our ideas of pleasure and pain might give us information about the world, just not about the way things benefit and harm us. On this line, ideas of pleasure and pain would be akin to our ideas of secondary qualities. Though these ideas do not represent (by resemblance) qualities of objects, we can gain some information about the world if we could adduce the causal relation between primary qualities and our ideas of secondary qualities. Similarly, this line would go, if we properly understood the causes of our pains and pleasures, those ideas could give us information about primary qualities. Locke seems to be suggesting this in a comparison he draws between ideas of secondary qualities and our ideas of pain in ECHU 2.8.16 and ECHU 2.8.18:

> And yet he that will consider that the same fire, that at one distance produces in us the sensation of warmth, does at a nearer approach produce in us the far different sensation of pain, ought to bethink himself what reason he has to say, that his idea of warmth, which was produced in him by the fire, is actually in the fire; and *his idea of pain, which the same fire produced in him the same way, is not in the fire*. Why are whiteness and coldness in snow, and pain not, when it produces the one and the other idea in us; and can do neither, but by the bulk, figure, number, and motion of its solid parts? (ECHU 2.8.16, emphasis mine).
>
> Besides, manna, by the bulk, figure, texture, and motion of its parts, has a power to produce the sensations of sickness, and sometimes of acute pains or gripings in us. *That these ideas of sickness and pain are not in the manna, but effects of its operations on us, and are nowhere when we feel them not*; this also every one readily agrees to. And yet men are hardly to be brought to think, that sweetness and whiteness are not really in manna; which are but the effects of the operations of manna by the motion, size, and figure of its particles on the eyes and palate; as the pain and sickness caused by manna are confessedly nothing but the effects of its operations on the stomach and guts, by the size, motion, and figure of its insensible parts (for by nothing else can a body operate, as has been proved). (ECHU 2.8.18, emphasis added).

These passages, interestingly, raise a question about how we are to distinguish between our ideas of secondary qualities and pleasures and pains. Hume, in his *Treatise on Human Nature*, raises just such a question in his discussion of scepticism with regard to the senses (THN 1.4.2.12), a section I take to be largely about the empiricist prospects for an account of object perception. This would seem to bear on my discussion later in this paper.

point of equivocation in Locke's account of our simple ideas of pleasure and pain, one that bears on whether Locke can be taken to have a conception of immersed experience. For Locke, each of our simple ideas, whether they be of sensation or reflection, are conveyed into the mind independently of one another, and it is central to his account that these simple ideas be unanalyzable. Given this assumption it certainly seems as if our ideas of pleasure and pain ought to be distinct from our other simple ideas. And often Locke's discussion of pleasure and pain does take this line. In ECHU 2.7.2 Locke writes,

> Delight or uneasiness, one or other of them, join themselves to almost all our ideas, both of sensation and reflection: And there is scarce any affection of our senses from without, any retired thought of our mind within, which is not able to produce in us pleasure or pain. By pleasure and pain I would be understood to signify whatsoever delights or molests us ...

Locke here makes two distinct claims. First, he notes that almost all of our ideas are joined to an idea of pleasure and pain, and in doing so he certainly suggests that each of the ideas that are joined together are distinct from one another. Second, he suggests that our ideas of pleasure and pain are *caused* by other ideas. While this second claim also presupposes that ideas of pleasure and pain are distinct ideas, it is puzzling with respect to ideas of sensation. Presumably, our sensory ideas of pleasure and pain derive directly from the workings of the world on our bodies, and not from the workings of our mind on itself. But in any case, it seems clear that Locke takes our ideas of pleasure and pain to be distinct from our other ideas.

This clear distinction, however, becomes somewhat murky in the very next paragraph. There Locke maintains that our creator,

> having also given a power to our minds in several instances, to choose, amongst its ideas, which it will think on, and to pursue the enquiry of this or that subject with consideration and attention, to excite us to these actions of thinking and motion that we are capable of; has been pleased *to join to several thoughts, and several sensations, a perception of delight. If this were wholly separated from all our outward sensations and inward thoughts, we should have no reason to prefer one thought or action to another; negligence to attention, or motion to rest* .... It has therefore pleased our wise Creator to *annex to several objects*, and the ideas which we receive from them, as also to several of our thoughts, a concomitant pleasure, and that in several objects, to several degrees; that those faculties which he had endowed us with might not remain wholly idle and unemployed by us. (ECHU 2.7.3, emphasis added)

While in this passage Locke still talks of pleasure and pain as joined to our thoughts, in almost the same breath he denies that pleasures and pains are 'wholly separated' from our sensations and reflections. Along the same lines, it is hard to know what Locke means when he claims that pleasures and pains are 'annexed' to our ideas of objects. Does he take the annexation to result in a simple idea, with aspects only distinguishable by reason? Or does the annexation amount to forming a complex idea? It is hard to say. In ECHU 2.7.5 he maintains that pleasure and pain are "blended ... together in almost all that our thoughts and senses have to do with." Again we can ask how we are to understand this 'blending'.

It seems important to Locke to maintain that pain and pleasure *are* distinct ideas; he seems to be bending over backwards to preserve their distinctness. Yet he realizes that this position faces some explanatory challenges. Pleasures and pains, for him, seem to function as the vehicles through which our attention is directed to one idea or another. But, of course, they cannot serve this function if there is no explanation of how the two come be joined together. It is for this reason Locke seems to want to qualify the distinctness of *all* our simple ideas from one another and to deny that pleasure and pain are wholly separate.

While Locke says relatively little about pleasure and pain, what he does say is fraught with equivocations. Sometimes he takes pain and pleasure to be essentially contentless, simply feelings of delight or uneasiness, which, while they ground our evaluations and move us to action, tell us nothing about the world. At other times, Locke takes pleasure and pain to indicate the ways things benefit and harm us, or otherwise impact on our well-being. Insofar as Locke does sometimes suggest the latter view, his account of experience might not be as instrumentally embodied as it initially seems. However, Locke is far from taking our experience to be essentially immersed insofar as he takes pleasures and pains to be distinct from our other simple ideas. Insofar as we can separate the sensory contents from pains and pleasures, Locke must take it that sensory information is gained *independently* of any information afforded by pleasure and pain, that is, independently of concern with self-preservation. This separability of sensory experience from any ends we might have is at the heart of the account of experience that takes our bodies as detecting instruments.

4 Pleasure and Pain and Our Ideas of Particular Substances

It might be tempting to note these equivocations in Locke's account of pleasure and pain but to discount them as on the whole unimportant. However, Locke's equivocation regarding the place of pleasure and pain in our *cognitive* economy does extend to other aspects of his account that philosophers have taken to be central. In this section, I focus on the implications of Locke's mixed mind regarding pleasure and pain, and so instrumental and immersed models of experience, on his account of particular substances.[13]

According to Locke, we form our ideas of objects, or what he terms particular substances,[14] in virtue of our noticing that "a certain number of these simple ideas

[13] In an extended treatment of this problematic, I would also want to consider in more detail Locke's epistemology, for it seems to me that we can think about the importance of the distinction there between intuitive and demonstrative knowledge, on the one hand, and sensitive knowledge on the other, as importantly dependant on the separability of pleasure and pain from our other simple ideas, and so dependent on an instrumentalist conception of embodiment.

[14] It might well be that for Locke our ideas of particular substances are ideas of *kinds* of objects, rather than of individual objects. Nonetheless, those ideas would presuppose, for him, ideas of individuals, and Locke owes his readers an account of our perceptions of objects. I don't see that that account of our ideas of individuals would be very different from his story about our ideas of particular substances.

go constantly together" (ECHU 2.23.1). We receive an ever changing set of simple ideas in the course of our experience, and, so the story goes, we notice that some of those simple ideas always seem to come into our mind together. We "collect … such combinations of simple ideas" together and form a complex idea. Locke takes it to be natural to seek an explanation of why those simple ideas do come into the mind together, and he posits that the qualities which cause those simple ideas all inhere in one underlying "something, I know not what" or pure substance in general. While readers usually focus on Locke's entitlement to this supposition, or how best to understand pure substance in general, or if Locke's account allows for particular substances to have real essences, these lines of inquiry take for granted that Locke can make sense of our ability to notice that some simple ideas come together in experience.

But how is that supposed to happen on Locke's view? As William James puts it, albeit two centuries after Locke, our first experience of the world is as of "one great blooming, buzzing confusion."[15] James is far from being the first to conceive of our experience in this way. We have already seen the shades of a similar position articulated in Descartes's Sixth Meditation, where Descartes presents us as perceiving various colours, sounds, smells, tastes, and so on, that must be organized in some way.[16] It would seem that Locke conceives of our experience in a similar way, at least at the beginning of his account. Locke makes no mention of any filter to control the conveyance of ideas into the mind, and a simple introspection reveals that the simple ideas I do sometimes collect together to form an idea of an object not only are not always joined together but also are sometimes joined with other ideas. Indeed, this seems to be part of the point of Descartes's meditator's consideration of the piece of wax. We cannot, according to this account, know by sensation the piece of wax – or, one might say, take there to be a single piece of wax, an object, before us – because the sensible qualities we perceive are constantly changing, and we cannot imagine the uncountable changes it might further undergo. The meditator requires and does not find within sensation itself a principle for grouping sensible qualities together into an object. The problem is not simply one of building up a complex idea from simple ideas of sensation, but also one of breaking down the manifold of our experience into objects. In peering into my dining area, I not only see my kitchen table, I also see the table cloth covering it, the light

[15] James 1950/1918, 462.

[16] Descartes's account of the three grades of sensation in his *Replies to the Sixth Objections* fits this reading well. On that account, the first grade of sensation consists simply in the physiological changes in our sense organs. The second and third grades both concern our sense perceptions, that is, sensations as mental states. The second grade of sensation consists of ideas of the various sensible qualities that are available to us. These sense perceptions do not yet allow us to form ideas of object, though there may be patterns to our perceptions. Descartes describes the third grade of sensation as our judgements of something before us – that we see a bent stick in water, for instance. (See 7:436ff; 2:294ff) It is hard to know how to read 'judgement' here. It seems clear that Descartes does not intend the sort of judgement that is the focus of the Fourth Meditation. I am inclined to read him as accounting for the way we reify the patterns we perceive into objects.

streaming through the window, the chairs around it, the floor it sits on, and so on, and one might say similar things about my tactile and auditory experiences of the table as well. One wants an account of how the simple ideas I receive all at once get grouped together into an array of complex ideas that constitute my idea of objects. To my knowledge, Locke does not explicitly offer any such account.

Descartes does not explicitly offer such an account either, but his remarks in the Sixth Meditation are suggestive. Recall there were two aspects to Descartes's account of sense perception. First, variations in our sense perceptions were taken to indicate real variation in the world, and second, our sense perceptions themselves were taken to be intrinsically agreeable or disagreeable, that is, pleasant or painful. While I cannot fully argue for it here, I want to suggest that these two aspects work together to afford us perceptions of objects. That our perceptions admit of variation in and of itself is insufficient to get us an idea of an object. Someone with poor vision need only take off her glasses to get this point: perceiving sensible qualities, even if we can perceive variations in those qualities is not the same thing as having an idea of an object. Rather we delimit objects in the varied landscape of our sensory experience by the pleasures and pains we experience. These pleasures and pains, for Descartes unequivocally, indicate the way things benefit and harm us and with this information we can, through a kind of judgement, form an idea of a well-delimited object.[17]

It is particularly noteworthy that Locke mentions nothing about pleasure and pain in his discussion about particular substances. Locke's equivocations regarding pain and pleasure preclude his telling a Cartesian story about how we manage to collect our simple ideas together in the first place, so as to form ideas of particular substances. If Locke takes ideas of pain and pleasure to be primitive simple ideas, providing no information about benefit and harm, they are just like any other of our many ideas of experience that need to be organized in some way. And insofar as they are not information bearing it is not clear what sort of role they could possibly play in any account of our complex ideas of objects. If he takes our pains and pleasures to provide us with information about benefits and harms, Locke might have a resource to draw on in providing an explanation of how we form ideas of objects similar to that of Descartes. Indeed, Locke's account of the role pleasure and pain play in focusing our mind's attention on one idea or another would seem to provide the beginning of such an account. However, this path is only open to him if he takes our sensations to be intrinsically pleasant or painful, that is, so long as he takes our bodies to be immersed in the world. If he takes our bodies instrumentally, that is, if he maintains that pleasure and pain are distinct and separable simple ideas, he cannot appeal to any information regarding benefit and harm as attention-focusing until he accounts for how those pleasures and pains come to be joined with our other simple ideas. But of course this is the problem for which we are seeking a solution in the first place.

[17] Again, the sense of judgement in play here cannot be that of the Fourth Meditation. (See n.16 above.)

Locke himself, even if he does explicitly mark a problem here, does recognize he must say something about how we form ideas of particular substances from simple ideas. In ECHU 3.6.28, he claims that

> the Mind, in making its complex *Ideas* of Substances, only follows Nature; and puts none together, which are not supposed to have a union in Nature. No body joins the Voice of a Sheep, with the Shape of a Horse; nor the Colour of Lead, with the Weight and Fixedness of Gold, to be the complex *Ideas* of any real Substances; unless he has a mind to fill his Head with *Chimeras*.

Here Locke simply asserts that our simple ideas are packaged into complex ideas in the mind in a way that conforms to the way the properties they represent are packaged in nature. While if this were so, it would solve the problem facing his account of particular substances, it is not clear how he is entitled to the claim. Even if we concede that our ideas of primary qualities do resemble the actual primary qualities of bodies, our holding these qualities in mind does not entail that we also hold the relations between them. Indeed, Locke's atomism about simple ideas suggests that they do *not* contain this relational content that is crucial to the story Locke is trying to tell here. Locke's professed realism here might well solve his problems but it is far from clear whether this strong a realist claim conforms to his empiricist methodological commitments.

It is intriguing that the problem of object perception in Locke seems analogous to what contemporary cognitive scientists call the Binding Problem. Consideration of Locke's mixed mind about the place of pleasure and pain in sense perceptions on his account of object perception can perhaps lend some insight to current discussion. The Binding Problem at its most basic level is the problem of how to understand our ability to represent a thing as having certain properties. As such, any solution to this problem must explain both how we come to represent a particular thing (i.e., have an idea of a particular substance, in Locke's parlance) and how we take that thing to be possessed of certain properties.[18] Clearly, different answers will turn on whether one takes representation to involve conscious awareness, how one models the ways in which properties can come to be 'bound' with one another, and more generally, the models of information processing available, as well as other factors.[19] Interestingly, standard answers to the Binding Problem, like Locke, ignore any affective dimension to our experience. Now there might well be pragmatic reasons for this. Models of neuroprocessing are still very much in development, and so merely meant to approximate the way our minds do work. Nonetheless, cognitive scientists might well want to evaluate the assumptions they tacitly make about the character of the information being processed. They seem to assume a kind of instrumental conception of experience, in a way similar to the reading of Locke I have been problematizing here. However, as I have shown, Locke himself

---

[18] For a helpful overview of this problem see Plate 2007. Other recent philosophically informed work includes Pylyshyn 2007 and Bermùdez 2007.

[19] Of course, if one does not begin from a Lockean perspective, assuming that sensory information is simply brought into the mind to be processed, one might reject the Binding Problem as a genuine problem. See for instance, O'Reagan and Noë 2001.

seems to have wavered between conceiving of our experience as instrumental and immersed in the world. Contemporary investigations of human understanding might well benefit from rethinking the models of sensory input, so that they include the affective dimension proper to an immersed conception of experience. Doing so might well lead to progress in solving the Binding Problem.

5 Resolving the Tensions in Locke: a Brief Overview of Empiricist Responses

Interestingly, at least some central figures in eighteenth century empiricist thought do seem to take our experience to be essentially immersed in the world. That is, they resolve the equivocations that I have pointed to in Locke's account of pleasure and pain in a particular way: they take our sensory experience to contain essentially an affective dimension. I here focus on Berkeley and Condillac to illustrate this point, but other figures to consider include Hutcheson, Diderot and Rousseau.

5.1 Berkeley

Berkeley's *A New Theory of Vision* is principally concerned with arguing against a geometrical, so rationalist, account of visual perception of distant objects and proposing an empiricist alternative, and in it he says very little about pleasure and pain. Nonetheless there we can see the beginnings of a view that would, unlike Locke, take at least some of our sense perceptions to involve some sense of benefit or harm. In a.59, in beginning to lay out the relation between ideas of sight and those of touch, he writes:

> We regard the objects that environ us in proportion as they are adapted to benefit or injure our own bodies, and thereby produce in our minds the sensation of pleasure or pain. Now bodies operating on our organs, by an immediate application, and the hurt or advantage arising therefrom, depending altogether on the tangible, and not at all on the visible, qualities of any object… (NTV, a.59)

Here Berkeley maintains that through our sense of touch we are able to sense the various benefits and harms the world might afford us. We do not, he here maintains, gain similar information directly through vision, though from past correlations between visual sensations and tactile ones, we can use our sense of sight to anticipate things we come across benefiting or harming us.[20]

[20] It is important to note that the distinction between touch and vision does not correspond to a distinction between ideas of secondary and primary qualities. This is certainly the case for Berkeley who rejects the distinction, but it is also the case for Locke. Touch, after all, allows us to have ideas of shape just as well as vision, not to mention solidity.

In the *Dialogues between Hylas and Philonous*, however, Berkeley seems to extend his position on touch to *all* our sensory ideas. In reading the *Dialogues*, commentators have typically focused on Berkeley's criticisms there of Locke's distinction between primary and secondary qualities, and on his argument following from the premise of both that criticism and Locke's own methodology – that what we immediately perceive are ideas – against the inference to the existence of any extra-mental cause of our ideas. However, with Locke's equivocation regarding the place of pleasure and pain in sensation at the fore, Berkeley's commitment to a central tenet of an immersed conception of experience comes out clearly. It begins to emerge as Berkeley's alter-ego Philonous prompts Hylas to recognize that all sensations of great heat or cold are "nothing distinct" from a "sensible pain":

> Phil. Seeing therefore they are both immediately perceived at the same time, and the fire affects you only with one simple or uncompounded idea, it follows that this same simple idea is both the intense heat immediately perceived, and the pain; and, consequently, that the intense heat immediately perceived is nothing distinct from a particular sort of pain.
> Hyl. It seems so.
> Phil. Again, try in your thoughts, Hylas, if you can conceive a vehement sensation to be without pain or pleasure.
> Hyl. I cannot.
> Phil. Or can you frame to yourself an idea of sensible pain or pleasure in general, abstracted from every particular idea of heat, cold, tastes, smells? &c.
> Hyl. I do not find that I can.
> Phil. Doth it not therefore follow, that sensible pain is nothing distinct from those sensations or ideas, in an intense degree?
> Hyl. It is undeniable; and, to speak the truth, I begin to suspect a very great heat cannot exist but in a mind perceiving it. (*Dialogues*, 12–13)

Note that Berkeley here makes a point of treating our sensation of heat as a simple idea, and from there concludes that the heat must be a kind of pain. As the discussion continues it becomes clear that Berkeley applies the same line of reasoning to all our sensations: they are all species of pleasure and pain. Here Hylas tries to recoil from that conclusion, but to no avail:

> Hyl. Hold, Philonous, I now see what it was deluded me all this time. You asked whether heat and cold, sweetness at were not particular sorts of pleasure and pain; to which I answered simply, that they were. Whereas I should have thus distinguished: – those qualities, as perceived by us, are pleasures or pains but not as existing in the external objects. We must not therefore conclude absolutely, that there is no heat in the fire, or sweetness in the sugar, but only that heat or sweetness, as perceived by us, are not in the fire or sugar. What say you to this?
> Phil. I say it is nothing to the purpose. Our discourse proceeded altogether concerning sensible things, which you defined to be, the things we immediately perceive by our senses. Whatever other qualities, therefore, you speak of as distinct from these, I know nothing of them, neither do they at all belong to the point in dispute. You may, indeed, pretend to have discovered certain qualities which you do not perceive, and assert those insensible qualities exist in fire and sugar. But what use can be made of this to your present purpose, I am at a loss to conceive. Tell me then once more, do you acknowledge that heat and cold, sweetness and bitterness (meaning those qualities which are perceived by the senses), do not exist without the mind? (*Dialogues*, 16)

Philonous insists that the only way of escaping that conclusion that our sensations are all species of pain and pleasure is to claim that something exists other than what we perceive immediately, a claim that Philonous thinks is wholly without warrant.

In the passages I have pointed to thus far, Berkeley adverts only to the so-called secondary qualities, and so can be seen as playing up Locke's willingness to draw an analogy between those qualities and pain and pleasure.[21] However, Berkeley does go on to suggest that even our ideas of extension, figure and motion are also pleasures and pains, albeit less vivid ones:

> Phil. It is not my business to account for every opinion of the philosophers. But, among other reasons which may be assigned for this, it seems probable that pleasure and pain being rather annexed to the former than the latter may be one. Heat and cold, tastes and smells, have something more vividly pleasing or disagreeable than the ideas of extension, figure, and motion affect us with. And, it being too visibly absurd to hold that pain or pleasure can be in an unperceiving substance, men are more easily weaned from believing the external existence of the Secondary than the Primary Qualities. You will be satisfied there is something in this, if you recollect the difference you made between an intense and more moderate degree of heat; allowing the one a real existence, while you denied it to the other. But, after all, there is no rational ground for that distinction; for, surely an indifferent sensation is as truly a sensation as one more pleasing or painful; and consequently should not any more than they be supposed to exist in an unthinking subject. (*Dialogues*, 27)

Berkeley thus resolves the equivocations in Locke in two ways. First, he admits that our sensations of pain and pleasure inform us about benefits and harms; they are information bearing states. Second, he denies that pleasure and pain are distinct simple ideas. Rather, all our simple ideas of sensation are pleasant or painful in themselves. In taking on these two positions, Berkeley clearly situates us as perceivers immersed in the world.

5.2 *Condillac*

Condillac was just as astute a reader of Locke as was Berkeley; his *Essay on the Origin of Human Knowledge* is a reworking of Locke's *Essay*. For my purposes here, however, I am interested in his *Treatise on Sensations*. Interestingly, there Condillac starts from the assumption that all our sensations essentially involve pleasure and pain. His self-described aim is to explain how all our cognitive functions can be explained from this starting point:

> Nature has given us our sense organs to alert us to what we have to seek out by pleasure, and to what we have to flee by pain. But our nature stops there. Experience has the job of giving us habits and completing the job that nature started. This object is new, and shows

[21] See n.13 above.

all the simplicity of the ways of the author of nature. Cannot one but wonder that to give birth to ideas, desires, habits, and all kinds of talents, it is only necessary to make human being sensible to pain and pleasure? (*Dessein;* Condillac 1754/1984, 12)

In many ways, then, Condillac's project is identical to Locke's: to explain human understanding without appeal to innate ideas and with only the resources of experience. However, there is one crucial difference. Whereas Locke is equivocal regarding whether pleasures and pains are distinct simple ideas, Condillac is clear they are not. For Condillac, they are integral aspects of our sensations. Moreover, that this is so is critical to the empiricist psychology Condillac wants to go on to develop.

Condillac proposes to defend his hypothesis that all our cognitive functions derive from our sensations through a thought experiment. Indeed, the whole of the work consists in just this exercise. Condillac asks us to consider a statue that is given one sense modality at a time and to determine when she comes to have a consciousness like our own. What Condillac seems to mean by consciousness here is simply the ability to perceive and identify objects as distinct from ourselves in the way we do. This work thus aims to provide a direct solution to the challenge Locke faces in explaining object perception in empiricist terms.

While I cannot give a full accounting of Condillac's argument here, it is useful to recognize just how the fact that our sensations are pleasant and painful figures into the account. As the statue is given one sense modality at a time, she finds her awareness completely consisting in the sensory quality of that mode she is experiencing. She first identifies herself with the smell she is experiencing, of a rose; there is nothing in her awareness but that smell. Then she identifies herself similarly with the color she experiences, the red of the rose; again she is not aware of anything but that color. Insofar as the sensory qualities each constitute her awareness, she has no consciousness of anything else. While in these cases, Condillac is clear that the experience of rose smell and of rose colour are pleasant, it is not clear what function this pleasure is to play in our cognitive life, other than to induce us to want to continue in having that pleasant experience.

It is not until that the statue acquires a sense of touch that she is able to form an idea of an object distinct from a perceiving subject. At its most fundamental level, touch, for Condillac, seems to be equivalent to proprioception, and it is this fundamental feeling, as Condillac calls it, through which we get a sense of ourselves.[22] Equally, we are only able to learn that there is something existing outside of us through the sense of touch. Pleasure and pain are essential to the statue's coming to understand that there are other things existing distinct from it. For with the sense of touch, understood first as a fundamental feeling of one's body, and so of oneself, pain and pleasure are what lead us to move our limbs:

In accord with its organization, therefore, pleasure and pain, or the passage from one to the other, cause it to have movements; it cannot then but happen that among these movements

[22] TS II.1.3; Condillac 1754/1984, 158.

> some will cut off or suspend a sensation which is hurtful, and others will procure it a sensation which is pleasurable. (*TS* II.5.2; Condillac 1754/1984, 101–2)[23]

Through these motions, motions prompted by feelings of pain and pleasure, the statue comes to discover that it has a body distinct from other bodies. In moving its limbs along itself and around in the world, it comes to distinguish when it feels itself, and when it does not:

> Placing its hands on itself it will discover that it has a body, but only when it has distinguished the different parts of it and recognized in each the same sentient being. It will discover there are other bodies when it touches things in which it does not find itself. (TS II.5.2; Condillac 1754/1984, 102)

Through touch, then, we begin to form ideas of objects. We are able to locate ourselves in a world populated by other things existing apart from us.

With an idea of an object existing distinctly in place, Condillac's next task is to explain how we are able to collect together our other simple ideas, ideas of color, smell, taste and so on, and attribute them to the objects our sense of touch has identified. Doing this, for Condillac, once again involves motion.

> Placing its hands by chance upon the objects it meets, it grasps a flower and holds it in its hand. Its arm moving aimlessly, is brought now towards its face, now away from it; it is conscious with more or less vividness of a particular mode of being. Surprised, it now repeats the experiment designedly. It lifts the flower and lets it fall several times, and it comes to think that it exists or ceases to exist in a certain manner according as the flower is near or distant. Finally, it begins to suspect that it owes to the flower the feeling which is a modification of its own being. (TS III.1.4; Condillac 1754/1984, 158)

We combine smell with touch, for instance, by first having two discrete sensations, and then noticing that there is correlation between the changes in the one (smell) while we move the other. Touch provides us with a fixed point through which we can measure changes in the other sensations we have, and in particular changes in pleasures and pains. In correlating those changes with our own motions, we come to attribute various sensory qualities to the objects defined through touch.

Condillac, thus, would seem to resolve the equivocations from Locke in a way similar to Berkeley. For both, pleasure and pain are an intrinsic aspect of all our sensations, and in particular those of touch. For both, pleasure and pain contain information about benefits and harms. Condillac ingeniously argues that with this conception of immersed experience, an empiricist has all the resources that are needed to resolve a problem facing Locke – the problem of explaining our ability to perceive objects.[24]

---

[23] See also TS II.8.1: "Without pleasure our statue would never have the wish to move: without pain it would move with confidence and infallibly perish. It must then always be exposed to pleasant and unpleasant sensations" (Condillac 1754/1984, 119).

[24] Condillac's account of how we come to perceive objects should be of particular relevance to cognitive scientists interested in the Binding Problem.

6 Locke's Mixed Mind: Possible Explanations

It is particularly interesting that two empiricists as committed as Berkeley and Condillac adopt the model of immersed experience, taking our sensations to be essentially affective, even though the model does seem to find its roots in the early modern period with canonical rationalists like Descartes and Spinoza. Thus, while it might perhaps be tempting to explain Locke's equivocations by claiming that immersed experience is somehow at odds with empiricist principles, this cannot be correct. So then why did Locke himself not resolve the equivocations in his account? Of course, one might well claim that he prioritized other issues, but for the sake of argument let me conclude by offering two other less pragmatic and more philosophical answers.

First, consider the status of animals in the early modern period. It was certainly highly contentious whether animals were capable of either pleasure or pain,[25] and if animals were taken to be incapable of these basic affective responses, this would certainly preclude their having immersed experience as such. But surely no one would want to deny that animals share a world with us, a world populated by objects with properties. If animals are incapable of feeling pleasure and pain, then their sensory perception must not involve this affective dimension. Are we then to understand humans and animals as experiencing the world in fundamentally different ways? A principle of simplicity would suggest that human and animal experience are not different in kind, but rather at different points on a spectrum. So long as animals are denied feelings of pain and pleasure, it would make sense then to adopt an instrumental conception of experience. We do share sense organs with animals, and so it does make sense that we would be able to take in similar bits of information. But equally, if we admit animals might feel pleasure and pain, then the possibility of a model of immersed experience opens. Locke's mixed mind about human experience thus parallels mixed views about animal experience.

Second, it is worth noting that the conception of human embodiment was intimately intertwined with theological issues, and in particular that of the immortality of the soul. The controversies around Locke's accounts of substance and personal identity – in particular his exchange with Stillingfleet – demonstrate just how vexed a topic the immortality of the soul was, and Locke does take care in to affirm his belief in the Resurrection and the immortality of the soul, distinct from its immateriality in the *Essay*.[26] In many ways, a model of immersed experienced is more threatening in this regard to Locke than it might be to someone like Descartes who was clearly committed to the view that the mind, or soul, was able to subsist without the body. For Descartes, whatever thoughts the soul might be able to have

---

[25] Descartes, for instance, initially denies animals are capable of pain and pleasure, just insofar as they are incapable of thought (see 3:85; 3:148 and 4:573; 3:302ff). Later, in correspondence with More, he backpedals a bit, simply suggesting that we cannot know whether animals experience pleasures and pains (5:276f; 3:365f).

[26] See for instance ECHU II.27.15.

in virtue of being united with a body, and so immersed in the world, the soul is still capable of independent existence and thereby persists after death; and so Descartes preserves the Christian doctrine of the immortality of the soul. Locke's skepticism about pure substance in general, his lack of commitment to dualism, and his denial of innate ideas put pressure on the model of immersed experience in the face of concerns about the soul's immortality. If all our ideas come from experience, and that experience is essentially situated in the world, what can possibly be retained once our body dies? Without our bodies, we would presumably lose the capacity to retrieve those ideas whose content depended on our having a body at all. And without its thoughts, what could the soul possibly be? Despite the advantages of a model of immersed experience, these kinds of considerations might well have moved Locke towards a more instrumental conception of our experience. If the content of our ideas does not depend on the body in which we find ourselves, then the mind can at least in principle retain those thoughts after the body dies, and so preserve some semblance of the doctrine of the immortality of the soul. Easing of various political pressures around this issue might well have allowed later empiricists the liberty to fully adopt the model of immersed experience. Certainly, these sorts of considerations need not constrain contemporary cognitive scientists in their efforts to address the question of object perception.

References

Anstey, Peter R., and Stephen A. Harris. 2006. Locke and Botany. *Studies in History and Philosophy of Biological and Biomedical Sciences* 37(2) (June): 151–171.

Aquinas, Thomas. 2002. *The Treatise of Human Nature: Summa Theologiae 1a75–89*. Ed. and trans. Robert Pasnau. Indianapolis: Hackett.

Berkeley, George. 1713/1979. *Three Dialogues Between Hylas and Philonous*. Ed. R.M. Adams. Indianapolis: Hackett. Cited internally as *Dialogues,* followed by edition page number.

Berkeley, George. 1784/1910. *A New Theory of Vision and Other Writings*. Ed. Lindsay, A.D. London: Everyman's Library. Cited internally as NTV followed by paragraph number.

Bermùdez, J.L. 2007. The Object Properties Model of Object Perception: between the Binding Model and the Theoretical Model. *Journal of Consciousness Studies* 14(9–10): 43–65.

Condillac, Étienne Bonnot de. 1754/1984. *Traité des Sensations, Traité des Animaux*. Paris: Fayard. Cited internally by TS, followed by Part, Chapter, Paragraph.

Condillac, Étienne Bonnot de. 1903. *Treatise on Sensations*. Trans. Geraldine Carr. Los Angeles: USC Press.

Descartes, René. 1985–1991. *The Philosophical Writings of Descartes, Vol I and II*. Eds. and trans. Cottingham, Stoothof and Murdoch, and *Vol III*. Eds. and trans. Cottingham, Stoothof, Murdoch and Kenny. Cambridge: Cambridge University Press.

Descartes, René. 1996/1908. *Œuvres de Descartes*. Eds. Charles Adam and Paul Tannery. Paris: Vrin.

Downing, Lisa. 1997. Locke's Newtonianism and Lockean Newtonianism. *Perspectives on Science* 5(3) (Fall): 285–310.

Downing, Lisa. 1998. The Status of Mechanism in Locke's *Essay*. *Philosophical Review* 107(3) (July): 381–414, www.csa.com (accessed July 22, 2009).

James, William. 1950/1918. *Principles of Psychology*. New York: Dover.

Jones, Jan-erik. 2007. Locke vs Boyle: The real essence of corpuscular species. *British Journal for the History of Philosophy* 15(4) (November): 659–684, www.csa.com (accessed July 22, 2009).

Keating, Laura. 2002. Linking mechanism and subjectivity in Locke's *Essay*. *Locke Studies: An Annual Journal of Locke Research* 2: 53–79.

Locke, John. 1975. *Essay Concerning Human Understanding*. Ed. P.H. Nidditch. Oxford: Clarendon. Cited internally as ECHU, followed by Book, Chapter, Paragraph.

Meynell, Guy. 2003. Locke's Corpuscularianism and Boyle's Corpuscular Philosophy. *Locke Studies: An Annual Journal of Locke Research* 3: 133–145.

Milton, J. R. 2001. Locke, Medicine and the Mechanical Philosophy. *British Journal for the History of Philosophy* 9(2) (June): 221–243.

O'Regan, J. K., Noë, A. 2001. A Sensorimotor Account of Vision and Visual Consciousness. *Behavioral and Brain Sciences* 24: 939–1031.

Plate, J. 2007. An Analysis of the Binding Problem. *Philosophical Psychology* 20(6): 773–792.

Pylyshyn, Z. W. 2007. *Things and Places: How the Mind Connects with the World*. Cambridge, MA: MIT Press.

Sanchez-Gonzalez, Miguel A. 1990. Medicine in John Locke's Philosophy. *The Journal of Medicine and Philosophy* (December): 675–695.

Simmons, Alison. 1999. Are Cartesian Sensations Representational? *Noûs* 33(3): 347–369.

Spinoza, Baruch. 1985. *The Collected Works of Spinoza*, Vol.1. Ed. and trans. E. Curley. Princeton: Princeton University Press.

Stein, Howard. 1990. On Locke, 'the Great Huygenius, and the incomparable Mr. Newton'. *Philosophical Perspectives on Newtonian Science*. Eds. Philip Bricker and RIG Hughes. Cambridge, MA: MIT Press.

Walmsley, Jonathan. 2008. Sydenham and the Development of Locke's Natural Philosophy. *British Journal for the History of Philosophy* 16(1) (February): 65–83.

Wilson, Catherine. 1995. *The Invisible World: Early Modern Philosophy and the Invention of the Microscope*. Princeton: Princeton University Press.

Empiricism and Its Roots in the Ancient Medical Tradition

Anik Waldow

Abstract Kant introduces empiricism as a deficient position that is unsuitable for the generation of scientific knowledge. The reason for this is that, according to him, empiricism fails to connect with the world by remaining trapped within the realm of appearances. If we follow Galen's account of the debate ensuing among Hellenistic doctors in the third century B.C., empiricism presents itself in an entirely different light. It emerges as a position that criticises medical practitioners who stray away from the here and now by indulging in theory-driven a priori forms of reasoning. In so doing empiricism remains at all times committed to the world and its agents. In this paper Galen's account of empiricism will serve me as a means to unravel the dynamics of a discussion that aims to reassess the standards of a dogmatic scientific practice. By looking at Bacon's and Gassendi's perception of the ancient medical tradition I will furthermore show that the understanding of what empiricism is crucially depends on the understanding of what scepticism is.

1 Introduction

In the *Critique of Pure Reason* Kant presents empiricism as a form of idealism. In the *Fourth Paralogism* he writes:

> The term '*idealist*' is not ... to be understood as applying to those who deny the existence of external objects of the senses, but only to those, who do not admit that their existence is known through immediate perception, and who therefore conclude that we can never, by way of any possible experience, be completely certain as to their reality.[1]

Kant thereby turns empiricism into a sceptical position that is sceptical for its admission of the claim "that, the existence of which can only be inferred as a cause of given perceptions, has a merely doubtful existence."[2] These considerations are

[1] Kant 1970, 345/A 369.

[2] Kant 1970, 344/A 367.

A. Waldow
Department of Philosophy, University of Sydney

C.T. Wolfe and O. Gal (eds.), *The Body as Object and Instrument of Knowledge: Embodied Empiricism in Early Modern Science*, Studies in History and Philosophy of Science 25,
DOI 10.1007/978-90-481-3686-5_14, © Springer Science+Business Media B.V. 2010

part and parcel of his attack on the scepticism that Kant regards as a consequence of Berkeley's and Hume's attempt to deal with Cartesian dualism by accepting that the *representation* of a physical object is genuinely different from the object *itself*, the so-called Ding an sich. Kant's announced strategy consists in tracing the limits of reason for the purpose of showing that knowledge is possible, *although* objects independent of our experience indeed lie beyond our grasp.[3]

In this paper I wish to show that the Kantian conception of empiricism is deeply mistaken. This conception suggest that thinkers before Kant were primarily concerned with the problem that real knowledge can be obtained only if precisely those causes are known that are taken to be unknowable. Kant thereby creates the wrong impression that early modern empiricism is a purely epistemological position that can but fail in its attempt to provide a genuine understanding of the world on the basis of appearances. I will oppose this Kantian conception of empiricism by examining the writings of Galen (129–216 A.D.) that provide us with an account of the debate between ancient empiricist and rationalist doctors, a debate that ensued among Hellenistic doctors in the third century B.C. and that was known to many early modern writers. I will argue that Galen's writings present ancient empiricism as an essentially pragmatic position which is marked by its hostility towards any sort of speculative science. Ancient empiricism thereby challenges the very metaphysical model that Kant attributes to empiricism in general, that is, the model that assumes the existence of unknown hidden causes and that renders empiricism a hopeless epistemological position. To correct the Kantian picture of empiricism is crucial as it may help us to understand that those whom we today identify as the great classic empiricists, that is, early modern philosophers who promoted experimental methods, did not understand themselves as the losers of a game that reaches for hidden causes, but as the avant-garde of a new generation of (natural) philosophers who most enthusiastically assumed that they can discover things that render the world of our everyday lives more comprehensible.

I will begin by giving a brief outline of the particularities of the Kantian account of the relation between empiricism, science and epistemology. This will help us to understand why Kant casts empiricism in terms of a purely epistemological theory. I will then turn to Galen's account of ancient empiricism in order to reveal the empiricist's commitment to the world of the here and now and his rejection of metaphysical hidden causes. I will substantiate this claim by examining empiricism with respect to its fierce opposition to theory-laden forms of reasoning and thereby create a link to the ancient sceptical tradition of Pyrrhonism. The remainder of the essay will focus on the perception of ancient empiricism and scepticism by thinkers such as Bacon, Gassendi and Montaigne. I will examine the reasons why empiricists became associated with that sort of scepticism that detaches appearances from the real things. I will conclude by suggesting that Galen's account of empiricism is able to provide a cure for the Kantian tendency to epistemologize empiricism.

[3] "Thus the critique of reason, in the end, necessarily leads to scientific knowledge; while its dogmatic employment, on the other hand, lands us in dogmatic assertions to which other assertions, equally specious, can always be opposed – that is in *scepticism*" (Kant 1970, 57/B 23).

2 The Kantian Turn

Kant has often been presented as the Copernicus of philosophy, that is, as the one who revolutionized the entire discipline. Apart from his followers, Kant himself even alludes to Copernicus in order to clarify his approach:

> We should then be proceeding precisely on the lines of Copernicus' primary hypothesis. Failing of satisfactory progress in explaining the movements of the heavenly bodies on the supposition that they all revolve around the spectator, he tried whether he might not have better success if he made the spectator revolve and the stars to remain at rest.[4]

According to Kant, Copernicus deserves merit because he understood that once a hypothesis fails to explain the phenomena, it needs to be replaced. The Kantian Kuno Fischer refers to Kant's admiration for Copernicus in slightly different terms; he adds a twist which reinforces the picture that Kant *overcame* the deficiencies of empiricism as much as Copernicus *overcame* the "first perspective that derives from sense perception."[5] The Copernican move is thus presented as a move away from the sensory level of experience toward a transcendental perspective. This apparently ignores Kant's own description of Copernicus' extraordinary achievement. For Kant, as pointed out above, it is one's readiness to adjust one's hypothesis if it fails to explain the observable data that deserves merit; it is not Copernicus' insight into the deficiencies of a perspective based on sense perception. But although Fischer transforms Kant's relation to Copernicus, it needs to be noted that he keeps in line with Kant's general opinion that empiricism is a mistaken epistemological approach. Fischer does so by suggesting that true insight is possible only if one abandons the perspective inherent in sense perception, because sense perception is unable to transcend the realm of the appearances.

Fischer continues his interpretation of pre-Kantian philosophy with an analysis of the situation of philosophy in the wake of the scientific revolution. According to him, early modern philosophy was faced with a dilemma: on the one hand, the success of the experimental sciences pressured it to acknowledge the importance of experience and sense perception; on the other hand, philosophy needed to show that experience is insufficient for the acquisition of knowledge, unless it wanted to render itself entirely otiose. Fischer's analysis thus proposes that philosophising is no more than a self-justificatory exercise: philosophers are those who need to show that analytic forms of reasoning are to rule any scientific enterprise. He writes: "If the empire of philosophy were not to die, it needed to regain a position that the empirical sciences respected and could not challenge."[6] One of the consequences of this analysis is that empiricism once more appears to be altogether untenable as a philosophical position, not merely because it fails to perform the Copernican move

---

[4] Kant 1970, 22/B xvii.

[5] "Gesichtspunkt der ersten, sinnlich nächsten Betrachtung" (Fischer 1909, 8).

[6] "Wollte ihr Reich [das Reich der Philosophie] nicht zugrunde gehen, musste sie sich eine neue, stete, von Seiten der Erfahrungswissenschaften anerkannte und unbestreitbare Stellung erobern" (Fischer 1909, 4).

and, therefore, remains trapped within the sphere of the senses. What's more, according to Fischer, empiricism lacks respect for analytic reasoning, although it is precisely this form of reasoning which puts philosophy in a position to defend its status as a scientific discipline.

As pointed out before, Kant himself presents empiricism as a form of half-hearted idealism that understands "all outer appearances" as being of "such a nature that their existence is not immediately perceived" but only inferred "as the cause of given perceptions."[7] In characterising empiricism along these lines, he ascribes to it a form of Cartesian dualism. He implicitly takes it for granted that the empiricist is aware of the existence of causes different from those things that appear before his mind. Thus Kant tells us that the dualist is ready to "admit the existence of matter without going outside his mere self-consciousness,"[8] while the empiricist refrains from such judgement, because for him the causes of appearances have "a merely doubtful existence."[9] Kant thereby suggests that the dualist *and* the empiricist agree that there is a reality that duplicates the reality of appearances and that they only disagree on whether or not one is able to gain knowledge about it. One consequence of this interpretation is that the empiricist turns into a sceptic, who "after wrongly supposing that objects of the senses, if they are to be external, must have an existence by themselves, and independently of the senses … finds, from this point of view, all our sensuous representations are inadequate to establish their reality."[10] Hence, for Kant himself, empiricism fails to offer a scientific approach, not so much because it opposes a priori forms of reasoning, as Fischer claims; it fails to do so because it is unable to create insight into the true being of things that the empiricist allegedly locates in the hidden beyond.

Note that, according to Fischer and Kant, the questions of what science is, how philosophy can be pursued scientifically and how empiricism fits into all this relate to the question of whether philosophy is able to achieve knowledge. The rejection of empiricism as a philosophical position, and as a scientific approach, thus relates to its failure to provide a framework that enables the philosopher to achieve true insight into the being of things, while true insight is defined in terms of knowledge of underlying causes. Kant himself proposes that knowledge is possible only if mind and world are not seen as separate from each other but as united by principles required for the possibility of experience. My aim in this paper is not to go into details about Kant's approach and its purported ability to solve the outlined epistemological problem the empiricist encounters, for I am concerned with his conception of what empiricism is. At this point, I only wish to draw attention to the fact that on Kant's account, he naturally turns out to be the genius who solved one of the major problems, if not *the* problem of philosophy that has troubled thinkers since antiquity: that is the problem of how we can achieve knowledge.

---

[7] Kant 1970, 344/A 367.

[8] Kant 1970, 346/A 370.

[9] Kant 1970, 429/A 474/B502.

[10] Kant 1970, 346/A 370.

As we will now see, Galen's discussion of ancient empiricism has a very similar strategic aim. It establishes Galen as the genius who reunites the advantages of the medical schools of rationalism and empiricism. His account of the disputes between the two schools is interesting not so much because it shows that the talk about what empiricism is needs to be received with care and evaluated against the relevant historical background. The comparison between Kant and Galen is primarily interesting because Galen's account reveals a dimension of empiricism that Kant entirely ignores. Galen presents empiricism as a position that challenges a hopelessly dogmatic and theoretical scientific practice. He thereby relates empiricism to a context that negotiated the aims and goals of science, without prioritising the question of whether and how knowledge is possible.

3 Empiricism à la Galen

In the Essay *On the Sects for Beginners* Galen introduces the concept of empiricism by its opposition to rationalism:

> Some say that experience alone suffices for the art [of medicine], whereas others think that reason, too, has an important contribution to make. Those who rely on experience [*empeiria*] alone are accordingly called empiricists. Similarly, those who rely on reason are called rationalists. And these are the two primary sects in medicine. The one proceeds by means of experience to the discovery of medicine, the other by means of indication. And thus they have named their sects empiricist and rationalist.[11]

In using the terms 'empiricism' and 'rationalism' Galen refers to the names that the medical sects gave themselves.[12] The standard objection that Kantians introduced the terms in order to refer to early modern philosophers who did not understand themselves in this way therefore does not hold.[13] And although it must be taken into account that Galen, similarly to Kant, provides us with a simplified story of empiricism that lets his own approach appear in a certain light – Nutton for instance emphasises that Hellenistic medicine was far more diverse than Galen tells us[14] – it is worthwhile going into details about some of the characteristics Galen attributes to the school of empiricism, because Galen's discussion notably influenced the medieval and early modern perception of ancient medicine and thereby the pre-Kantian perception of what empiricism is.

Empiricists, as Galen puts it, aim for the knowledge of the right medicine by studying the symptoms; they make use of their memory in order to recall the different occasions and circumstances that obtained when a certain treatment succeeded or failed. In so doing the empiricist opts for inductive forms of reasoning:

[11] Galen 1985, 3.

[12] Galen 1985, 26.

[13] For this kind of criticism see Woolhouse 1985 and Loeb 1981.

[14] Nutton 2005.

he generalizes upon his previous experiences and thereby commits himself to the so-called epilogism, a form of reasoning that is defined "solely in terms of what is apparent" and can be used for the "discovery of things which are not manifest *temporarily*."[15] The means by which the empiricist can gain knowledge here turn out to be rather limited. If he remains true to his own method, he cannot infer to something never seen or experienced before. Galen therefore rightly complains that the empiricist's treatment is based on the trial and error of cures. Why the empiricist limits himself to this procedure – which is particularly worrisome for the one who serves as his guinea pig – becomes clear if we look at the various respects in which epilogism is considered a useful weapon:

> The epilogism also is useful, if one wants to refute those who dare to argue against what is manifest. It is also of use to point out that some phenomenon has been overlooked and to counter sophistical arguments. In reasoning this way, one never departs from what is clear but throughout stays within its limits.[16]

It here seems that the empiricist understands the limits of his method as a virtue that he purposefully chooses in order to counterbalance the rationalist's speculations and sophistical arguments.

One could perhaps be tempted to interpret the ancient empiricist's commitment to the limits of reason in Kantian terms. That is, one could argue that the ancient empiricists were mainly interested in the epistemological question of how one can achieve knowledge. And one could point out that their answer is that one has to respect the limits of reason by restricting one's reasoning to the systematisation of the previously experienced. Galen's account of ancient empiricism would thus nicely tie in with the Kantian story that traces the battle between empiricists and rationalists back to Plato and Aristotle.[17] What this interpretation ignores, however, is that the context of the debate between ancient empiricists and rationalists was entirely different from the context to which Kant refers when developing his own critical philosophy as the long awaited solution to the problem of how knowledge is possible. Galen was a doctor and practitioner; and his discussion of the leading medical traditions pursues the aim of assessing the most promising way of treating diseases successfully. His discussion thus has a genuinely practical focus. In other words, the question is not that of how one can know, merely for the aim of knowing the truth, but that of how one can find a remedy that is efficient enough to cure the ill patient. Galen states in the opening line of *On the Sects for the Beginners*: "The aim of the art of medicine is health, but its end is the possession of health."[18]

Of course, practicing medicine involves the knowledge "of what is healthy and what is not healthy"[19]; and it is in this respect that empiricists object to rationalists

[15] Galen 1985, 9 (my emphasis).

[16] Ibid.

[17] Kant 1970, 667/A 854/B 882.

[18] Galen 1985, 3.

[19] Ibid.

for relying on forms of reasoning unwarranted by experience, the so-called *analogism*. Galen writes: "The analogism starts from what is apparent but then proceeds to matters which are entirely unclear and that is why it takes so many forms."[20] The rationalist would thus turn observable phenomena into signs, namely, into indicative signs – a class of signs the ancient empiricist vehemently rejects[21] – because he takes phenomena to indicate things which are not accessible in themselves. In so doing the rationalist would not only rely on the unproven assumption that the observable is analogous to the unobservable, but also make use of unwarranted theories that inform the practitioner about the precise way in which the two relate to each other.[22] The problem with this approach would be, Galen continues his presentation of the empiricist's attack on the rationalist, that the phenomena do not allow the rationalist to rule out that it is one cause rather than another that produces the observable effect:

> For, starting from the same phenomena, it arrives now at one and now at another unobvious conclusion. And, at this point, they [the empiricists] bring up the problem of the discord which cannot be settled and which they claim is a sign of the incomprehension of things. This is the language they use: 'comprehension' for true and certain knowledge, 'incomprehension' for the opposite of the first. And they say that the incomprehension is the cause for the discord which cannot be settled and that the discord, in turn, is a sign of the incomprehension. They point out that it is the disagreement concerning matters which are not manifest which cannot be settled, not the disagreement concerning matters which are manifest. For, in their case, everything once it is apparent what it is like, confirms those who are right about it.[23]

Galen's empiricist here mocks the rationalist for aspiring to "comprehension" if all he can achieve is "incomprehension" and "discord." His more to general point seems to be that a medical practice that is built on as insecure a foundation as the knowledge of hidden causes is entirely unsatisfactory in its tendency to create nothing but contradiction and speculation.[24] Jonathan Barnes characterises this problem as arising from the underdetermination of theory: "The symptoms of fever can be explained in many ways; yet the different doctors stick each to his own explanation. In general there are many ways of explaining the fact that *q*; but most aetiologists cite just one of those ways."[25] Barnes furthermore emphasises that this attack

---

[20] Galen 1985, 9f.

[21] The empiricist, as much as the Pyrrhonist, only allows for commemorative or recollective signs which are used to draw inferences on previously observed causal chains. See Barnes 1983, 158f.

[22] Lolordo cites an Epicurean example to illustrate the nature of analogism: "The void must exist, because motion could not exist unless the void existed. Thus, motion is a sign of the void. How do we come to know that motion is impossible without an absolute void? Epicurus' answer is that we know this by analogy with evident things. In ordinary cases, a (relatively) solid body cannot move somewhere unless that place is (relatively) empty" (Lolordo 2007, 96).

[23] Galen 1985, 9f.

[24] Barnes remarks that the same point has been made by Cicero, namely, when he presented the sceptic Carneades as someone who turned useless forms of logic, the so-called sorites, against the Stoic Chrysippus who, according to Carneades, explained nothing. See Barnes 1982, 45.

[25] Barnes 1983, 167.

on the search for hidden causes (*aiteologia*) lay at the heart of the empiricist's concern. This is certainly true, for, as we have seen, Galen's empiricist opposes the idea that we need to dig deep into the metaphysical realm in order to find successful cures and he recommends to base treatment on the experiences with the patients.[26] The empiricist thereby seems to take it that phenomena have an explanatory value in themselves and do not merely function as signs for a truly explanatory cause, because for him the phenomena provide sufficient information for determining the right cure.

The empiricist's commitment to a medical practice that results in an effective cure rather than unfounded speculation and contradiction seems to find Galen's approval. This becomes clear three chapters later when Galen – after having given a short outline of a third group of practitioners beside those of the rationalists and empiricists, namely the group of the so-called methodists – sets out to specify the term 'apparent'. According to him, this specification is necessary because the methodists would misinterpret 'apparent' in order to deny the need for theory and learning: "If you do say", Galen objects to the methodist in the voice of the empiricist, "as I also have heard from you in the beginning, that all that is not manifest is useless and if you agree to follow what is obvious, then, perhaps, I can point out to you what it is that you are overlooking, reminding you of what is apparent."[27] To illustrate his point Galen considers the case of two men who were bitten by a mad dog. Both men received medical treatment, the first by someone who considers solely what is apparent, in this case the open wound, the other by an empiricist doctor who also pays attention to background information, in this case the fact that the dog is mad. While the first man dies, the second survives. Galen finishes this story by asking: "Do you think that, in such cases, one inquires in vain into the antecedent cause and that the man died for any other reason than the negligence of the doctor, who failed to ask at all about the cause and to apply the treatment observed in this case?"[28] Note that the way Galen describes empiricism in this passage presents it as the recommended position, because the empirical doctor allows for the consideration of circumstances that previous experience has shown to be crucial to the success of the treatment (further circumstances that Galen names are for instance seasons, the location and age of the patient). Galen's comparison between the empiricist and the methodist thus gives the characteristics of a desirable medical practice in terms of its success: it underlines that in order to secure the success of a certain treatment, it is necessary to consider not only the momentary apparent circumstances, but also the so-called antecedent causes, that is, causes that were once apparent. What does not seem to matter is which sort of reality – obvious or hidden causes – the practitioner is to know in order to achieve this success.

---

[26] Experience comprises sources other than one's own experience: history, reports and testimony of other practitioners can be consulted in order to identify the needed remedy. See Galen 1985, 27.

[27] Galen 1985, 13.

[28] Galen 1985, 14.

The comparison of empiricism with the school of the methodists shows how Galen gradually introduces new dimensions of empiricism that turn it into a more and more refined position.[29] In the same context Galen points out – again in the voice of the empiricist who keeps arguing against the methodist – that anatomy is essential to a successful medical practice: "That you do not even consider the parts of the body seems to me to be a rather strange thing for you to say and indeed quite absurd."[30] Galen does not mention at this point what he explains on the opening page of *On the Sects for Beginners* when opposing empiricism to rationalism, that is, that the empiricist rejects the use of theory for its tendency to lead the practitioner away from the realm of the experienceable to an underlying reality. Of course, if anatomy is understood as a theory that emerges as a result of the experimental study of corpses and living bodies, there is no conflict between Galen's earlier and later account of empiricism, because the theory in question would be derived from the study of the observable. Unfortunately, Galen does not specify in which respect anatomy is to play a part within the empiricist's practice. This suggests that he plainly understands it in the traditional sense as *phusiologia*, that is, as a form of theory that reaches for principally unexperiencable causes.[31]

Why does Galen present the empiricist as someone who is open to theory? It is clear that Galen himself held rather traditionally that science necessarily involves theory: for him the best doctor is a philosopher who is trained in logic, physical theory and ethics. In defence of Hippocrates and against the methodist he claims:

> They scold the man who has said that life is short and the art long. Quite the contrary, they say that the art is short and life is long. For, if one does away with all those things which have been wrongly taken to further the art and if we attend only to the communities, then medicine is neither long, nor difficult, but rather is easy and clear and can be learned as a whole in a matter of six months. (Sects, 11)

Galen's point in this passage is that the methodists would deprive the 'art of medicine' of its status as a science. They would do so by introducing the simplistic model of communities or generalities, which describe diseases in terms of manifest states of constriction, dilation, or a combination of both.[32] In raising this point Galen clearly reveals the traditional view that "a true … science has to be based on truly general knowledge, which only reason not experience can provide us with"[33]; and he subscribes to the orthodox claim that experience does not give us any explanations but only facts.[34] Despite his conservatism, it needs to be noted,

[29] Frede points out that the methodists took an equally mediating stance in the debate between empiricists and rationalist as Galen, an aspect that Galen would purposefully conceal. See Frede 1982.

[30] Galen 1985, 14.

[31] Barnes notes that Stoics and Aristotelians agreed with their Epicurean rivals that "*phusiologia* is 'to give a precise account of the causes of the most important things' (Ad. Hdt.78)" (Barnes 1983, 150).

[32] Galen 1985, 11.

[33] Frede's introduction in Galen 1985, xxiv.

[34] See Dear 2006.

however, that Galen often sympathises with the empiricist, for instance, when claiming that experience is the most reliable criterion[35] for truth and that empirically unwarranted theories need to be abandoned.[36]

If we combine these characteristics of Galen's own position with his interpretation of empiricism as a position that assimilates theory in form of *phusiologia*, it seems that Galen uses his discussion of the school of empiricism for a specific purpose. By discussing the methodists in opposition to the empiricists, he emphasises that theory is needed if medicine is to qualify as science. Thus, Galen is convinced that some persons can acquire the "art of medicine" by experience alone,[37] although he insinuates that this art is limited in its scope and fails to qualify as a true science.[38] Conversely, by discussing the empiricist's criticism of the rationalist Galen draws attention to the problem of overemphasising theory and alerts his readers to the problem that theory easily leads to useless speculations about entities about which nothing definite can be said. He thereby reveals the need for new scientific methods, namely, methods that are more reliable and useful and less speculative which can be achieved through a robust commitment to experience. Galen's purpose of presenting us with the methods and beliefs of the empiricist school of medicine in opposition to other schools thus turns out to consist in arguing for an experience-sensitive medical science that relies on theories that enable the practitioner to treat his patient successfully.[39]

It might here strike us as a parallel between Kant and Galen that both discuss empiricism in order to show which standards a truly scientific discipline needs to respect. Kant wants to lead philosophy on the highway of science; and his disciple Fischer points out that the scientific status of philosophy can be preserved only if it succeeds in showing the need for analytic forms of reasoning. Galen surely takes a different stance. For him it is clear that philosophy is science, because science just is logical reasoning. Hence, for him the question is not so much that of how philosophy can preserve its status as a science, but that of how a medical practice that pays due respect to experience can qualify as scientific. A further difference between Kant's and Galen's discussion of empiricism is that the latter seems to use this discussion to advocate some sort of practically-oriented conception of science that opposes speculation and contradiction and thereby ensures the success of the treatment.

---

[35] Galen *De simpl. med.* I, 40, Kühn, XI 456.

[36] Galen *De fac. Nat.* I, 13, 501 III, 132; II, 8, 500 III, 186.

[37] Galen 1985, 44–45.

[38] Galen, *De methodo medendi* V, 1, X, 306; XIV, 5, Kühn X, 962, in Galen 1821–33/1965. See Tieleman 2009, 53.

[39] This characterisation applies to Galen's own position rather than Galenism as it had developed as an intellectual system by 600 A.D. Nutton describes the different between the two positions as follows: "His [Galen's] empiricism, his observational genius and his willingness to think on his feet found little place in Galenism, for its central texts were those that emphasized his conclusions rather than the means by which he had researched them. Anatomical dissection for the purpose of investigation, so much stressed by Galen, seems to have vanished almost entirely, although both the Byzantines and the Arabs were extremely proficient in surgery" (Nutton 2008, 363).

If so, Galen's discussion of the various medical schools is to be read as an attempt to transform established science by bringing into play considerations about the practical value of scientific enquiry. Kant and Fischer certainly do not share this conception of science: their enterprise is driven by the quest for knowledge, while practical considerations play no role. But this is not all that distinguishes Galen's conception of empiricism from the Kantian conception: as Galen's discussion of the various medical traditions has shown, ancient empiricism is a position that attacks the model of a twofold reality and abandons the idea that unreachable causes need to be known in order to provide an appropriate understanding of the world. Galen's discussion is thereby able to challenge the Kantian concept of empiricism as a form of Cartesian dualism. It can do so because it shows that the ancient empiricist does not construe reality as something which is detached from the world of the appearances. He takes reality to be the reality of the experienceable world and opposes all attempts to search for a deeper level.

4 Empiricism cum Scepticism

I will now try to answer the question of why empiricism is usually associated with the twofold metaphysical model that construes reality as something which lies hidden beyond the appearances and that we cannot access. One obvious answer to this question is that empiricism puts forward a weak concept of knowledge, that is, a concept that is weak, one could think, precisely because it is based on experiences, rather than the things that cause our experiences. The conclusion that seems to follow from this is that empiricism can merely offer suggestions about what is most likely to happen, but not about what needs to take place given that certain causes are inevitably at work. Although it is true that empiricism is a position that admits that it cannot predict the course of nature infallibly, it is not true that empiricism needs to be committed to the view that there is a realm of reality that, if it were accessible, would allow us to make such infallible claims. To render this point more evident let us take a closer look at the relation between ancient empiricism and scepticism (I will focus on Pyrrhonism and leave Academic Scepticism unconsidered).

Galen draws a direct comparison between the empiricist and the sceptic Pyrrho:

> The empiricist will not be a man of many words or of long speeches but will talk little and rarely, just like Pyrrho the Sceptic. Pyrrho had looked for the truth and, not finding it, was in doubt about all things nonevident, but in his daily activities he followed what is evident, whereas in everything else he remained in doubt. The empiricist's attitude towards medical matters is like the sceptic's attitude towards the whole of life. He does not lack in reputation, but he also is not arrogant; he is unassuming and not boastful, just as Timon claimed Pyrrho to have been.[40]

If we follow this characterisation, empiricism needs to be understood as a specialised form of Pyrrhonism: while the latter marks an attitude which is relevant to *all* areas

[40] Galen 1985, 42.

of life, the former describes an attitude that is relevant with respect to a *certain medical practice*. Ancient scepticism and empiricism thus turn out to be similar in spirit: they are both marked by a readiness to overthrow previously formed opinions. Although this understanding of scepticism is Galen's and not Sextus Empiricus',[41] it seems advisable to take the link between empiricism and scepticism seriously. Barnes for instance emphasises that "it was the doctors, not the philosophers, who began the debate about causation and who first turned the weapon of skepticism against *aitiologia*. Some of the arguments which Sextus directs against causation can still be traced back to Herophilus and to Erasistratus, and it is scarcely to be doubted that many more of the arguments had a medical origin."[42]

One point that is remarkable about Galen's interpretation of the nature of Pyrrho's scepticism is that it opposes the standard view that Pyrrhonism is a position that rejects belief *tout court*. According to Galen, both empiricism *and* scepticism admit of beliefs that are based on the evident. Thus he points out that Pyrrho naturally held beliefs while interacting with the world and other people in his ordinary life. Michale Frede has defended Galen's interpretation of Pyrrhonism as a position that involves beliefs by drawing attention to Sextus Empiricus' own words:

> In P.H. I 13 ff, Sextus explains in what sense the skeptic is not dogmatic. What is not in question, at least if we follow Sextus, whether the skeptic has no dogmas, no beliefs at all but whether he has no beliefs of a certain sort. Sextus distinguishes between a wider (koinoteron) and a narrower sense of 'belief'; and only beliefs in a narrower sense count as dogmatic. Hence, there can be no doubt whatsoever that, according to Sextus, a serious Pyrrhonean skeptic can have beliefs.[43]

According to Frede's interpretation of Sextus, the Pyrrhonist merely rejects beliefs resulting from theoretical reasoning, but not beliefs that he comes to hold naturally and which are necessary for the mastery of everyday life. The reason for this is not so much that theoretical reasoning is taken to be suspicious in itself; rather it is because the Pyrrhonist understands that reason, once set into action, is able to undermine any one of its previously formed claims by presenting equipollent arguments on both sides of the issue.

Galen draws a similar distinction when reflecting on the status of medicine. He distinguishes between beliefs resulting from a natural process of reasoning and those that require reflection and theoretical reasoning. Thus he declares that there is a form of reason that "men have by nature"[44] and that it is this form that provides us with insight into necessary truths. Of course, Galen here clearly distances himself from the sceptic, for he claims that natural reason can reveal things that are necessarily true, while the sceptic may accept beliefs for their naturalness but not for their truth.

[41] There is no evidence that Galen knew Sextus' writings. See Annas' and Barnes' introduction of Sextus Empiricus 2007, xii.

[42] Barnes 1983, 153.

[43] Frede 1987, 186.

[44] Galen 1985, 45.

At the same time, however, Galen agrees with the sceptic (and the empiricist) that natural and theory-driven belief-forming processes form two separate categories of belief. If regarded against the background of Galen's characterisation of empiricism as a position that rejects theory-dependent beliefs, that is, beliefs that Galen also criticises if unwarranted by experience, it becomes clear that the issue at stake is not so much the question of what is required in order to have true belief, but that of *how much theory* is needed for a scientific approach to the world.[45] This issue is certainly one of importance, especially for those who challenge established orthodoxies, such as the sceptics, empiricists and, to a certain extent, even Galen. After all, theories are man-made and often enough infested by dogma.

One way of opposing dogma certainly consists in raising doubts about the appropriateness of established positions, another in abstaining from the mode of judgement that establishes dogma. I will now show in which way ancient Pyrrhonists made use of the second strategy. Pyrrhonists are famous for their practice of suspending judgement: that is for the practice of claiming that things appear to them in this or that way, while abstaining from claims to knowledge. Barnes distinguishes four senses of "to appear":

> A proposition is of *type* (A) if it contains a term purporting to refer to something "by nature nonevident"; for example: (1) The tower is composed of atoms – where atoms are those nonevident corpuscles hypothesized by some schools of belief. Propositions of *type* (B) refer to evident objects and describe their evident characteristics; for example: (2) The tower is square. Propositions of *type* (C) again refer to evident objects, but report on how they seem (how they look, feel, etc.); for example: the tower looks round. Finally propositions of *type* (D) make no reference to any object, but merely state how things seem to be; for example: (4) It looks as though there's a round tower.[46]

In short, by saying that something appears in a certain way one can either form claims about things which are non-evident as in case (A), about things which are evident as in case (B) and (C) or about no things at all. Furthermore, it seems that only case (A) can be interpreted as a sense of "to appear" in which phenomena are taken to indicate a hidden reality, because in cases (B) and (C) the things to which we refer are evident, while in case (D) the appearance is not taken to refer to an object at all. Barnes ends this passage by suggesting that the "Sober Skeptic" has a problem only with propositions of type (A), because these alone would refer to principally nonevident objects: "Faced with an example of type (A), the appropriate answer is a sceptical shrug. As with (B), (C) and (D), the Sober Skeptic will make room for them all."[47] More generally speaking, one could say that (sober) sceptics are those who form statements that remain concerned with the observable reality without reaching for non-evident objects. Note that this conclusion sits particularly well with the above-developed account of empiricism as a position that does not admit of a second layer of reality. If it holds that empiricists are specialised

[45] For similar interpretations see Burnyeat 1984, 230f.

[46] Barnes 1983, 159.

[47] Ibid.

sceptics, and if sceptics are permissive towards claims about the evident, while merely rejecting claims about the non-evident, the interpretation of empiricism as a position that concentrates on the world of the experiencable as the *only* world finds confirmation.[48]

The close link between ancient empiricism and scepticism now allows us to redefine our understanding of the empiricist's fallibilism. As we have seen, ancient scepticism and empiricism are describable in terms of an attitude that challenges a worldview based on theory and pure reason. Furthermore, it has been suggested that ancient empiricists as well as ancient sceptics happily accept beliefs that are useful and necessary for the conduct of our lives if these beliefs are based on the evident. Both, ancient sceptics and empiricists thus remain bound to the world that is precisely the world we experience when acting in our everyday lives.[49] If we take this into consideration, it becomes clear that the empiricist's motivation for operating with a fallibilist concept of knowledge does not seem to derive from the disappointing news that the mind is unable to reach the real being of things; it derives from the empiricist's commitment to the reality of our everyday lives that all too often teaches us that we were wrong. What we gain here is a concept of empiricism that is characterisable by a general openness to revision and the replacement of beliefs. Empiricism thus-understood is sceptical, if by sceptical we understand a general unwillingness to formulate incorrigible and dogmatic claims to the truth and falsehood of things; however, it is non-sceptical, if scepticism is taken to result from the inability to reach causes that would need to be known in order to predict things infallibly.

5 Perception of Empiricism and Scepticism

At this point one could ask which developments needed to take place in order to create the impression that empiricism is genuinely sceptical in its adherence to a twofold metaphysical model that both ancient empiricism and scepticism originally ventured to attack. That is, the metaphysical model that treats appearances as things that stand in between mind and world and thereby block access to the true being of things. As we have seen, Kant takes it for granted that empiricism operates with this metaphysical model. He characterises empiricism as a form of idealism or Cartesian

---

[48] Burnyeat 1984, 232 points out that even the distinction between evident and non-evident is the distinction of the dogmatist but not that of the Pyrrhonist; Perler 2006, 21 attributes the sort of scepticism that results from a twofold metaphysical model to the Academics rather than the Pyrrhonists.

[49] Caluori argues that it is this practical commitment to the everyday world that qualifies the Renaissance doctor Sanchez as a Pyrrhonist: "Although … Sanchez claims that our sense perception and our reason often go astray, this critique is primarily a problem for a *theoretical* approach that is based on sense perception and reason. The usefulness of senses and reason, however, for daily life is not being questioned" (Caluori 2007, 43).

metaphysical dualism that separates appearances from their underlying real causes. In characterising empiricism along these lines he suggests that the understanding of what empiricism is must have changed with Descartes. Let us now have a look at Bacon and Gassendi in order to find out whether this impression is correct.

In the *New Organon* (1620) Bacon writes:

> Those who have treated of the sciences have been either empiricists or dogmatists. Empiricists, like ants, simply accumulate and use; Rationalists like spiders, spin webs from themselves; the way of the bee is in between: it takes materials from the flower and of the garden and the field; but it has the ability to convert and digest them.[50]

Bacon furthermore advocates "a closer and more binding alliance (which has never been made) between these faculties (the experimental and the rational)"[51] for the purpose of improving the established scientific practice.

This passage once again gives a good example of how empiricism and rationalism have been repeatedly presented as one-sided in order to introduce one's own more sophisticated position as a combination of the best of the two. Bacon first explains that empiricism is a position that prioritises experience and fails to integrate the use of reason, before suggesting that a mix of both, experience and reason, is the solution to all problems. He thus proposes what Galen has offered before him and Kant would repeat one and a half centuries later.[52] Despite these broad similarities, however, it becomes clear that all in all Bacon's discussion of empiricism stands closer to Galen's than to Kant's: Bacon neither introduces the epistemological twist so characteristic for the Kantian analysis of empiricism, nor does he impose a metaphysical model that detaches the realm of appearances from the real world. Bacon raises the topic of empiricism within his general attack on Aristotelian science that he takes to rely on useless forms of argument. He thereby pays tribute to the pragmatic considerations of the ancient debate and resumes its anti-dogmatic spirit: "We see remote and superficial generalities do but offer knowledge to scorn of practical men; and are no more aiding to practice than Ortelius' universal map is to direct the way between London and York."[53] Bacon's presentation of the opposition between the two schools thus exhibits the same function as Galen's. It triggers reflections on method and urges a reassessment of the aims and goals of science.[54] Bacon thereby clearly transcends the realm of purely epistemological considerations.

We will now see that Gassendi, Descartes' contemporary and opponent, introduces a perspective that somewhat alters the picture by offering a conception of appearances that detaches them from the real things. In the *Syntagma* Gassendi writes that the "Academic Sceptics" are those who

---

[50] Bacon 2000, 79.

[51] Ibid.

[52] It is interesting to note that Kant inserts a passage from Bacon's *Great Instauration* as the motto in the B edition of the Critique.

[53] Bacon 1950, 145.

[54] For further discussion of these aspects of Bacon's empiricism see Van Fraasen 2002, 32f.

> say that the appearance of things, or what things appear to be on the outside, is one thing and the truth, or the inner nature of things, namely what the things are in themselves, is another matter, and that when they say that nothing can be known certainly and that there is no criterion, they are not speaking of what things appear to be and of what is revealed by the senses as if by some special criterion, but of what things are in themselves, which is so hidden that no criterion can disclose it.[55]

Pyrrhonists would be those "who make a habitual distinction between *tê phantasia*, "appearances", or what things seem to be, and *tê alêtheia*, "the truth", or what things are, and as those "who swear that they will raise no quarrel concerning the appearance of things, but only concerning their truth."[56] Gassendi finally adds that Pyrrhonists only "assent to appearances" and "engage in dispute with the dogmatists only over the claim they make to know not only how things appear but also what they are like in themselves"[57] Gassendi thus presents scepticism, i.e. Pyrrhonism and Academic scepticism, as a position that is committed to the view that there is an "inner nature of things" that cannot be known, neither by the senses nor by reason. He thereby provides us with an interpretation of scepticism that assumes precisely the metaphysical model Kant attributes to empiricism, for Gassendi clearly distinguishes the appearance of things not only from their unknown essence, but also from the things themselves. From what has been said so far, it is clear that the sceptic certainly does not rely on such a metaphysical model. If the sceptic opposes belief in everything non-evident, how can he base his scepticism on a metaphysical model of a hidden real? If considered from a sceptical point of view, Gassendi's analysis therefore appears questionable and likely to provoke a shrug, at least if we follow Barnes, because a shrug is precisely the reaction the sober sceptic offers to statements that refer to principally non-evident objects.

In this context it is interesting to note that Antonia Lolordo claims that Gassendi is as much of a direct realist as Sextus Empiricus: "Gassendi understands Sextus as denying knowledge of essences or inherent qualities, but accepting knowledge of appearances, so that we can, for instance, assent to the claim 'honey tastes sweet to me' but not 'honey is sweet in itself'. For Gassendi as for Sextus, appearances are not mental entities but ways external objects appear."[58] This may be true about Sextus own account as it may be true about Gassendi's more general account of perception. However, it can barely be denied that Gassendi's theory of signs, and especially his admission of the notion of the indicative sign, conflicts with his alleged direct realism. Gassendi points out that if "the truth in question is hidden, lying concealed beneath appearances" we must inquire "whether it is still possible to know it through some sign and whether we have a criterion by which we may recognize the sign and judge what the thing truly is."[59] These considerations show

---

[55] Gassendi 1972, 294.

[56] Gassendi 1972, 304.

[57] Ibid.

[58] Lolordo 2007, 64; also see 76.

[59] Gassendi 1972, 329.

that Gassendi, at least when it comes to the use of indicative signs, endorses the metaphysical model that he attributes to the ancient sceptics. He understands appearances as signs that provide information about an underlying reality.

If we go back to Montaigne and his perception of ancient scepticism in "An Apology for Raymond Sebond" we gain valuable hints about what may have influenced Gassendi to think of appearances as indicators of a deeper level of reality. Montaigne is commonly seen to be the first to provide an extensive discussion of ancient scepticism after the publication of Sextus Empiricus' writings in Western Europe in the sixteenth century.[60] When discussing the deceitfulness of the senses Montaigne writes:

> Our mental faculty of perception directly in touch with the outside objects – which are perceived via the senses, and the senses do not embrace an outside object but only their own impressions of it; therefore the thought and the appearance are not properties of the object but only the impressions and feelings of the senses. Those impressions and the objects are different things. So whatever judges from the appearances judges from something quite different from the object itself.[61]

This is clearly a statement about the difference between the thing and its appearance: Montaigne here distinguishes between the inner world of the perceiving subject, "the impressions and feelings of the senses" and the "outside object"; and he tells us that judgement based on sense perception is based on something "different from the object itself."

Craig Brush draws our attention to the fact that in this passage Montaigne notably alters the original arguments of Sextus Empiricus by exclusively concentrating on "the separation that exists between the senses and the external reality"[62] and by omitting the discussion of the criterion that attacks syllogism and inductive reasoning. As an explanation Brush offers that Montaigne "is simply not interested in being a logician"[63] but rather in the role of the judge, "for he is more concerned with the subjective of the question than with the objective."[64] The overall conception of Montaigne's work seems to endorse Brush's claim. Again and again Montaigne discusses change and alteration, choice of perspective and the influence of non-cognitive factors that influence processes of judgement. Even the above-cited passage in which Montaigne discusses the impossibility of reaching the outer reality ends with the words:

> There is no permanent existence either in our being or in that of objects. We ourselves, our faculty of judgement and all mortal things are flowing and rolling ceaselessly: nothing certain can be established about one from the other, since both judged and the judging are ever shifting and changing.[65]

---

[60] MacLean 2005; Popkin 2003; Schmitt 1983.

[61] Montaigne 2003, 679.

[62] Brush 1966, 17.

[63] Brush 1966, 16.

[64] Ibid.

[65] Montaigne 2003, 680.

But if it is change, perspective and subjectivity that cause problems for the generation of certain judgement, why, one could ask, does Montaigne introduce the inner-outer distinction and with it a model that supports the idea that appearances are detached from the reality of things?

One answer to this question arises if we consider that Montaigne, similar to the ancient sceptics and empiricists, ventures to attack the established philosophical opinion of his time, in his case Aristotelian Scholasticism, and more specifically, Aristotle's account of perceptual cognition.[66] This becomes particularly clear if we pay attention to the fact that in the discussion of the fallibility of perceptual judgement, he challenges the Aristotelian view that in processes of perception the mind is able to perceive the external object by forming sensibles (*aistehon*), that is, mental entities that are taken to *resemble* the external thing and which allow the perceiver to extract the "essence" or "form" of the perceived object. Montaigne questions this account by asking: "How can our rational soul make sure that they [sense-impressions] are resemblances, since it has no direct contact of its own with the outside objects?"[67] It here appears that the reason for which Montaigne casts processes of perception in terms of an inner-outer distinction is that he wants to show the flaws of this particular perceptual theory and the metaphysical model going along with it. Montaigne thus proceeds in a similar vein as the ancient empiricists: he attacks a position that detaches the realm of the appearances from the real being of things by questioning how we can be sure that appearances indeed tell us something about the things from which they are detached. Of course, this does not entail that Montaigne commits to the view that he criticises, that is, to the metaphysical model that detaches the apparent from the real.[68]

Having clarified some of the background considerations that could have influenced Montaignes' decision to refer to appearances as something separate from the things themselves, let us now return to the question of why Gassendi holds on to the notion of indicative signs and the model of a twofold reality, although he openly sympathises with the sceptics who reject both. Gassendi proceeds in a very similar way to Montaigne when using sceptical arguments as a means to attack established Aristotelian metaphysics. Thus Gassendi claims that Aritotelian science is founded on the assumption that we can achieve a grasp of inner natures; and he argues that science pursued along these lines is a hopeless enterprise, because we are plainly incapable of grasping hidden essences. Gassendi exclaims: "Yes, Aristotle, dear Peripatetic! Yes indeed the fundamental elements of reality are matter, form and privation.

[66] Gaukroger characterises the Aristotelian position as follows: "Aristotle's account had involved a commitment to the most obvious and natural epistemology for a theory of vision, resemblance: what you saw resembled what was there in the world." See Gaukroger's introduction in Arnauld 1990, 3.

[67] Montaigne 2003, 679.

[68] This interpretation seems to be endorsed by the concluding paragraphs of the Apology where Montaigne suggests that reality itself is as changeable and unsteady as appearances: "There is nothing in Nature, either, which lasts or subsists; in her, all things are either born, being born, or dying." Montaigne 2003, 682.

All I ask is that, using these, you explain to me the essence of just one single thing."[69] But although Gassendi happily engages in sceptical arguments when attacking Aristotelian metaphysics, he opposes Sextus by citing the Epicurean claim that "the sign is a thing of the senses."[70] If we believe in Gassendi's frequently expressed enthusiasm for Epicurean thought, it here becomes clear that the purpose of using sceptical arguments is not so much that of downgrading our knowledge claims for the very reason that all we can reach are appearances, but not the things themselves or their essences. By defending the notion of the indicative sign he urges us to use appearances as signs that indicate an underlying reality. So Gassendi's problem is obviously not the problem that appearances block access to the real being of thing.[71] His – as well as Montaigne's – problem seems to be that Aristotelianism offers an unacceptable approach towards the world, while he, similar to Montaigne, uses sceptical arguments to reveal this.

6 Kant's Concept of Empiricism Revised

What does all this have to do with Kant's interpretation of empiricism as a form of idealism or Cartesian metaphysical dualism? At first glance it seems that Gassendi's and Montaigne's interpretation of ancient scepticism reveals that Kant was not altogether mistaken in attributing the metaphysical model of a twofold reality to early modern empiricists and rationalists. It indeed looks as if in response to scholasticism and with the various interpretations of Sextus Empiricus' scepticism a gap between the inner and the outer of the perceiving subject opened up and created epistemological problems with which philosophers subsequently tried to cope. If so, Kant's story of empiricism that presents empiricism as an account of how to solve sceptical problems would be vindicated. It would be vindicated because empiricism turned out to be no more than a strategy by which the sceptical threat was to be brought under control.

This is surely a rather simplistic reading. As has been pointed out, Montaigne and Gassendi cite the model of a twofold reality in order to attack orthodox philosophical positions; and it is in this context that they consider scepticism a useful weapon. From this it does not follow that both thinkers understood scepticism as a

[69] Gassendi 1972, 82–6.

[70] Gassendi 1972, 333.

[71] This interpretation opposes Walker's claims that Gassendi would have benefited from Kant's insight that knowledge of the Ding-an-sich is impossible: "It would certainly be a bonus from Gassendi's point of view if the result were such as to exclude the possibility of our knowing about things as they are in themselves. Kant, in fact, was to put forward a line of thought which at least in his own opinion met the relevant requirements, and which I cannot help thinking Gassendi would have found rather congenial" (Walker 1983, 333). According to my reading, Gassendi was plainly not troubled by the Kantian question; for further discussion of this point see Waldow 2010.

problem that rendered any intellectual and scientific activity entirely otiose. Both, after having exploited sceptical arguments in order to show the deficiency of the established opinions, provide positive accounts of what sort of knowledge one is to seek: Montaigne recommends the study of oneself, while Gassendi advocates an experience-based examination of the world. Whether these recommendations can be seen as an attempt to solve the sceptical problem is questionable. It seems that both plainly stop bothering with sceptical arguments, once the adversary philosophical position has been defeated. Gassendi repeatedly states that it is entirely absurd to believe that there is no world and that we are just dreaming when actually perceiving;[72] and Montaigne does not seem to be overly troubled by the fact that even the knowledge of ourselves can never gain the status of objective knowledge, for it is the self that becomes the subjective judge of herself.[73]

If we consider the spirit Galen's discussion of empiricism exhibits, we can understand why scepticism need not necessarily involve the fear that nothing can be known.[74] Galen presents empiricism as a pragmatic position that employs sceptical arguments with the aim of scrutinising the theory-laden practice of the medical dogmatists. Empiricism thus-understood is a position that casts doubt on the very belief that it is legitimate to transcend the realm of the experienceable by inferring to underlying causes. It is not a position that can be characterised by its struggle with the idea that knowledge is impossible. In drawing our attention to this use of sceptical arguments, Galen enriches our understanding of empiricism *and* scepticism at the same time. He explains that both positions reject the unconstrained quest for knowledge that fails to impact on our lives. Empiricism and scepticism here present themselves as positions which are deeply committed to the world and its agents, a commitment that goes beyond purely epistemological considerations. If we add that, according to Ian MacLean, "Galen's *De optimo modo docendi* as well as others of his texts" were "sources for sceptical and antisceptical thinking"[75] of the Renaissance, it becomes clear why for early modern thinkers, such as Gassendi and many others, it was natural to make use of sceptical arguments *and* to believe in the possibility of gaining insight into the world. These thinkers were presumably well aware of the fact that sceptical arguments are perfect tools for replacing established orthodoxies and for advocating an experience-based and practically-oriented form of philosophical investigation, even if this sometimes entailed their adherence to a metaphysical model that ancient empiricists originally rejected.

To conclude it can be maintained that Kant may have a point – at least as it concerns early modern forms of empiricism – when claiming that empiricists are quasi-metaphysical Cartesian dualists who differ from them only in their denial that it is possible to know the causes of appearances. Thinkers such as Gassendi

[72] See for instance Gassendi 1972, 328 and Gassendi's reply to Descartes in Descartes 2005, 180.

[73] Hartle defends a non-sceptical reading of Montaigne in Hartle 2005.

[74] Popkin 2003.

[75] MacLean 2006, 253.

implicitly endorse the view that inferences based on appearances can lead us to an understanding of their non-apparent causes. However, Kant's account of empiricism does not only fail to do justice to ancient empiricism which clearly rejects the twofold model of reality. What is more, Kant is impermissibly reductive in his attempt to present empiricism as a purely epistemological position. In so doing he clearly departs from the course of authors like Galen, Bacon and even Gassendi who all refer to ancient empiricism and scepticism in order to advocate new forms of useful, experience-based and non-dogmatic science.

References

Barnes, Jonathan.1983. Ancient skepticism and causation. In *The Skeptical Tradition*, ed. M. Burnyeat, 149–204. Berkeley/CA: University of California Press.

Arnauld, Antoine. 1990. *On True and False Ideas*. Trans. and ed. Stephen Gaukroger. Manchester/New York: St. Martin's Press.

Bacon, Francis. 1950. *The Advancement of Learning*, London: J.M. Dent & Son, New York: E.P. Dutton.

Bacon, Francis. 2000. *The New Organon*. Cambridge: Cambridge University Press.

Barnes, Jonathan. 1982. Medicine, experience, and logic. In *Science and Speculation*, eds. J. Barnes, J. Brunschwig, M. Burnyeat, M. Schofield, 24–68. Cambridge/New York: Cambridge University Press.

Brush, Craig. 1966. *Montaigne and Bayle. Variations on the Theme of Skepticism*. The Hague/The Netherlands: Nijhoff.

Burnyeat, Miles. 1984. The skeptic in his place and time. In *Philosophy in History*, eds. R. Rorty, J.B. Schneewind, Q. Skinner, 225–254. Cambridge: Cambridge University Press.

Caluori, Damian. 2007. The scepticism of Francisco Sanchez. *Archiv für Geschichte der Philosophie*, 89: 30–46.

Dear, Peter. 2006. The meaning of experience. In *The Cambridge History of Science: Early Modern Science*, eds. K. Park, L. Daston, vol. 3, 106–131. Cambridge: Cambridge University Press.

Fischer, Kuno. 1909. *Immanuel Kant und seine Lehre*, Heidelberg: Carl Winter.

Frede, Michael. 1982. The method of the so-called methodical school of medicine. In *Science and Speculation*, eds. J. Barnes, J. Brunschwig, M. Burnyeat, M. Schofield, 1–23. Cambridge: Cambridge University Press.

Frede, Michael. 1987. The sceptic's beliefs. In *Essays in Ancient Philosophy*, 179–200. Oxford: Clarendon.

Galen. 1821–1833. *Galeni Opera Omnia*, ed. C.G. Kühn. Leipzig (reprint, Hildesheim, Olms 1964–65.

Galen. 1985. *Three Treatises on the Nature of Science*. eds. M. Frede, R. Walzer. Indianapolis: Hackett.

Gassendi, Pierre. 1972. Syntagma. logic. In *Selected Works of Pierre Gassendi*, ed. C. Brush. New York: Johnson Reprint Corporation.

Gassendi, Pierre. 2005. Fifth set of objections. In *The Philosophical Writings of Descartes*, trans. J. Cottingham and others, vol. 2, 179–240. Cambridge: Cambridge University Press.

Hartle, Ann. 2005. Montaigne and scepticism. In *Cambridge Companion to Montaigne*, ed. U. Langer, 183–206. Cambridge: Cambridge University Press.

Kant, Immanuel. 1970. *Critique of Pure Reason*. Trans. Norman Kemp Smith. London: MacMillan, St. Martin's Press.

Loeb, Louis. 1981. *From Descartes to Hume: Continental Metaphysics and the Development of Modern Philosophy*. Ithaca: Cornell University Press.

Lolordo, Antonia. 2007. *Pierre Gassendi and the Birth of Early Modern Philosophy*. Cambridge: Cambridge University Press.
MacLean, Ian. 2005. Montaigne and the truth of the schools. In *Cambridge Companion to Montaigne*, ed. U. Langer, 143–162. Cambridge: Cambridge University Press.
MacLean, Ian. 2006. The 'skeptical crisis' reconsidered: Galen, rational medicine and the *Libertas Philosophandi*. *Early Science and Medicine*, 11(3): 247–274.
Montaigne, Michel de. 2003. *The Complete Essays*. London: Penguin.
Nutton, Vivian. 2005. Galen and the history of ancient medicine. *Science in Context* 18(1): 11–121.
Nutton, Vivian. 2008. The fortunes of Galen. In *The Cambridge Companion to Galen*, ed. R.J. Hankinson, 355–390. Cambridge: Cambridge University Press.
Perler, Dominik. 2006. *Zweifel und Gewissheit*. Frankfurt am Main: Klostermann.
Popkin, Richard. 2003. *The History of Scepticism. From Savonarola to Bayle*. Oxford: Oxford University Press.
Schmitt, Charles B. 1983. The rediscovery of ancient skepticism in modern times. In *The Skeptical Tradition*, ed. M. Burnyeat, 225–252. Berkeley: University of California Press.
Sextus Empiricus. 2007. *The Outlines of Scepticism*. ed. J. Annas, J. Barnes, Cambridge: Cambridge University Press.
Tieleman, Teun. 2005. Methodology. In *The Cambridge Companion to Galen*, ed. R.J. Hankinson, 49–65. Cambridge: Cambridge University Press.
Van Fraasen, Baas. 2002. *The Empirical Stance*, New Haven/London: Yale University Press.
Waldow, Anik. 2010. The Pretence of scepticism and its non-epistemological relevance in early modern philosophy. *History of Philosophy Quarterly* 27(1): 35–55.
Walker, Ralph. 1983. Gassendi and skepticism. In *The Skeptical Tradition*, ed. M. Burnyeat, 319–336. Berkeley: University of California Press.
Woolhouse S. Roger 1985. *The Empiricists*. Oxford: Oxford University Press.

Embodied Stimuli: Bonnet's Statue of a Sensitive Agent

Tobias Cheung

Abstract In this paper, I focus on the relations between sensibility, organic fibres and the soul–body-interface in the second half of the eighteenth century in France. The process of the mediation and transformation of embodied stimuli in a "human statue" is the main theme of Etienne Condillac's *Traité des sensations* (1754) and Charles Bonnet's *Essai analytique sur les facultés de l'âme* (1760). Both call embodied stimuli "sensations" (*sensations*) and the faculty to receive sensations "sensibility" (*sensibilité*). Like in the writings of the Montpellier medical doctors Théophile Bordeu and Paul Joseph Barthez, the term "sensation" refers to a wide range of organic and reflexive phenomena beyond the Hallerian dualism of muscular "irritability" and nervous "sensitivity." While Condillac is mainly interested in the transformation of sensations into "experiences" and "judgments" of the soul, Bonnet develops a detailed scheme of the fibre œconomy of sensitive agents that transform outer stimuli into "physical sensations" (*sensations physiques*). Bonnet thus compares the organic differentiations of physical sensations and the relation of "physical ideas" (*idées physiques*) to the "organization" of living "organized bodies" (*corps organisés*).

1 Introduction

In the second half of the eighteenth century, the process of the mediation and transformation of embodied stimuli in a "human statue" becomes the main theme of Etienne Condillac's *Traité des sensations* (1754) and Charles Bonnet's *Essai analytique sur les facultés de l'âme* (1760). Both call the embodied stimulus "sensation" (*sensation*) and the faculty to receive sensations "sensibility" (*sensibilité*). Like in the writings of the Montpellier medical doctors Théophile Bordeu, Jean-Joseph Menuret, Henri Fouquet and Paul-Joseph Barthez, the term "sensation" refers to a wide range of organic and reflexive phenomena beyond the Hallerian dualism of muscular

T. Cheung
Max-Planck Institute for the History of Science, Berlin

C.T. Wolfe and O. Gal (eds.), *The Body as Object and Instrument of Knowledge: Embodied Empiricism in Early Modern Science*, Studies in History and Philosophy of Science 25, DOI 10.1007/978-90-481-3686-5_15,

"irritability" and nervous "sensitivity."[1] Bonnet and Condillac use the term "sensation" for mechanical processes of stimulation, muscular, vascular, glandular and nerve-related movements, feelings, perceptions, passions and memories.

The sensation discourse is closely related to discussions about fibre models and soul–body-interactions. From around 1750, the fibre becomes through its passive and active properties the main operative building block and at the same time the first unifying principle of function–structure-complexes and stimuli-reaction-schemes of organic bodies.[2] It occupies the role that the cell takes up in the cell œconomies of the second third of the nineteenth century.[3] After Giorgio Baglivi's pathology of fibre types and before Xavier Bichat's catalogue of tissue types, the bodies of plants, animals and humans are for Bonnet all constituted through fibres,[4] and their differences mainly rely on different fibre types, structures and properties – plants feed through fibres, animals move and sense through fibres, and humans think through fibres.

While Julien Offray de La Mettrie and Claude-Adrien Helvétius already outline the framework of a material anthropology in which reflexive faculties of humans belong to their physical organization,[5] Condillac and Bonnet still distinguish between the sensitive "faculties" of a thing called "soul" and a thing called "body." However, Condillac and Bonnet have different interests. Condillac is mainly interested

[1] In the fifteenth volume of the *Encyclopédie* 1765, Henri Fouquet (who is like Bordeu a medical doctor at Montpellier) defines "sensibility" as a general organic property that comprises both "irritability" and "sentiment" (*Encyclopédie*, vol. 15 1765, 50): "L'*irritabilité* n'est autre chose que la *mobilité* ou *contractilité* dont il a été question au commencement de cet article, & que nous avons dit être une des deux actions comprises dans l'exercice de la *sensibilité*; c'est toujours l'expression du sentiment; mais une expression violente, attendu qu'elle est le produit de la *sensibilité* violemment irrité par des *stimulus*; aussi est-elle quelquefois désignée sous le nom même de *stimulus* chez les Phisiologistes, ou sous celui de *fibre motrice*, &c." In the article 'Spasme' of the *Encyclopédie*, Menuret refers to the human as a "machine" in which each part is sensible or irritable: "Qu'est-ce que l'homme? ou pour éviter toute équivoque, que la méchanceté & la mauvaise foi sont si promptes à faire valoir; qu'est-ce que la machine humaine? Elle paroît à la premiere vûe; un composé harmonique de différens ressorts qui mûs chacun en particulier, concourent tous au movement général; une propriété générale particulierement restreinte aux composés organiques, connue sous les noms d'*irritabilité* ou *sensibilité*, se répand dans tous les ressorts, les anime, les vivifie & excite leurs mouvemens; mais modifiée dans chaque organe, elle en diversifie à l'infini l'action & les mouvemens; par elle les différens ressorts se bandent les uns contre les autres, se résistent, se pressent, agissent & influent mutuellement les uns contre les autres; cette commixture réciproque entretient les mouvemens, *nulle action sans réaction*. De cet antagonisme continuel d'actions, résulte la vie & la santé" (*Encyclopédie*, vol. 15 (1765), 435b; thanks to Charles Wolfe for this hint).

[2] Cf. Le Camus 1769, 8: "Toutes les substances créées sont organisées ou sans organisation. Les premières sont composées de fibres, jouissent de la vie, et sont connues sous les noms d'animaux et de végétaux. Les dernieres sont massives, n'ont que des particules appliquées les unes contre les autres et sont inertes. Elles constituent le regne minéral."

[3] Cf. Berg 1942, 336.

[4] Fibres can be themselves composed of smaller organic units, like Buffon's organic molecules.

[5] For the larger context of material anthropologies in the seventeenth and eighteenth centuries, debates about sensibility, and the so-called ideologists, see Picavet 1891; Baruzzi 1968; Moravia 1978; Lawrence 1979; Baasner Baasner 1988; Mullan 1988; Henry 1989; Van 1993; Duchet 1995; Bourdin 1998; Wolfe 1999; Riskin 2002; Audidière et al. 2006; and Thomson 2008.

in the transformation of sensations into "experiences" (*expériences*) and "judgments" (*jugements*) of the soul, while Bonnet develops a detailed scheme of the fibre œconomy of sensitive agents that transform outer stimuli into "physical sensations" (*sensations physiques*). Bonnet thus discusses the organic differentiations of physical sensations and the relation of "physical ideas" (*idées physiques*) to the "organization" of living "organized bodies" (*corps organisés*). I will refer to both "spheres" of "operation" of Bonnet's model of a sensitive agent.[6] In the first part, I reconstruct Bonnet's "œconomy of fibres" of organized bodies. In the second part, I focus on the relation between the œconomy of fibres and the soul–body-interface.

2 Bonnet's Œconomy of Fibres of Organized Bodies

After his observations of the parthenogenesis of aphids and Abraham Trembley's experiments on the complete regeneration of artificially detached parts of the fresh water polyp *Hydra* in 1744, Bonnet worked on a germ-fibre-theory to explain the development of organic bodies. If pieces of something living can again grow to entire organic bodies with heads and arms or stems and branches, then, Bonnet argued, there have to be small organized units with reproductive potentials within the visible units. These small, point-like units are germs that extend their contracted fibre architecture like "mini-machines" (*machinules*) into the "organization" of adult bodies.[7]

In the first decades of the eighteenth century, web-like representations of fibre architectures and canal systems were already a standard reference for anatomists of plant and animal bodies. In his manual *Corporis humani anatomia* (1693), which was reedited at least thirteen times in Latin and German until 1739, Philip Verheyen portrays the human body as a complex system of different types, dispositions and combinations of fibres (Fig. 1)[8]:

Beside the representation of web-like canal systems and a great variety of catalogues of different geometrical shapes and dispositions of fibres from Vesalius to Verheyen, the problem of self-moving properties of organic units characterizes

[6] Most of the secondary literature about Bonnet focuses either on his germ theory (within the framework of preformationist debates) or on his "psychology." For a discussion of both aspects, see e.g. Marx 1976; Anderson 1982; Mazzolini et al. 1986; Cheung 2005a, c. For Bonnet's germ-fibre-theory, see also Rieppel 1988; Buscaglia 1994; Cheung 2004, 2005b.

[7] Cf. Bonnet, *Contemplation de la nature*, in Bonnet 1779, vol. 4/1, 360–361. In *De Motu Animalium* (1680–1681), Giovanni Alfonso Borelli often referred to the Latin expression *machinulae* to characterize mini-machines of fibre complexes in organic parts as muscles.

[8] Cf. Berg 1942, 390–392; Ishizuka 2006, 72–75; and Suy 2007. Boerhaave combines the notion of a hollow "nerve fibre" as the basic building-block of all organic structures with Frederik Ruysch's visualization of canal systems through various injection techniques. In Boerhaave's hydraulic model of organic bodies, "canals" or "tubes" (*canales*) are differentiated into smaller tubes that "finally form, branched within themselves, the structure of a net (*retis structura*)." Cf. Boerhaave 1703, 10: "... canalis est conicus, elasticus, inflexus, divisus in similes minores eodem trunco ortos, qui ultimo ... retis structura in se mutuo patent."

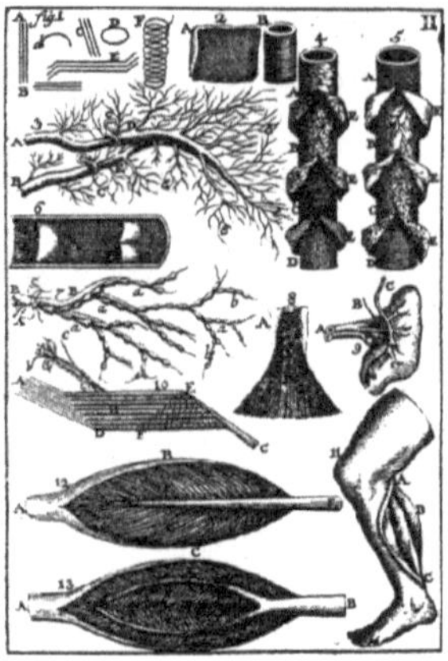

Fig. 1 Fibre architecture of the human body. Verheyen 1712–1713, vol. 1 (1712), 30, plate II

fibre debates from René Descartes, Francis Glisson and Thomas Willis to Giorgio Baglivi and Albrecht von Haller. Their explanatory patterns refer to mechanistic or chemical explanations of elasticity, contractility, and irritability and to so-called vital explanations of these properties through an inherent force.[9]

In *Primae lineae physiologiae* (Haller 1747), Haller clearly distinguished between three types and general functions of fibres that constitute all "tissues" (*tela*) and "membranes" (*membranae*) of human and animal bodies: the fibres of the "cellular tissue" (*cellulosa tela*), the muscle fibres, and the nerve fibres.[10] Living beings are "cellular fibre fabrics" (*cellulosae fabricae fibris*).[11] The fibres of the cellular tissue serve as the basic stabilizing material for organic structures, irritable muscle fibres can contract themselves through a specific *vis insita*, and nerve fibres transmit sense data or instructions from a regulating, soul-like agent.

For Bonnet, the germ-fibre-unit of an individual organized body is first of all a preexisting system of parts with basic regulating and structural properties:

> … I understand in general by the word germ all these preordinations or preformations of parts that are themselves able to determine the existence of a Plant or an Animal (*de parties capables par elle-même de déterminer l'existence d'une Plante ou d'un Animal*).[12]

---

[9] Cf. Temkin 1964; Jaynes 1970; and Ishizuka 2006.

[10] Cf. ibid., 11 (§ 23): "Summam dignitatem huius cellulosae telae adgnoscet, qui cogitabit, ab ea sola pendere omnium arteriam, nervorum, fibrarum muscularium, adeoque compositarum inde carnum viscerumque legitimam firmitatem, stabilitatem … Ea cum vasi, nervis, fibris muscularibus & tendinieis … omnia viscera, omnes musculos, glandulasque & ligamenta, & capsulas componit, ab ea sola, eiusque varia longitudine, tensione, copia, proportione, glandularum viscerumque diversitatem oriri, & ab ea denique longe maximam partem corporis ipsius effici certum est, si quidem non totum ex huiusmodi fillis cellulolis componitur."

[11] Cf. Haller 1747, 7 (§ 17).

[12] Bonnet, *Tableau des Considérations sur les corps organisés* in Bonnet 2002, 82. Cf. Bonnet 1985, 63 (§ 83): "L'organisation primitive des germes détermine l' arrangement que les atômes nourriciers doivent recevoir pour devenir parties du tout organique."

In *Considérations sur les corps organiques* (1762), Bonnet introduces his model of a germ-fibre-unit as an automaton that is composed of indefinitely small fibres. These fibres interact as "mini-machines" (*machinules*) to "animate" the *tout organique*:

> A fibre, as simple as it might appear, is nonetheless a *tout organique* that feeds, growths and vegetates.[13]

The *tout organique* can differentiate into nine thresholds of organized units: the germ, the fibril, the fibre, the "mesh" (*maille*) or the "cell," the "fascicle" or "fibre bundle" (*faisceau*), the tissue, the organ, the apparatus, and the entire organism. Each higher organized unit is composed of lower units.

The fibre that is "supposed to be simple is itself composed of a multitude of fibrils (*fibrilles*) that are, as for themselves, composed of a multitude of more or less homogeneous particles (*molécules*) as their first elements, while the fibrils are the second elements."[14] Each fibre represents in general a basic fusion point of the external particles with the inner order of the organic body. It possesses assimilative properties:

> ... the fibre determines through the Mechanics of its Structure the arrangement of its nutritive Atoms ...[15]

From this fusion result the characteristics of the "primordial layers" of higher organs in the form of specific "patterns" (*trames*). These patterns can appear as "meshes" (*mailles*) or "cells." They belong again to various types of "fibre bundles" (*faisceaux*) and "tissues."

Lower units, as for example the parenchymatic or cellular tissue,[16] are less determined in their operative effects than higher units (organs or apparati). Cells, bundles of strings and tissues represent the structural or "solid" physiological basis of the organic "network" (*réseau*). The organs and the apparati operate through the co-existing and co-ordinated "meshes" of the "network":

> ... we learn from physiology that there is no organic part that is not covered externally and internally with a cellular or parenchymatic tissue. It is so universally extended that it enfolds the whole system of fibres. One could thus think of it as the principal instrument of growth. The diverse incrustations or incorporations that determine the consistence [of organic parts], the [process of] growth and the most essential modifications of each part are situated within its nearly infinitely varied meshes or pores.[17]

The organic "network" thus is a "networking system" (*ouvrage à réseau*)[18] of organized units on various levels, a system of a "nearly infinite" number of intermediate "chain-links" (*chaînons*) that appear in certain subordinated levels of order.

---

[13]Bonnet 2002, 257. Bonnet sometimes also describes "fibrils" (*fibrilles*) as "mini-machines." Cf. Bonnet, *Contemplation de la nature*, in Bonnet 1779, vol. 4/2, 361; and Bonnet 2002, 232 and 262.

[14]Ibid., 258.

[15]Bonnet 1973, 68 (§ 101).

[16]Cf. Bonnet 2002, 288.

[17]Ibid., 296.

[18]Bonnet 1973, 65 (§ 97).

This networking system depends on the "enchained" microscopic structures of "bended," "stretched" and "hooked" fibres or "threads" (*fils*) that form "meshes" and "tissues." To describe the growth of the organic "network," Bonnet often refers to the process of weaving and to the mechanism of looms. The folded skins of organs become thus different kinds of weaving patterns or "cloths" (*étoffes*):

> The organized bodies are more or less fine tissues, they are networks, kinds of cloths of which the chain forms itself the pattern of an art that we would have to admire even if it would be known to us ...The bodies of plants and of animals are thus sorts of looms, more or less composed machines that convert the diverse materials that are used for the activities of their mechanisms (*ressorts*) and their fluids.[19]

Bonnet's loom analogy reflects the invention of new machines of spinning and weaving processes for the fast growing market of the weaving industry, especially in England and France.[20] It was not before 1785 that Edmond Cartwright patented the first fully mechanized loom.[21] However, the automatization of spinning machines and looms already begun in the first half of the eighteenth century. In 1733, John Kay invented the so-called flying shuttle.[22] The weaver could move Kay's shuttle, itself loaded with cotton, only with a flick of the wrist of one hand from one side of the loom to the other and change at the same time the pattern of the warp threads.[23] In 1745, Jacques Vaucanson developed a completely automated process of the production of woven patterns through punch cards.[24] But Vaucanson's machine remained a prototype until 1805 when Joseph-Marie Jacquard combined Cartwright's power loom with an automated punch card system.

In his *Lettres philosophiques* (Bourguet 1729), Louis Bourguet already compared the mechanism of looms with the mechanism of organic bodies. However, Bourguet focused on the automated process of weaving itself and not on the fibre architecture of the product:

> The Mechanic Arts still reveal a couple of other good examples, and there is one among all the others that I have to mention because it seems to be truly accurate for a better understanding of the subject that we examine [i.e. organic bodies]. I would like to talk about the vertical loom[25] that is used to produce *Rubans*, *Galons*, and all kinds of *Cloth*s with gold, silver, silk, wool and linen. This work of art is such that a simple Worker, who understands nothing, can produce all theses nice Works in passing one or a couple of shuttles along the Chain and in putting the feet on a certain number of Pedals.[26]

---

[19]Bonnet, *Contemplation de la nature*, in Bonnet 1779, vol. 4, 359–360. Cf. Bonnet 1985, 63 (§ 83): "Un solide non-organisé est un ouvrage de *marquetterie*, ou de pièces de rapport. Un solide organisé est une étoffe formée de l' entrelacement de différents fils. Les *fibres élémentaires* avec leurs *mailles*, sont la *chaîne* de l'étoffe; les atomes nourriciers qui s'insinuent dans ces mailles, sont la *trame*."

[20]Cf. Baum 1913, 25–42.

[21]Four years later, a steam engine powered the movements of the loom.

[22]In the same year, John Wyatt invented an automated spinning machine.

[23]Cf. Bohnsack 2002, 154–156.

[24]Cf. Barlow 1879, 141.

[25]*Haute-Lice*. The threads of this loom are disposed vertically. In the "*basse lice*," they are disposed horizontally.

[26]Bourguet 1729, 145.

For Bonnet, it is both the automated production process and the product that are similar in looms and living beings, although the comparison should not be pushed too far.[27] Bonnet explains the order of organic bodies through the assimilative "reproduction" of the order of a regulating "plan" within a "system" of mini-machines that produce, sustain and repair "tissues." These regulated processes are executed by a "secret force" (*force secrete*) that acts like a self-moving flying shuttle. The shuttle "chases the nutrition into the meshes (*mailles*)."[28]

However, in difference to the mechanism of looms, organic systems react to outer "stimuli." The regulating activity of the preexisting "plan" depends on a continuous initiation of movements through various "impulse" series.[29] Such "impulses" emerge in muscles and nerves. While Haller's irritability of isolated muscles served Bonnet as a standard reference for an experimental proof of the general existence of this property in organic units, he tried to combine Haller's irritability with the "impulse" series of nerve fibres to describe a general property of the natural automaton to initiate and repeat coordinated movements. Such a system of impulses and movements is for Bonnet an expression of the "vital principle" (*principe vital*) of living beings.[30]

Bonnet localizes the natural spinning unit of tissues in an "evolving" germ that "contains"[31] ever smaller and smaller germs, an image that is similar to Malebranche's example of an apple tree that is contained in its seeds.[32] But every imaginative approach to the process of *emboîtement* risks cutting short a proper understanding of parts of matter that could be indefinitely small, and yet organized.[33] Such an indefinitely small, but organized matter is not simply an "abridged sum" (*abrégé raccourci*) of its "evolved," visible body. Rather, Bonnet emphasizes, it is necessary "to give a much larger signification to the notion of germ."[34]

In his *Principes philosophiques* (1754), Bonnet thought of the preexistence of a similar, though smaller organic body in the germ that "evolves" during its development.[35] The publication of Haller's research on the development of embryos in the egg yolk of chickens in 1757 seemed to confirm Bonnet's viewpoint.[36] But from the *Considérations sur les corps organisés* (1762) to the *Palingénésie philosophique* (1769), Bonnet changes his position towards a more dynamic, regulative process that characterizes organic development:

---

[27] Cf. Bonnet 1985, 63 (§ 83).

[28] Bonnet, *Contemplation de la nature*, in Bonnet 1779, vol. 4/1, 259.

[29] Cf. Bonnet, 508–509 (§ 797).

[30] Cf. Bonnet 2002, 267.

[31] Beside the *emboîtement* theory of germs, Bonnet also discussed a dissemination theory in *Considérations sur les corps organisés*. He finally preferred the *emboîtement* theory. Cf. Savioz 1948, 75–78.

[32] Cf. Bonnet, *Mémoires sur les germes*, in Bonnet 1779, vol. 5, 10. Savioz 1948, 76–77 and Marx 1976, 57–58 discuss the influence of Malebranche on Bonnet's preformation theory.

[33] Cf. Bonnet, *Mémoires sur les germes*, in Bonnet 1779–1783, vol. 5, 2.

[34] Bonnet 2002, 257.

[35] Bonnet, *Principes philosophiques*, in Bonnet 1978, 210.

[36] Cf. Roger 1993, 724.

> When the chicken is still in the state of a *germ*, all its parts have forms, proportions and positions (*situations*) that differ extremely from those in which evolution will cover them. This goes so far that, if we could see this small germ in larger form, it would be impossible for us to recognize it as a chicken ... In brief, the germ is composed of a sequence of points that afterwards form a line. These lines prolong and multiply themselves, and they produce surfaces.[37]

Besides the discussions on developmental processes, Bonnet's model of organic bodies and living looms is also part of debates on the soul–body-interface. From Willis's *Cerebri anatome* (1664) and William Croone's *De ratione motus musculorum* (1664) to Baglivi's *De fibra motrice et morbosa* (1700) and David Hartley's *Observations on Man* (1749), the fibre fabric of organic bodies has been associated with the transmission of stimuli and sensations through vibrations that originate in a soul-agent and in the senses.[38] Bonnet refers to such vibrations. However, in his later texts, Bonnet rejects – like Louis de Lacaze in his *Idée de l'homme physique et moral* (1755) – a mechanical, string-based model for the soul–body-interface. In the nerve system, the impulse series seems for Bonnet to depend on an "electric fluid," instead of being the result of minimal "vibrations":

> The nerves that have been illustrated as strings of a music instrument are not at all under tension like the strings of such an instrument. They are not, like them, made to oscillate; they are not, like them, stretched in a straight line; they appear to have a lot of inflections. Finally, there are neither elastic nor irritable. Their proper substance is malleable and mushy ...[39]

---

[37] Bonnet 1985, 481 (§ 351). Cf. Bodemer 1964, 25–27. However, in the *Palingénésie*, Bonnet still argues that a similar, though smaller organic body preeexists in the germ. Cf. Bonnet 2002, 260: "Mais quand il est question d'expliquer la reproduction d'un tout organique dissimilaire, il me paroît, que je suis dans l'obligation philosophique d'admettre, que ce tout préèxistoit dans un germe proprement dit, où il étoit dessiné très en petit et en entier. J'admets donc, qu'une tête, une queue, une jambe préèxistoient originairement sous la forme de germe, dans le grand tout organique où elles étoient appellées à se développer un jour."

[38] Cf. Glassman et al. 2007 and Whitaker et al. 2007. Besides models of vibrations and strings, the movement of fluids in a canal system played a crucial role since Descartes for the soul–body-interface. Jean-Paul Marat refers in *De l'homme* (1775–1776) to the "soul" as an agent that moves a "fluid" in the "nerves."

[39] Bonnet, *Méditations sur l'origine des sensations et sur l'union de l'ame et du corps*, in Bonnet 1779, vol. 8, 391. Cf. Bonnet 1973, 21: "Mais les *Nerfs* sont mols, ils ne sont point tendus comme les Cordes d'un *Instrument*: les Objets y exciteroient-ils donc des vibrations analogues à celles d'une Corde pincée? ces vibrations se communiqueroient-elles à l'instant au Siège de l'ame? la chose paraît difficile à concevoir. Mais si l'on admet dans les *Nerfs* un *Fluide* dont la subtilité & la mobilité approchent de celui de la *Lumière*, on expliquera facilement par le secours de ce Fluide, & la célérité avec laquelle les impressions se communiquent à l'Ame, & celle avec laquelle l'Ame exécute tant d'Opérations différentes." See also Haller, Letter to Bonnet of the 6 September 1754; in Sonntag 1983, 49. Newton combined theories of electricity and vibration for the transmission of stimuli in organic bodies. Cf. Wallace 2003. For theories on animal electricity in the later half of the eighteenth century, see Bernardi 1992. Clericuzio 1994 focuses on the development of chemical explanations of the actions of animal spirits.

In a letter of 25 June 1771, Bonnet challenges Haller, the "famous experimenter," to search a single explanatory scheme for the initiation of movements in nerves and muscles through the combination of irritability and electricity:

> How does thus the contact of some fluid or solid bodies, of a sharp spine or any kind of acid, excite in the *Touts organiques* a more or less strong and continuous *Irritability*? ... I would like to have *Electricity* applied to *Irritability*, & and that much attention is paid to the effects of this application. It would be enough to bring an *electric fluid* on the recently detached limbs (*membres*) of their Subject. These experiments could be varied in thousands & thousands of different manners.[40]

The organic "network" thus is an electric "networking machine" (*ouvrage à réseau*)[41] of organized units on various levels, a system of a "nearly infinite" number of intermediate "chain-links" that appear in certain subordinated levels of order and that are regulated by a force that executes a preexisting plan. This networking system depends on the "enchained" microscopic structures of fibres or "threads" (*fils*) that form meshes, tissues, membranes, and canals for the transport of liquids and the constant assimilation and excretion of outer and inner parts.

3 Organized Fibre Bodies and the Soul-Body-Interface

After the *Essai de psychologie* (1754), Bonnet develops in the *Essai analytique sur les facultés de l'âme* (1760) a detailed explanatory framework of the soul–body-interface of a sensitive "statue" and its relation to the "œconomy of fibres" (*œconomie des fibres*)[42] of organized bodies. The soul–body-interface relies on two categorically different "substances": the "immaterial substance" of the soul and the "material substance" of the body.[43] However, Bonnet does not give a detailed explanation of the ontological status of the two "substances." Rather, he wants to discuss the "actions" (*actions*) of an "agent" (*agent*) that are "observable" through "sensations."

"Sensations" are for Bonnet the basis of all kind of knowledge.[44] In this perspective, his explanatory framework of the agency of organic bodies is situated within the

[40]Letter from Bonnet to Haller of 25 June 1771; in Sonntag 1983, 946. Bonnet mentioned this idea already in the very beginning of his correspondence with Haller in a letter of 16 August 1754: "If you suppose that the *Esprits Animaux* have a Nature that is similar to the one of the *electric Fluid*, then the effects of the *irritations* of fibres correspond to the rubbing of electric Tubes" (ibid., 44). See also Bonnet 1973, 20–21 (§ 31).

[41]Ibid., 65 (§ 97).

[42]Ibid. 75.

[43]Cf. ibid., 3: "*Je suppose* que l'*Homme* eft un *Composé* de deux *Substances*, l'une *immatérielle*, l'autre *Corporelle* ...."

[44]Cf. ibid., 13–14: "L'EXPÉRIENCE démontre que la *privation* d'un *sens* emporte avec elle la *privation* de toutes les *idées* attachées à l'*exercice* de ce *sens*: la privation de *tous* les *sens*, ou, ce qui revient au même, leur inaction *absolue* emporteroit donc avec elle une privation *totale d'idées*. JE ne m'arrêterai point ici à combattre l'opinion des Idées *innées*: elle a été trop souvent & trop

context of Lockean sensationalism and its French reception.[45] Bonnet combines Lockean sensationalism with Malebranche's Cartesian fibre body, in which various fibre structures connect the brain and the senses through vibrations or the movements of so-called animal spirits within tiny "nets" (*filets*).[46] Bonnet focuses on the material conditions of the "mechanics of the operations of our soul" (*Méchanique des opérations de nôtre Ame*),[47] the "physics of the ideas" (*physique des idées*)[48] and the "physics of reflection" (*physique de la reflexion*).[49] His model of a sensitive "statue" thus is part of a general shift of interest from the soul-body-interface to the mechanisms of fibre movements that mediate between physical impressions and the regulating activitites of a mover.[50] However, Bonnet also thinks that certain self-experiences of sensitive agents cannot be adequately explained by mechanic models. The experience of the "I" as the author of various volitional acts points for him to the existence of two "substances" that "act on each other.[51]

---

solidement refutée. Je ne m'arrêterai pas non plus à prouver que nos idées les plus *abstraites* ont une origine *corporelle*: il suffira de dire que nous n'avons ces idées qu'à l'aide des *Signes* qui les représentent; & ces *signes* sont figures, sons, mouvemens, *corps*. TOUTES nos idées dérivent donc originairement des *sens;* ... Je prends ici le mot *d'idées* dans le sens le plus étendu, pour toute *manière d'être* de l'Ame dont elle a la *conscience* ou le *sentiment.*" See also ibid., 174–175: "Nos Idées les plus *abstraites*, les plus *spiritualisées*, si je puis employer ce mot, dérivent donc des Idées *sensibles,* comme de leur source naturelle. L'Idée de DIEU, par exemple, la plus spiritualisée de toutes nos Idées, tient manifestement aux *Sens.*"

[45]For the influence of Locke, see Marx 1976, vol. 1, 105–108; Yolton 1991; and especially Wellman 1992, 149–163 for La Mettrie's Lockeanism.

[46]Cf. Malebranche 2006, vol. 1, 192–199 and 246–250. For a comparison between Locke's and Malebranche's physiology of sensation, imagination, feeling and thinking, see Yolton 1983, 153–189.

[47]Bonnet 1973, Preface, XIV.

[48]Cf. ibid., 46.

[49]Cf. ibid., 176.

[50]Cf. Tripp 1986, 45: "Pourtant il faut se rendre compte du changement – inauguré par Borelli – dans la discussion scientifique concernant la correspondance entre l'âme et le corps. Non seulement l'application stricte de la géometrie au corps vivant, mais aussi le développement et la différenciation de l'analyse fournirent les moyens de restreindre de plus en plus la fonction de l'âme traditionnelle et de favoriser l'extension de la pensee mécanique, meme sous la forme de la machine"; and Des Chene 2005, 255: "As with Borelli, we see [in Perrault's model of organic order] a definite demarcation between the 'mechanics' of animals and what we might call their 'energetics'. The source of active power is the soul, about which very little is said. What remains is to explain the transmission and application of that power to the end of locomotion, and that is a matter of applying mechanical knowledge. Perrault acknowledges that the animal-machine resembles 'pure machines' in the manner of its operation. But every machine requires a mover, and this the mechanism itself is incapable of supplying."

[51]Cf. ibid., Préface, XIV: "Ce n'est point parce que je crois l'Ame un Etre plus excellent que la Matière, que j'attribue une Ame à l'Homme: c'est uniquement, parce que je ne puis attribuer à la Matière tous les Phénomènes de l'Homme." See also ibid., 5: "Je reçois donc l'*Union* de l'Ame & du Corps & leur *influence réciproque,* comme un *Phénomène* dont j'étudie les *Loix,* & dont je fais profession d'ignorer profondément le *comment.*"

Bonnet calls the interactions between the supposed substances "relations" (*rapports*) or "liaisons" (*liaisons*).[52] Sensations are the "result" or the "effect" of these interactions. The human as a sensitive agent is "not a *certain* soul or a *certain* body," but "the result of the union of a *certain* soul and a *certain* body."[53] The sensitive agent is a "mixed being" (*être mixte*).[54] Mixed beings are "humans" and "animals," maybe also "plants."[55] The unity of mixed human beings, the "human system" (*Système Humain*), is always both, a "machine"-like body and a regulating or intervening soul.[56] Taken for themselves, neither the body nor the soul is the "whole human being" (*tout l'Homme*).[57]

The "organized" body of the human system is capable to process "impressions." During its development from a germ-fibre-unit into an adult human, its soul, that is in the beginning a mere "puissance" or a mere "potential," specifies its "faculties" through its "reactions" to physical impressions.[58] Thus, before the soul-part of the sensitive agent possesses specific faculties, the organized body has to "sense" or to receive impressions. These bodily impressions or sensations always follow movements, and the movements can be, in general, observed or experienced[59] – although the "mechanique of the sensible fibres" (*méchanique des fibres sensibles*) is unknown.[60]

Through the a priori sensibility of the organized body, the soul can have experiences – especially the experience of itself as an entity or an "I" (*moi*) in which all sensations are united. The "I" is "modified" through sensations and

---

[52] Cf., ibid., 93–95.

[53] Ibid., 5: "*L'Homme* n'est pas une *certaine Ame*; il n'eft pas un *certain Corps*, il eft le *resultat* de l'*union* d'une *certaine Ame* à un *certain Corps*."

[54] Cf. ibid., 3; and ibid., Préface, XIII: "L'Homme eft un Etre *mixte*; il n'a des Idées que par l'intervention des sens, & ses Notions les plus abstraites dérivent encore des Sens. C'est sur son Corps, & par son Corps que l'Ame agit. Il faut donc toujours en revenir au Physique, comme à la première origine de tout ce que l'Ame éprouve."

[55] Cf. ibid., 16: "Si les Plantes sont insensibles, ce qui n'est point démontré, la Statue [qui n'a pas encore des sensations] est immédiatement au dessus de la *Plante*: elle est entre la *Plante* & *l'Animal*."

[56] Ibid., 18.

[57] Ibid.

[58] Cf. ibid., 93: "Ailleurs j'ai défini *l'Ame* une *Force*, une Puissance, une *Capacité d'agir* ou de produire *certains* effets. C'était tout ce que je pouvais dire de l'Activité de l'Ame en la considérant sous ce point de vue général." See also ibid., 29–30 and 100: "Mais, cette Activité, que je suppose que l'Ame exerce sur les Fibres, eft *en soi* une Force *indéterminée*; c'est un simple *pouvoir d'agir*, ou de produire certains *effets*; & ce n'est point tel ou tel effet en particulier."

[59] Cf. ibid., 94: "Je vois une *Sensation* suivre un *Mouvement:* j'ignore ce que le Mouvement & la Sensation sont en eux-mêmes; mais j'étudie ce qu'ils sont par rapport à moi, c'est à dire par rapport à ma manière de concevoir. Cette étude me conduit à reconnoître que chaque Sensation a un mouvement qui lui correspond; & que ce mouvement est aussi distinct de tout autre mouvement, que cette Sensation est distincte de tout-autre Sensation."

[60] Cf. Bonnet, *Contemplation de la nature*, Bonnet 1779, vol. 4/1, 151.

also capable to act upon the body. The soul-related events occur without any observable spatial movement or extension,[61] and the body-related events seem just to "give occasion" (*donner occasion*) to the actions of the soul.[62] It is an "inner sentiment" (*sentiment intérieur*) that "persuades" Bonnet that he is the "author" of these events as "actions" of a sensitive, immaterial soul that executes volitional acts:

> In the same way that I sense that I exist, because I am conscious of the actual modification of myself, I sense that I possess a will to move certain parts of my body, and that this will performs an act. I thus admit that my soul is capable of an activity that occurs in different modes. I understand that this activity is a force (*capacité*) of my soul to produce inside & outside of itself, or on its body (*en elle & hors d'elle,* ou *sur son Corps*), certain effects.[63]

Such experiences are for Bonnet "facts" (*faits*) of the "inner sentiment." However, Bonnet admits that he knows nothing about the way how an "I"-agent acts.[64] He can only reconstruct the organic "milieu" (*milieu*)[65] within which it receives, processes and transforms sense data. The order of the milieu depends on the "œconomy of fibres." And from this œconomy emerges also the order of "ideas":

> … we know that ideas are attached to the play (*jeu*) of certain fibres. We can examine these fibres through that what we can see of them. We can study to some degree their movements, the results of their movements, and the relations that exist between them.[66]

Within and through the "milieu" of the "œconomy of fibres," events are mediated from the body to the soul and from the soul to the body. The soul produces in the sensible fibres "the same or analogue impressions (*impressions semblables ou analogues*) as those that the activity of objects would produce in her, or the corpuscules that emanate from them."[67] Like the organized body itself, the soul acts according to a stimulus-reaction-scheme. For Bonnet, the

---

[61] Cf. Bonnet 1973, Préface, XVIII–XIX and ibid., 3.

[62] Cf. ibid., 5: "Je suppose que le Corps *agit* sur l'Ame, ou, si l'on aime mieux, qu'à l'*occasion* des *mouvemens* que les *Objets* excitent dans les *sens, l'activité* de l'Ame *se déploye* d'une certaine manière, d'où naissent les *Sensations* & les *Volitions.*"

[63] Ibid., 4. Cf. ibid., 18 ("l'*action* suivit *constamment* la *décision* de la *Volonté,* comme la *Volonté* suit *constamment* la *décision* de l'*Entendement*") and 7: "Nous sommes constitués de manière que nous nous croyons *Auteurs* de nos *actions* … "

[64] Cf. ibid., 93: "Nous nommons *Agents* les Etres dans lesquels nous pensons qu'eft la *Raison* de ces changemens, & cette Raison nous est aussi inconnue que les Essences *réelles.* Le mot d'*Action* qui revient si souvent dans nos discours n'emporte donc point la *connoissance* de la *manière* dont les *Agents* opèrent, mais, simplement celle de ce qu'ils opèrent. Nous voyons des *Faits;* & tout ce qui est au delà des Faits n'est pour nous que ténèbres plus ou moins épaisses. Toutes nos Théories de *Causes & d'Effets* se bornent au fond à connoître l'*Ordre* dans lequel les Choses se succédent; ou les *Rapports* suivant lesquels l'*Existence,* ou les *Modifications* des unes, paroissent *déterminées* par l'Existence, ou les Modifications des autres."

[65] Cf. ibid., Préface, 20.

[66] Ibid., Préface, XIII.

[67] Ibid., 100.

"perceptions" (*perceptions*) of the soul are at the same time its "sensations" (*sensations*).[68] The knowledge about the "liaison and the reproduction of Ideas," and the progress of "psychology" in general, thus depends on the knowledge about the "relations" between the fibres.[69] The movement and the "order of fibres" (*Ordres de Fibres*) displays the "production" (*génération*) and the order of ideas.[70] They are the "natural signs" (*Signes naturels*) of the ideas that can be, in principle, observed:

> I do not know how the movement of certain fibres of my brain produces ideas in my soul, but I know at least very well that I only have ideas because of the movements (*en consequence des mouvemens*) that excite certain fibres in my brain. I thus examine these fibres and their movements. I think of them as natural signs of ideas, and I study the results of their possible combinations. From these analyses, I can deduce the order of the generation of ideas in my soul.[71]

Bonnet admits that the order of fibres in organized bodies is still quite unknown. However, he supposes that this order resembles the structure of a tree. He thinks that the interface of the soul and the body (the "seat of the soul") are surrounded by a "bundle" (*faisseaux*) of fibres as a "trunk" (*tronc*) of a tree in which the fibres of the "branches" (*branches*) and the "smallest ramifications" (*des plus petits Rameaux*) are united.[72] It is through this organized and complex net of fibres that the "undetermined force" (*force indeterminate*)[73] of the soul exposes a certain tendency to act upon the fibers, after it has received certain patterns of impressions. These tendencies become her own "faculties." To retrace the development of these faculties, Bonnet focuses, like Condillac, on the minimal conditions of the soul-body-interface and of soul-body-interactions within a fibre "automaton" that he calls a human "statue" (*statue*).[74]

The statue becomes a "sensing being" (*Etre sentant*) in the very moment of the first sensation.[75] Its first sensation has to be "initiated" through an outer "object." Without the initiation event from the outside, the statue simply has no sensation at all and

---

[68] Cf. ibid., 98.

[69] Cf. ibid., 53.

[70] Ibid.

[71] Ibid., Préface, 21–22.

[72] Ibid.

[73] Cf. ibid., 100.

[74] Cf. ibid., 7: "Ne considérons point un Homme fait, placé au milieu d'une Campagne, & environné de mille Objets divers: l'examen des opérations du Cerveau d'un tel Homme deviendroit pour nous infiniment trop compliqué. Allons par degrés: Simplifions; pouvons-nous trop simplifier dans un sujet si composé, & si singulierement composé ? N'ENTREPRENONS pas même d'étudier les Enfans: ils sont encore trop difficiles à observer." For Bonnet's critique of Condillac's *Essai*, see ibid., 10–11.

[75] Cf. ibid., 24: "J'APPROCHE donc une *Rose* du *Nez* de la Statue: au même instant elle devient un *Etre sentant*. Son Ame est *modifiée* pour la première fois: elle est *modifiée* en odeur de Rose; elle *devient* une odeur de Rofe; elle se *reprèsente* une odeur de Rose."

does not know what a sensation is.[76] In a series of thought experiments that rely on observations and experiences, Bonnet sets the fibre automaton step by step into motion:

> Let us thus have recourse to a fiction (*fiction*). It is not Nature, but it has its foundation in Nature. We separate things that are, in their natural condition, united. However, we do this only to understand them better. We will reunite them afterwards step by step and get again closer to Nature. Imagine a Human whose senses are all intact, but who did not begin to use them. Suppose that we have the power to restrain the senses of this Human or to let them free in the order, in the time and in the manner that pleases us. Offer successively to each sense, and later on also simultaneously to different senses, the objects that are proper to affect them. Let us observe what results from these impressions … This Human will be a kind of *Statue*, and we will give him this name.[77]

Bonnet begins with the sense of smell, but he does not successively introduce the other four senses. Rather, he focuses on the different steps that connect the first bodily impressions of the statue with its first cognitive reactions. He thus reconstructs a process that begins with the emanation of specific particles from an object – a rose – that act upon the surface-fibres of the skin, and that continues with the mediation of these impressions to "olfactory nerves," their transport to the soul–body-interface in the "brain," and their transformation into the first "sensation or perception" of the soul.

In general, impressions that have been transformed into data of olfactory nerves move the net of fibres that covers like a rhizome the zone of the soul–body-interaction in the "brain" – the receptive and regulating center with which all nerves "communicate."[78] These movements occur in the form of certain patterns of simultaneously moved fibres or fibre bundles. The soul has the faculty to sense or recognize these impressions as "modifications" (*modifications*) of itself.[79]

---

[76]Cf. ibid., 15: "Déjà les mouvemens *vitaux* s'opèrent dans la Statue; les Liqueurs y circulent & portent à toutes les Parties la nourriture qui leur est nécessaire. Les *sens* sont prêts à jouer; mais, ils ne jouent point encore: le *Sentiment* n'est pas né."

[77]Ibid., 8.

[78]Cf. ibid., 20: "Un Organe qui communique avec tous les *Sens*, & par lequel l'Ame agit sur toutes les Parties de son *Corps* soumises à son empire est, sans doute, un Organe prodigieusement composé. Il est en quelque sorte l'abrégé de tous les Organes, un *Systême Nerveux* en raccourci. Les ramifications de tous les *Nerfs* doivent aller aboutir à cet Organe ou avoir avec lui la communication la plus étroite. Le *Siège de l'Ame s*eroit ainsi un *Centre* où tous les *Nerfs* iroient *rayonner*." For Bonnet's discussion of the "seat of the soul," see ibid., Préface, XVIII, and ibid., 18–19: "La découverte de l'origine des *Nerfs*, a conduit à placer l'*Ame* dans le *Cerveau*. Mais comme il n'y a que les *Corps* qui ayent une *relation* proprement dite avec le *Lieu*, nous ne dirons pas que l'Ame occupe un *Lieu* dans le *Cerveau*; nous dirons que l'Ame est présente au *Cerveau*, *& par le Cerveau* à son *Corps* d'une manière que nous ne pouvons définir … Quoiqu'il en soit de cette décision de l'*Anatomie*, que l'on ne prendra si l'on veut que pour la décision d'un Anatomiste, j'admets qu'il est quelque part dans le *Cerveau* une Partie que je nomme le *Siège de l'Ame*, & que je regarde comme l'instrument *immédiat* du *Sentiment*, de la *Pensée*, & de *l'Action*. Il est indifférent à mon but que cette Partie soit le *Corps Calleux*, ou tout autre Corps."

[79]Cf. ibid., 29: "Le *mouvement que* la *Rofe* imprime au *Nerf olfactif*, & que celui-ci transmet à l'*Organe du Sentiment* donne lieu à cette *modification* de l'*Ame* que nous exprimons par les termes d'*Odeur de Rose*. Cette *modification* est une *manière d'Etre* de l'*Ame*, un *état* distinct de tout autre *état*."

"Modifications" are more or less distinct perceptions with the help of which the soul can process epistemic acts. In a similar way as in David-Renaud Boullier's transformation of Malebranche's model of the soul-body-interface (first discussed in his *Essai philosophique sur l'âme des bêtes* (1728)[80]), Bonnet's soul reacts to modifications according to a pleasure principle and its volitional decisions. "Pleasure" (*plaisir*) and "pain" (*douleur*) depend for Bonnet on the "degree" of the impact of physical impressions on the structure of fibers.[81] Fast movements or "vibrations" can cause "painful" structural changes in fibres, while "smooth" movements are often "pleasant."[82] Pleasant sensations reinforce a certain tendency of the living agent to repeat the event or to actively try to repeat certain acts. Bonnet calls this effort "attention" (*attention*).[83] The repetition of pleasure results into passions for something that has been sensed or experienced. "Decisions" of the will "satisfy" these passions.[84]

---

[80]Cf. Cheung 2008, 113–117; and Boullier 1985, 359–360: "Supposons un Agent immatériel capable de remuer la matiére, uni pour cela à une portion de matiére organisée; supposons un Esprit uniquement susceptible de perceptions confuses qui auroient pour objet les petits mouvemens excités dans cette Machine à laquelle il est uni; soit que ces mouvemens naissent du différent choc que les Corps extérieurs produisent sur ces organes, soit qu'il naisse de l'interieur de la Machine même. Supposons outre cela, que par la constitution essentielle de ce principe spirituel, quelques-unes de ces perceptions soient agréables, et quelques autres affligeantes. ... Il ne faut plus autre chose, si ce n'est que le Créateur ait tellement ajusté les ressorts de cette Machine faite pour l'Ame de la Bête, que les desirs confus qui correspondent aux Sensations douloureuses ou agréables, produisent dans le cerveau diverses impressions lesquelles, en vertu de la structure générale, feront mouvoir la Machine d'une maniére propre à éviter la cause de la douleur, et à s'unir à celle du plaisir. Il y aura dans tout cela une merveilleuse harmonie, et j'y vois point de difficulté."

[81]Cf. Bonnet 1973, 90: "Ainsi en supposant que l'action d'un Objet sur deux Individus soit précisément la même, celui-là sera le plus *sensible* à cette action, dont les Fibres seront les plus *mobiles.* Si cette *mobilité* est *excessive,* l'Individu aura une Sensation *désagréable*; les Molécules tendront à se désunir. Si les Fibres n'ont au contraire, que fort peu de mobilité, l'Individu ne sera affecté que très foiblement. Il le sera dans la proportion qui fait le *Plaisir* si les Fibres ont une mobilité *tempérée.*"

[82]Cf. ibid., 87: "Nous ne pouvons pas plus définir le *Plaisir* ou la *Douleur,* qu'une Sensation quelconque. Nous sçavons seulement que toute Sensation tient à un *mouvement,* & qu'un mouvement *plus* ou *moins* fort, plus ou moins accéléré fait naître la Douleur ou le Plaisir. La plus *légère* Sensation ne diffère du *Chatouillement* le plus vif, & celui-ci de la *Douleur* que par le *degré*; & c'eft au degré du mouvement que répond dans l'*Ame* ce *Sentiment* que nous exprimons par les termes de *Plaisir* ou de *Douleur,* comme c'est à l'*espéce* du mouvement ou de la Fibre, que répond la *Sensation* que nous exprimons par les termes *d'Odeur de Rose,* ou *d'Odeur d'Oeillet.* Ainsi la même Fibre qui produit le Plaisir lorsque ses vibrations sont accélérées dans un *certain* degré, fait naître la douleur lorsque ces vibrations sont accélérées au point de *séparer* trop les unes des autres les *Molécules* de la Fibre. La Douleur sera à son dernier terme, si cette séparation va jusqu'à la *Solution de continuité.*"

[83]Cf. ibid., 34: "J'entends ici, par l'*attention,* cette *réaction* de l'Âme sur les *Fibres* que l'Objet a mises en mouvement, par laquelle l'Ame tend à conserver, à fortifier ou à prolonger ce mouvement."

[84]Cf. ibid., 21–22: "L'*Action des* Objets, ou celle de l'Ame peut produire sur le *Fluide Nerveux* des effets analogues à ceux que la chaleur ou les frictions produisent sur le *Fluide, électrique* ... Nous ignorons la nature des *Esprits Animaux:* ils sont encore plus hors de la portée de nos sens & de nos Instrumens que les *Vaisseaux* qui les filtrent ou qui les préparent. Ce n'est que par la voye du raisonnement que nous sommes conduits à admettre leur existence, & à soupçonner quelqu'analogie entre ces *Esprits* & le *Fluide Electrique.* Cette analogie repose principalement sur certaines Propriétés très singuliéres de ce *Fluide*; en particulier sur la rapidité & la liberté avec lesquelles il se meut, le long d'une ou de plusieurs Cordes, au travers d'une masse d'Eau, même en mouvement."

Finally, knowledge about something that left an impression on the body emerges when various modifications have been "memorized" and "compared" with each other. Like in Condillac's statue, the first act of judgment of Bonnet's statue about the status of objects within its environment begins with "comparisons" of memorized or stored data. These data are "signs" which represent the sense-objects. After these first comparisons, more complex and abstract epistemic acts take place. The long chain of *impression – mediation – transportation – cognitive transformation – modification – pleasure & pain – attention – passion – memorization – comparison – abstraction* produces, in the last chapters of Bonnet's *Essai*, "thoughts" (*pensées*).

Bonnet compares the mediation of the object-impression to the soul with a musical performance or "play" (*jeu*) that is executed on an instrument (the fibre body).[85] The "original constitution" (*constitution originale*) of the body-automaton determines the quality of the sensation through its different fibre types that are moved like the keys of an organ, their spatial "arrangement" and their "coordination."[86]

In humans, sensations occur according to their five senses. These five sense "classes" can be again differentiated into an "indefinite number of genres and species."[87] Each sense is constituted through fibres that are made for "various kinds of sensations" (*diverses Espèces de Sensations*).[88] The sense of smell, for example, is composed of fibres that are apt to receive impressions from the "corpuscules" that "emanate from a rose," and others that are apt to receive impressions from a

---

[85]Cf. ibid., 15–16 and 52. Already in his *Essai de psychologie* (1755), Bonnet combined the image of a soul that plays on the body like an organist on an organ with the fibre fabric of the evolved germ. Bonnet 1978, 13–14: "Le siège de l'ame est une petite machine prodigieusement composé, & pourtant fort simple dans sa composition. C'est un abrégé très complet de tout le Genre Nerveux, une Neurologie en mignature. On peut se représenter cet admirable instrument des opérations de notre ame, sous l'image d'un clavessin, d'une orgue, d'une horloge, ou sous celle de quelque autre machine beaucoup plus composée encore … L'ame est le musicien qui exécute sur cette machine différens airs, ou qui juge de ceux qui y sont éxécutés, & qui les répète. Chaque fibre est une espèce de Touche, ou de marteau destiné à rendre un certain ton. Soit que les Touches soient muëes par les objets, soit que le mouvement leur soit imprimé par la force motrice de l'ame, le jeu est le même; il ne peut différer qu'en durée & en intensité." Cf. Bonnet 1973, 15–16 and 52. From Descartes, Pascal and Charles Perrault to Diderot, comparisons between clavichords or organs and the fibre mechanisms of organic order are frequently employed. Cf. Descartes, *Traité de l'homme*, 1996, vol. 11, 165; Perrault 1680, vol. 3; Perrault 1680, Avertissement, 1; and Diderot 1965, 50–51. For further references, see Proust 1963; Davies 1989, 108–150; Kassler 1995, 43–48; and Mayer 1959, 269–270 and 320–322.

[86]Bonnet 1973, 16: "La valeur *Physique & Morale* de nôtre Automate dépendra donc de sa *constitution originelle,* & de la *manière* dont nous aurons sçû *jouer* de cette *Machine.*"

[87]Cf. ibid., 23: "Nous avons cinq *Sens,* dont procèdent cinq *Classes* de *Sensations* qui ont sous elles un nombre indéfini de *Genres & Espéces.*"

[88]Cf. Bonnet, *Contemplation de la nature*, in Bonnet 1779, vol. 4/1, 150: "Chaque sens renferme donc probablement des fibres spécifiquement différentes. Ce sont autant de petits sens particuliers, qui ont leur maniere propre d'agir, et dont la fin est d'exciter dans l'ame des perceptions correspondantes à leur jeu."

carnation or a tuberose.[89] Bonnet highlights that the structural differences of most of the sense-specific fibres are still unknown.[90] However, he refers to the "papillae" (*papilles*) of the senses of taste and of touch. Their "pyramidal form" contains "fibres of different lengths that probably correspond to the diversity of impressions that they have to receive and to transmit" – just as the different "proportions" of "strings" of musical instruments correspond to various tones.[91]

Sensations of the soul are for Bonnet acts of representation of outer objects, and this representation is specifically coded through the organic fibre "constitution" of the statue in order to transmit suitable and exploitable sense data to the soul-body-interface.[92] For Bonnet, the specific interaction between the "corpuscles" of the rose and the "net" (*rameaux*) of nervous fibres of the sense of smell cannot be otherwise conceived than through an "impulsion" that results from the collision of bodies in the form of "vibrations" (*vibrations*), "undulations" (*ondolations*) or "pressure" (*pression*).[93] The impulsion-trace of the impression of the outer object is finally

89Cf. Bonnet 1973, 52: "LES Faits nous conduisent donc à penser que la *diversité* des Sensations ne dépend pas de la *diversité* des mouvemens imprimés par les Objets à des fibres *identiques* ... Ainsi, nous sommes acheminés à admettre qu'il est dans chaque *sens* des Fibres appropriées aux diverses *Espèces* de Sensations que le *sens* peut exciter dans l'Ame; qu'il y a, par exemple, dans l'Organe de *l'Odorat* des Fibres appropriées au jeu des Corpuscules qui émanent de la *Rofe,* d'autres au jeu des Corpuscules de l'*Oeillet,* d'autres à celui des Corpuscules de la *Tubereuse,* &c."

90Cf. ibid., 23 and 52–53.

91Cf. ibid., 52: "La forme piramidale des *Papilles* du *Goût* & de celles du *Toucher* semble confirmer cette Hypothèse. Il résulte de cette forme que chaque Papille contient des Fibres de différentes longueurs assorties, sans doute, à la diversité des impressions qu'elles doivent recevoir & transmettre. Personne n'ignore qu'en variant les *proportions* des Cordes d'un Instrument de Musique, on varie les *Tons*."

92Cf. Bonnet, *Contemplation de la nature*, in Bonnet 1779, vol. 4/1, 150: "Chaque sens renferme donc probablement des fibres spécifiquement différentes. Ce sont autant de petits sens particuliers, qui ont leur maniere propre d'agir, et dont la fin est d'exciter dans l'ame des perceptions correspondantes à leur jeu." In *Médicine de l'esprit* (1753), Le Camus refers to the impact of fibre properties on the "vivacity of movements" that excite "distinct ideas" (Camus 1753, vol. 1, 59): "La vivacité du mouvement qui excite en nous les idées distinctes, part aussi de quatre chefs. 1°. L'impétuosité des esprits, qui tire son origine de l'énergie des causes mouvantes nommées ci-dessus. 2°. La disposition des fibres, qui provient de leur structure, de leur sécheresse, de leur tension, de leur élasticité. 3°. La facilité qu'elles ont à se mouvoir à cause de certains mouvemens antécédens plusieurs fois répétés. 4°. Une seule, ou plusieurs de ces causes. Ce qui peut rendre compte de tous les degrés qui se trouvent dans l'intervalle d'un entendement médiocre à un génie heureux."

93Bonnet 1973, 27: "LA *manière* dont les *Corpuscules odoriférans* agissent sur les *Fibres nerveuses* m'est inconnue: je n'ai aucune voye pour parvenir à cette connoissance. Mais, comme dans l'ordre de mes idées, je ne conçois pas qu'un *Corps* puisse *agir* sur un autre *Corps* autrement que *par impulsion*; je pense que les *Corpuscules odorifèrans* étant doués d'un *certain mouvement,* & d'un *certain degré* de mouvement, communiquent ce *mouvement* dans une *certaine proportion* aux *Rameaux* du *Nerf olfactif* ... LA *nature* de ce *mouvement* est au nombre de ces *Déterminations* que j'ignore. Je ne sçais si c'edt un mouvement de *vibration, d'ondulation,* de *pression,* ou tout autre mouvement que je pourrois imaginer: je me borne donc à dire en général que les *Corpuscules odoriférans* impriment un *mouvement* aux *Rameaux du Nerf olfactif.*"

"transmitted" to the brain through a "nervous liquid" (*fluide nerveux*) and its various types of small particles, the animal spirits (*esprits animaux*). These animal spirits are still not observable. However, Bonnet suggests that they should have similar properties as the particles of "electric fluids."[94]

Physical sensations do not immediately end after an object stopped to "affect" a sensing agent. Rather, the length of a sensation depends on the "mobility" (*mobilité*) of the nerves and the "activity" (*activité*) of the outer corpuscles on the nerves which causes "vibrations" (*vibrations*) or "perturbations" (*ébranlements*) of the fibres.[95] But this is not the only effect of the impressions of outer objects. They also change the "original form" (*forme originelle*) of the affected fibres and the "position" (*position*) of their "molecules." Impressions can thus change the entire "structure" (*structure*) of the affected fibres.[96] The new structures possess new functional properties. These properties represent an adaptation of the fibres to the stimuli or impressions that they received. Bonnet calls the "capacity" of fibres to receive impressions their "temperament" (*tempérament*) and the capacity to acquire new dispositions their "mutability" (*mutabilité*).[97]

Fibres as mediators of sensations are "determined" through physical impressions and their various movements. Affected fibres have a "tendency" (*tendance*) or a "disposition" (*disposition*) to perform certain movements.[98] "Dispositions" thus represent the physical "memory" of the "physical ideas" that they received and that the soul can activate through its own movements:

---

[94]Cf. ibid., 21–22: "L'*Action des* Objets, ou celle de l'Ame peut produire sur le *Fluide Nerveux* des effets analogues à ceux que la chaleur ou les frictions produisent sur le *Fluide électrique* … Nous ignorons la nature des *Efprits Animaux:* ils sont encore plus hors de la portée de nos sens & de nos Instrumens que les *Vaisseaux* qui les filtrent ou qui les préparent. Ce n'est que par la voye du raisonnement que nous sommes conduits à admettre leur existence, & à soupçonner quelqu'analogie entre ces *Esprits* & le *Fluide Electrique.* Cette analogie repose principalement sur certaines Propriétés très singuliéres de ce *Fluide*; en particulier sur la rapidité & la liberté avec lesquelles il se meut, le long d'une ou de plusieurs Cordes, au travers d'une masse d'Eau, même en mouvement."

[95]Cf. ibid., 32: "AINSI quoique la Rose n'affecte plus *l'Odorat* de la Statue, elle peut continuer à sentir; mais plus foiblement. La *durée* de la Sensation est *proportionnelle* à la *mobilité* du Nerf, & à l'*activité* des Corpuscules qui ont agi sur le Nerf. Au même instant où l'ébranlement finira, la Statue cessera de sentir."

[96]Cf. ibid., 42: "L'*action* des Objets sur les Fibres y produit l'un ou l'autre de ces deux effets, & peut-être tous les deux ensemble: elle *modifie* la forme originelle de leurs Molécules, ou en *change* la position *respective*."

[97]Cf. ibid., 41: "LA *capacité* de recevoir ces déterminations, ou pour m'exprimer par un seul mot, la *mutabilité* des Fibres, a sa *raison* dans leur *Structure.*"; and ibid., 90: "JE définis le *Tempéramment* d'une *Fibre, l'aptitude* plus ou moins grande de cette Fibre à *céder* à l'impression de l'Objet. Cette aptitude tient en général, *aux proportions* de la Fibre, & à la *facilité* qu'ont ses *Molécules* de glisser les unes sur les autres, ou de s'écarter les unes des autres.

[98]This tendency determines also the "passions." Cf. Bonnet, *Contemplation de la nature*, in Bonnet 1779, vol. 4/1, 148: "Pourquoi les passions, qui ont leur source dans le tempérament, sont-elles si difficiles à maîtriser ? Elles tiennent fortement à la machine, et par la machine à l'ame. Les passions se nourrissent donc, croissent, se fortifient comme les fibres qui en sont le siege."

> In their first beginning, *Ideas* were only the movements that objects impressed on the fibres of the *Senses*. Therefore, the *conservation* of ideas through the memory depends on the disposition that the fibres of the *senses* possess to *follow* these movements and to repeat them (*à se prêter à ces mouvemens & à les répéter*).[99]

If the same impressions occur frequently, the "nutritive atoms" (*atomes nourriciers*) that are "assimilated" to the fibres of the senses, "fortify" their new dispositions. This fortification results into a "habit" (*habitude*) as a stable structure-function-complex of fibres that is capable to receive and to perform certain patterns of impressions and movements.[100] Embodied stimuli thus become part of an œconomy of fibres of living beings that sense their environment.

4 Concluding Remarks

In Bonnet's writings, the fibre is the main operative building block of structure-function-complexes in the œconomy of living automata that can receive, process and express sensations. His human statue represents a model of organized bodies in which merge (1) the Lockean concept of the genesis of ideas with Malebranche's fibre interface between the soul and the body, (2) anatomical, medical and physiological references to the fibre as the smallest organic building block, and (3) debates about the common order of humans and animals as living beings and machine-like entities with type-specific perfectibilities.

---

[99]Bonnet 1973, 40. Cf. Bonnet, *Contemplation de la nature*, in Bonnet 1779, vol. 4/1, 151: "Et puisque la réitération des mêmes mouvemens dans les mêmes fibres, y fait naître une disposition habituelle à les reproduire dans un ordre constant, nous pouvons en inférer que les fibres sensibles ont été construites sur de tels rapports avec la maniere d'agir des objets, qu'ils y produisent des changemens ou des *déterminations* plus ou moins durables, qui constituent le précieux fond de la mémoire et de l' imagination."

[100]Bonnet 1973, 69: "En se plaçant relativement à la disposition *actuelle* de la Fibre, les Atomes nourriciers maintiennent cette disposition, & si le même mouvement est répété de tems en tems dans la Fibre, & qu'il ne survienne point de mouvement contraire, ils la fortifient cette disposition, puisque leur incorporation dans la Fibre tend à augmenter sa Solidité. Voilà, la naissance de l'*Habitude*. Si l'on dit en général, que la répétition des Actes la fortifie, c'est que la répétition des actes est une répétition de mouvemens, & qu'une répétition de mouvemens augmente la *tendance* aux mouvemens." Cf. Bonnet, *Contemplation de la nature*, Bonnet 1973–1983, vol. 4/1, 152: "Mais les fibres sensibles se nourrissent comme toutes les autres parties du corps: elles *s'assimilent* ou s'incorporent les matieres alimentaires; elles croissent, et tandis qu'elles se nourrissent et qu'elles croissent, elles continuent à s'acquitter de leurs fonctions propres; elles demeurent essentiellement ce qu'elles sont. Leur méchanique est donc telle, qu'elles s'incorporent les matieres alimentaires dans un rapport direct à leur structure et à leurs déterminations acquises. Ainsi la nutrition tend à conserver aux fibres ces déterminations et à les y enraciner; car à mesure que les fibres croissent, elles prennent plus de consistance, et je crois entrevoir ici l'origine de l'habitude, cette puissante reine du monde sentant et intelligent."

For Bonnet, life is sensation, and the most important structure-function-complexes of the sensibility of living beings are the fibre apparati. Only living beings possess fibre apparati, and their life lasts as long as they are able to sense. All tissues and organs are composed of fibres, and the sensibility of the senses depends on the specific dispositions of their fibre bundles.

Like Condillac, Bonnet still applies a dualistic model of a soul–body-interface to explain the faculties of human sensitive agents. However, his notion of the sensitive agent as a mixed body relies not on the metaphysics of the substantial properties of the soul, but on an organized "milieu" of fibres as a transitory zone of the physical body that mediates both stimuli and cognitive data. This mediation takes place between the sense apparati that are covered by the skin and the rhizome-like nerve nets around the soul–body-interface in the brain. It is, according to Bonnet, observable in so far as movements of affected fibres are related to it. These movements establish not only causal connections between the affecting outer objects and their inner perception; they also occur in sequences and patterns that represent the specific appearance of the objects. These sequences and patterns can be memorized through the material "dispositions" of fibers. A similar fibre automaton is characteristic of Diderot's, Cabanis's and Destutt de Tracy's physical anthropologies.[101]

Bonnet's focus on stimuli-reaction-schemes and their visible movements in œconomies of fibres finally represents a general shift of interest from the ontology of the soul-body-interface and the mechanismis of movements to the processuality of the inside-outside-interface of living bodies. This shift, that is also indicated through other statue models in the writings of Buffon, Boureau-Deslandes and Condillac, takes place in the second half of the eighteenth century. In Bonnet's model of a human statue, the research object *living body* becomes a mediating physical entity that receives, transforms and expresses stimuli in the same way as it assimilates and excretes food particles. Both processes rely on the stimuli-reaction-schemes of sensitive fibres within organized bodies that function as adaptable, and yet stable containers for their performance.

[101] Some authors refer to Bonnet's fibre model of a human statue as a paradigmatic example for such a physical anthropology. See for example Mendelssohn 1784, 152–153: "Ein feines Gewebe von Fasern, welche ineinander verschlungen sind, und welche die Schwingungen und Bebungen, worin sie von äussern Gegenständen gesetzt werden, sich einander harmonisch mittheilen, dieses sind die Materialien, aus welchen sie [Condillac und Bonnet] eine ganze Geisterwelt erbauen wollen; und die Gesetze, nach welchen sich diese Schwingungen im Sichtbaren oder Fühlbaren erzeugen und mittheilen lassen, sollen dieselben sein, aus welchen sich die Gesetze der geistigen Verrichtungen alle herleiten lassen. Nachdem diese oder jene Fibern, mehr oder weniger, stärker oder schwächer, erschüttert worden; ist der Erfolg bald ein Urtheil des Verstandes, bald ein witziger Einfall, bald eine großmüthige Entschließung. Diese Denker müssen sich in der That haben aus der Acht kommen lassen, dass Faser, Gewebe, Schwingen, Beben, und alle Synonyma, deren sie sich zum Behuf ihrer Hypothese bedienen, ursprünglich aus der Finger- und Augensprache entlehnt sind, und also nichts anders beweislich machen können, als was sich betasten oder besehen lässt ... Allein ... dieses Uebersetzen und bildliche Vorstellen für eine Erklärung ausgeben, heißt den Witz mit Händen greifen oder die Vernunft durch die Brille sehen wollen." On the same subject, see also Engel 1784.

References

Anderson, Lorin. 1982. *Charles Bonnet and the order of the known*. Dordrecht, Boston, London: D. Reidel Publishing.

Audidière, Sophie, et al. eds. 2006. *Matérialistes français du XVIIIe siècle. La Mettrie, Helvétius, d'Holbach*. Paris: P.U.F.

Baasner, Frank. 1988. Der *Begriff 'sensibilité' im 18. Jahrhundert: Aufstieg und Niedergang eines Ideals*. Heidelberg: Winter.

Barlow, Alfred. 1879. *The History and Principles of Weaving by Hand and Power* (1st edn. 1878). 2nd edn. London: Sampson et al.

Baruzzi, Arno, ed. 1968. *Aufklärung und Materialismus im Frankreich des 18. Jahrhunderts*. München: List.

Baum, Gustav. 1913. *Die Baumwoll-Spinnerei und Weberei in ihrer bautechnischen und maschinellen Entwicklung*. Berlin: M. Krayn.

Berg, Alexander. 1942. Die Lehre von der Faser als Form- und Funktionselement des Organismus. Die Geschichte des biologisch-medizinischen Grundproblems vom kleinsten Bauelement des Körpers bis zur Begründung der Zellenlehre. *Virchows Archiv für Physiologische Anatomie und Physiologie und für klinische Medizin* 309: 333–460.

Bernardi, Walter. 1992. *I fluidi della vita. Alle origini della controversia sull'elettricità animale*. Florence: L. S. Olschki.

Bodemer, Charles W. 1964. Regeneration and the Decline of Preformationism in Eighteenth Century Embryology. *Bulletin of the History of Medicine* 38: 20–31.

Boerhaave, Hermann. 1703. *De usu ratiocinii Mechanici in Medicina. Oratio habita in auditorio magno. 24. September 1703*. Lugduni Batavorum [Leiden]: Apud Johannem Verbessel.

Bohnsack, Almut. 2002. *Spinnen und Weben. Entwicklung von Technik und Arbeit im Textilgewerbe* (1st ed. 1981). 2nd edn. Bramsche: Rasch Verlag.

Bonnet, Charles. 1978. *Essai de psychologie* (1st edn. 1755). Reprint. Hildesheim, New York: Olms.

Bonnet, Charles.1973. *Essai analytique sur les facultés de l'âme* (1st ed. 1760). Reprint. Hildesheim, New York: Olms.

Bonnet, Charles. 1779–1783. *Œuvres d'Histoire naturelle et de Philosophie*. 10 vols. Neuchâtel: Fauche.

Bonnet, Charles. 1985. *Considérations sur les corps organisés* (1st ed. 1762). Reprint of the *Œuvres* edition (1779–1783) vol. 3. Paris: Fayard.

Bonnet, Charles. 2002. *Palingénésie* (1st ed. 1769). Reprint. Paris: Fayard.

Boullier, Renaud-David. 1985. *Essai philosophique sur l'âme des bêtes* (1st ed. 1728). Reprint of the 2nd edn. (1737). Paris: Fayard.

Bourdin, Jean-Claude. 1998. *Diderot: Le matérialisme*. Paris: P.U.F.

Bourguet, Louis. 1729. *Lettres philosophiques sur la formation des sels et des crystaux et sur la génération et le méchanisme organique des Plantes et des Animaux*. Amsterdam: F. L'Honoré.

Buscaglia, Mario. 1994. Bonnet dans l'histoire de la méthode expérimentale. In *Charles Bonnet (1720–1793): Savant et Philosophe. Actes du colloque international de Genève*, ed. M. Buscaglia et al., 283–316. Genève: Editions Passé-Présent.

Camus, Antoine Le. 1753. *Médecine de l'esprit. Où l'on traite des Dispositions & des Causes Physiques qui, en consequence de l'union de l'ame avec le corps, influent sur les opérations de l'esprit; & des moyens de maintenir ces opérations dans un bon état, ou de les corriger lorsqu'elles sont viciées*. 2 vols. Paris: Chez Ganeau.

Camus, Antoine Le. 1769. *Médecine de l'esprit …* (1st ed. 1753). 2nd ed. Paris: Chez Ganeau.

Cheung, Tobias. 2004. Die Ordnung des Organischen. Zur Begriffsgeschichte organismischer Einheit bei Charles Bonnet, Spinoza und Leibniz. *Archiv für Begriffsgeschichte* 46: 87–108.

Cheung, Tobias. 2005a. Charles Bonnets allgemeine Systemtheorie organismischer Ordnung. *History and Philosophy of the Life Sciences* 26: 177–207.

Cheung, Tobias. 2005b. The Hidden Order of Preformation: Organized Bodies in the Writings of Louis Bourguet, Charles Bonnet and Georges Cuvier. *Early Science and Medicine* 11: 11–49.

Cheung, Tobias. 2005c. *Charles Bonnets Systemtheorie und Philosophie des Organischen.* Frankfurt am Main: Harri Deutsch Verlag.

Cheung, Tobias. 2008. *Res vivens. Agentenmodelle organischer Ordnung 1600–1800.* Freiburg im Breisgau, Berlin, Wien: Rombach.

Clericuzio, Antonio. 1994. The Internal Laboratory. The chemical reinterpretation of medical spirits in England (1650–1680). In *Alchemy and Chemistry in the 16th and 17th Centuries*, eds. P. Rattansi and A. Clericuzio, 51–83. Dordrecht et al.: Kluwer Academic.

Davies, Catherine Glyn. 1989. *Conscience as Consciousness: the Idea of Self-awareness in French Philosophical Writing from Descartes to Diderot.* Oxford: Voltaire Foundation.

Des Chene, Dennis. 2005. Mechanisms of Life in the Seventeenth Century: Borelli, Perrault, Régis. *Studies in History and Philosophy of Biological and Biomedical Sciences* 36: 245–260.

Duchet, Michèle. 1995. *Anthropologie et histoire au siècle des Lumières.* (1st ed. 1971) Reprint. Paris: Albin Michel.

Engel, Johann Jakob. 1784. Die Bildsäule. *Berlinische Monatszeitschrift* 4: 396–407.

Glassman, Robert B. & Buckingham, Hugh W. 2007. David Hartley's Neural Vibrations and Psychological Associations. In *Brain, Mind and Medicine: Essays in Eighteenth-Century Neuroscience*, eds. H. Whitaker et al., 177–190. Boston: Springer.

Haller, Albrecht von. 1747. *Primae lineae physiologiae in usum praelectionum academicarum.* Göttingen: Vandenhoeck.

Henry, John. 1989. The matter of souls: medical theory and theology in seventeenth-century England. In *The Medical Revolution of the Seventeenth Century*, eds. R. French and A. Wear, 87–113. Cambridge: Cambridge University Press.

Ishizuka, Hisao. 2006. Enlightening the Fibre-Woven Body: William Blake and Eighteenth-Century Fibre Medicine. *Literature and Medicine* 25: 72–92.

Jaynes, Julian. 1970. The Problem of Animate Motion in the Seventeenth Century. *Journal of the History of Ideas* 31: 219–234.

Kassler, Jamie C. 1995. *Inner Music. Hobbes, Hooke, and North on Internal Character.* London: Athlon Press.

Lawrence, Christopher. 1979. The nervous system and society in the Scottish Enlightenment. In *Natural Order. Historical Studies of Scientific Culture*, eds. B. Barnes and S. Shapin, 19–40. Beverly Hills, London: Sage.

Malebranche, Nicolas. 2006. *De la recherche de la vérité.* (1st ed. 1674–675) Présentation, édition et notes par J.-C. Bardout et al. 2 vols. Paris: J. Vrin.

Marx Jacques. 1976. *Charles Bonnet contre les Lumières 1738–1850.* 2 vols. Oxford: Voltaire Foundation.

Mayer, Jean. 1959. *Diderot. Homme de science.* Rennes: Imprimerie Bretonne.

Mazzolini, Renato G. & Roe, Shirley A. 1986. *Science Against the Unbelievers: The Correspondence of Bonnet and Needham, 1760–1780.* Oxford: The Voltaire Foundation.

Mendelssohn, Moses. 1784. Die Bildsäule. *Berlinische Monatszeitschrift* 4: 130–154.

Moravia, Sergio. 1978. From Homme Machine to Homme Sensible: Changing Eighteenth-Century Models of Man's Image. *Journal of the History of Ideas* 39: 45–60.

Mullan, John. 1988. *Sentiment and Sociability: The Language of Feeling in the Eighteenth Century.* Oxford: Clarendon.

Perrault, Charles. 1680–1688. *Essais de physique ou recueil de plusieurs traitez touchant les choses naturelles.* 4 vols. Paris: J.-B. Coignard.

Picavet, François. 1891. *Les idéologues. Essai sur l'histoire des idées et des théories scientifiques, philosophiques, réligieuses, etc., en France depuis 1789.* Paris: Félix Alcan.

Proust, Jacques. 1963. Variations sur un thème de l'«Entrétien avec D'Alembert». *Revue des sciences humaines* 112: 435–454.

Roger, Jacques. 1993. *Les sciences de la vie dans la pensée française au XVIIIe siècle* (1st edn. 1963). 2nd edn. Paris: Albin Michel.

Rieppel, Olivier. 1988. The Reception of Leibniz's Philosophy in the Writings of Charles Bonnet (1720–1793). *Journal of the History of Biology* 21: 119–145.
Riskin, Jessica. 2002. *Science in the Age of Sensibility: The Sentimental Empiricists of the French Enlightenment*. Chicago: University of Chicago Press.
Savioz, Raymond. 1948. *La philosophie de Charles Bonnet de Genève*. Paris: Vrin.
Sonntag, Otto, ed. 1983. *The Correspondence Between Albrecht von Haller and Charles Bonnet*. Berlin, Stuttgart, Vienna: Hans Huber Publishers.
Suy, Rau. 2007. Philip Verheyen (1648–1710) and his Corporis Humani Anatomiae. *Acta Chirurgica Belgica* 107: 343–354.
Thomson, Ann. 2008. *Bodies of Thought. Science, Religion, and the Soul in the Early Enlightenment*. Oxford, New York: Oxford University Press.
Tripp, Matthias. 1986. Le modèle méchanique comme paradigme epistémologique de la nature et de la pensée aux XVIIe et XVIIIe siècles. In *Epistémologie et Matérialisme*, ed. O. Bloch, 35–50. Paris: Méridiens Klincksieck.
Van Sant, Ann Jessie. 1993. *Eighteenth-Century Sensibility and the Novel: The Senses in Social Context*. New York: Cambridge University Press.
Verheyen, Philip. 1712–1713. *Corporis humani anatomiæ*. (1st ed. 1693) New, enlarged edition. 2 vols. Coloniae: Ab Egmond.
Wallace, Wes. 2003. The Vibrating Nerve Impulse in Newton, Willis and Gassendi: First Steps in a Mechanical Theory of Communication. *Brain and Cognition* 51: 66–94.
Whitaker, Harry & Turgeon, Yves. 2007. Charles Bonnet's Neurophilosophy. In *Brain, Mind and Medicine: Essays in Eighteenth-Century Neuroscience*, eds. H. Whitaker et al., 191–200. Boston: Springer.
Wellman, Kathleen. 1992. *La Mettrie. Medicine, Philosophy, and Enlightenment*. Durham, London: Duke University Press.
Wolfe, Charles. 1999. Machine et organisme chez Diderot. *Recherches sur Diderot et sur l'Encyclopédie* 26: 213–231.
Yolton, John W. 1983. *Thinking Matter. Materialism in Eighteenth-Century Britain*. Minneapolis: University of Minnesota.
Yolton, John W. 1991. *Locke and French Materialism*. Oxford: Clarendon.

Empiricist Heresies in Early Modern Medical Thought

Charles T. Wolfe*

Abstract Vitalism, from its early modern to its Enlightenment forms (from Glisson and Willis to La Caze and Barthez), is notoriously opposed to intervention into the living sphere. Experiment, quantification, measurement are all 'vivisectionist', morally suspect and worse, they alter and warp the 'life' of the subject. They are good for studying corpses, not living individuals. This much is well known, and it has disqualified vitalist medicine from having a place in standard histories of medicine, until recent, post-Foucauldian maneuvers have sought to change the situation (but for unrelated, contextualist reasons). What is perhaps more suprising is that if we consider the emergence of medical 'theory' as a whole, from Harvey through to Locke and Sydenham, is the presence of a sustained anti-experimentalist line of argument, and this from the 'empiricist' (not Cartesian or Boerhaavian rationalist) side. It would seem then that 'empiricks', medical empiricists and other protagonists of an 'embodied empiricism' are not Boylean experimentalists who seek to map out Nature in its transparency, but deliberately archaic, Hippocratic observers of living bodies.

It is known that empiricism is essentially a medical invention, dating back to the original 'Empirics' in the third century BC, with Serapion, a disciple of Herophilus of Alexandria (although 'empiricist' tendencies can be made out as far back as Hippocrates' treatise *On Ancient Medicine*).[1] We are familiar with the distinction

*Earlier versions of this paper have been presented at meetings of AAHPSSS (the Australasian Association for History, Philosophy and Social Studies of Science) at the University of New England, Armidale, and RMIT, Melbourne, in 2007 and 2008; HOPOS, at the University of British Columbia, Vancouver, in 2008 and especially at the Embodied Empiricism workshop in Sydney, 2009. Thanks especially to Peter Anstey, Hal Cook, Sean Dyde, and John Sutton for helpful comments and suggestions.

[1]See Hankinson 1995, 64, and more broadly Pentzopoulou-Vallas 1990. For medical 'empiricks' in the period of the Scientific Revolution see Cook 1990.

C.T. Wolfe
Unit for History and Philosophy of Science
University of Sydney

C.T. Wolfe and O. Gal (eds.), *The Body as Object and Instrument of Knowledge: Embodied Empiricism in Early Modern Science*, Studies in History and Philosophy of Science 25,
DOI 10.1007/978-90-481-3686-5_16, © Springer Science+Business Media B.V. 2010

between Empiricists, Methodists and Dogmatists from Galen's accounts in treatises such as *On Medical Experience* and *On Medical Sects*[2] (the key trait of the empirics is their emphasis on description rather than definition and the compilation of what might be called tables of induction: collections of instances where one thing (say an illness) is seen to follow from another, which then constitutes an "experience" [*empeiria*].) Indeed, if we fast forward to the early modern era, we find Mandeville, in his book on hysterick and hypochondriac diseases, quoting the basic empiricist credo that 'nothing is in the mind which was not first in the senses' (*nihil est in intellectu* [or: *in mente*] *quod non fuerit in sensu*) and attributing it to the physician Sylvius.[3] In the eighteenth century, if we consult Diderot and D'Alembert's *Encyclopédie*, the article "Empirisme" is much shorter than the article "Empirique," which is entirely devoted to medicine.[4]

In what follows I would like to briefly consider this uniquely medical or embodied nature of empiricism, in order to stress how it differs from a more general or generic 'experimentalist' picture of scientific practice, including the common view of empiricism as stemming from the new Royal Society culture of experiment.[5] I have discussed elsewhere how careful consideration of even the 'Epistle to the Reader' of Locke's *Essay* reveals a project very different from the atomistic, sense-data-oriented doctrine of empiricism we have come to be familiar with since at least the early twentieth century; how we should no longer think of Locke (and indeed of other empiricists) as the "underlabourer" of Newton, Boyle, Huygens and Sydenham, due to the uniquely moral, theological and political motivations of his doctrine.[6]

The specifically medical dimension I want to focus on here, if combined with these reflections on Locke, could lead us to significantly revise our picture of what empiricism was in the early modern period, leading up to the notion of an *embodied empiricism*. A good deal of robust revisionist work has changed our picture of 'British empiricism' in the past 10–20 years (from Michael Ayers and David Fate Norton onwards). Its 'Britishness', its relation to experiment, its relation to Cartesianism and generally its status as an epistemology have all come under fire; most relevant here is the idea that the thinkers we call empiricists were both (a) less empirically minded than their sixteenth and seventeenth-century forebears, and in fact (b) were critical of empiricism, at least in the case of Hume.[7]

---

[2]Galen's relevant treatises are available in Walzer and Frede 1985.

[3]Mandeville 1730/1976, Preface, vi. On the history of the phrase (which is both philosophical and medical), see Cranefield 1970.

[4]A very different, more 'Baconian–Boylean' point of view is found in the article "Expérimental" by D'Alembert (1756, 298f.). This analysis can be extended all the way to the picture of materialism found in the French eighteenth century, which unlike its present-day variants is a 'biologism' rather than a 'physicalism'. I try to sketch out some aspects of this in Wolfe and Aury 2008, Wolfe 2009a, b.

[5]Contrast the standard view of empiricism as tied to the Royal Society's experimental culture and its origins in Bacon and the Oxford Philosophical Club, e.g. in Kaplan 1993, 44. Analyses which contextualize this culture of experiment as one of 'trust' and 'civility' (Shapin et al.) do not alter the classic view, for present purposes, as also indicated in the Introduction to this volume.

[6]See Salter and Wolfe 2009.

[7]Norton 1981, 334, 340.

I suggest that there might at the very least be *different kinds of empiricism* operative in English thought in this period, such as – for the sake of discussion – the following:

– A 'Royal Society', *experimentalist empiricism*, which may be the context in which an actual 'philosophy of experiment' emerges (as in Boyle's "There is a big difference betwixt the being able to make Experiments, and the being able to give a Philosophical Account of them"[8] or Bacon's "Founding a real model of the world ... cannot be done without dissecting and anatomizing the world"[9]); this is summarized in 1728 by Ephraim Chambers, in his *Cyclopædia*, with the statement that "Experiments, within these fifty or sixty years, are come into such vogue, that nothing will pass in philosophy, but what is founded on experiment, or confirm'd by experiment, &c. so that the new philosophy is altogether experimental."[10]

– A *moral/practical empiricism* (Locke and Hume), in which themes such as anti-innatism (that is, the denial that there are innate ideas in the mind) are in fact not epistemological, that is, not primarily reducible to concerns about the nature of knowledge or the cognitive states of the knower, but are rather motivated by embedded concerns such as anti-authoritarianism (as in Locke's rejection of an innate notion of 'patriarchal' authority) and the desire to articulate a notion of toleration.[11]

– A *medically motivated, 'embodied' empiricism*, as found in such diverse figures as William Harvey (see the current work of Alan Salter), Pierre Gassendi, Thomas Sydenham; in a different sense, La Mettrie, especially when located as he meant to be within the tradition or trend of 'medical Epicureanism'[12], as in Gassendi-Lamy-La Mettrie[13]; and the Montpellier vitalists.

[8]Boyle 1661/1999, 221.

[9]Bacon, *Novum Organum*, I, § 124.

[10]"Experimental Philosophy," in Chambers 1728, vol. 1, 368.

[11]On the idea of a specifically moral or practically motivated empiricism, see Dario Perinetti's work on 'moral certainty', Waldow & Wolfe (ms.) and more generally Gaukroger 2005.

[12]I discuss the tradition of 'medical Epicureanism' – whether as something medical or as a polemical, rhetorical construct (as Olivier Bloch puts it, "une médecine assez littéraire en somme . . . qui se réclame de Démocrite et Lucrèce face à Aristote et Galien"; Bloch 1992, 79) – in Wolfe 2009b; see also Cook 2007, 385, on Boerhaave's Epicureanism. Mandeville is obviously an Epicurean in his social theory, including the theory of the passions, but not in his medico-philosophical work on hysterical and hypochondriac diseases (which contains extensive materialist reflections on 'pneumatology', but not of a particularly Epicurean sort).

[13]On the Paris physician Guillaume Lamy 1644–1683, an atomist Epicurean author of works such as *De Principiis rerum* (1669), *Discours anatomiques* (1675) and *Explication mécanique et physique des fonctions de l'âme sensitive* (1677), see Anna Minerbi Belgrado's extensive, informative introduction to her edition of Lamy's *Discours anatomiques & Explication méchanique et physique des fonctions de l'âme sensitive* (Lamy 1996). For additional biographical information, see Mothu 1993, which is also available online.

That Locke worked closely with Sydenham, Richard Lower, and to some extent Thomas Willis, does not however render his empiricism a *medical* one. Locke's *Essay* contains not a single medical example, and never makes a 'realist' ontological claim involving bodily states and their truth-value (nor any reductive claims). It is clear that Locke devoted a number of years of his life to medicine (in different forms and contexts), and that his collaboration with Sydenham does reflect an emphasis on the 'practical' dimension; but after 1689, he ceases to care about medicine.[14] Furthermore, he explicitly states that his 'historical, plain Method', his project of a logic of ideas, is not itself to be confused with "Physical consideration of the Mind"; any effort to "enquire philosophically into the peculiar Constitution of Bodies" is "contrary to the Design of [his] Essay," which is chiefly motivated by "practical" concerns (the moral project I alluded to above).[15] However, it is possible to see medicine as falling broadly within a practical project, if it is a medicine emphasizing observation and the history of diseases, as I indicate below.

Here, focusing on the medical side of empiricism, I will claim that this trend culminates in the figure of vitalism, which philosophers of science are accustomed to present as the stupid person's view (the belief in extra-causal vital forces and other otiose entities) but which was in fact the most sophisticated and diverse medical 'model' available in the Enlightenment (especially if we define vitalism so that it encompasses figures like Haller in addition to Bordeu, Barthez, Ménuret, Fouquet, etc., while conversely not applying the term 'vitalism' retroactively to a series of incompatible views including chimiatry and animism). A structurally similar claim was made a generation ago by Georges Canguilhem, in his study of the origins of the notion of reflex action; Canguilhem wanted to show that the true founder of neurophysiology was not the mechanist Descartes but the chimiatric, Helmontian Thomas Willis.[16]

In this specifically medical empiricism, the stress is on observation rather than on experiment, on bodily states rather than on de-personalized, quantitative measures. This explains the otherwise puzzling presence of polemics against anatomy, quantitative methods, and instruments such as the microscope (similarly to Harvey) – experiment and instruments go better with mechanism. The microscope

---

[14]I differ here from Peter Anstey (and from earlier, much more inflated claims such as Patrick Romanell's), and find myself in greater agreement with J.C. Walmsley's work on Locke and medicine; Milton 2001, 221 similarly comments that Locke no longer exhibited any interest in medicine after 1689.

[15]Locke 1975, I.i.2, II.xxi.73. This should also lead us to reject, or at least be very cautious with, the image of Locke as the 'underlabourer' of the Scientific Revolution – which is not at all the same as the image of a 'neurophilosophical' Locke (or Hume) concerned with relations between 'trains of animal spirits' and 'trains of thought' (as discussed in Sutton's contribution to this volume; the extent to which Sutton's new, physiologically embedded description of empiricism concurs with or challenges the picture of empiricism I am presenting here, is a matter for further discussion).

[16]Canguilhem 1955/1977. The same point is made in Martensen 2004, without any reference to Canguilhem. The present claim differs from Canguilhem's inasmuch as it does not seek to rehabilitate one figure at the expense of another, but rather to revise our picture of a construct: empiricism.

is much friendlier to the iatromechanist, or the 'expanded mechanists' like Malebranche or Leibniz, than to the type of empiricist who promotes unaided sense-perception.

Locke and Sydenham similarly reject the value of microscopy:

> All that Anatomie can do is only to shew us the gross and sensible parts of the body, or the vapid dead juices all which, after the most diligent search, will be noe more able to direct a physician how to cure a disease than how to make a man. ... How to regulate his dose, to mix his simples and to prescribe all in a due method? All this only from history and the advantage of a diligent observation of these diseases, of there beginning progress and ways of cure ... sugar in some stomachs turns to acidity and milk the most universal and innocent food in the world, is to some men as bad as poison. The anatomist will hardly be enabled to tell us, therefore, what changes any particular medicine either makes or receives in the body ...[17]

and especially,

> it is ... beyond controversy that nature perform all her operations in the body by parts so minute and insensible that I think noe body will ever hope or pretend even by the assistance of glasses or any other invention to come to a sight of them.[18]

For Mandeville, too, "Our shallow Understandings will never penetrate into the Structure of Parts of that amazing as well as mysterious Composition, the Mass of Blood."[19] It is apparent, then, that empiricists "rejected instrument-assisted sense perception," whereas the microscope and related technologies were "very much welcomed by the mechanical philosopher."[20] Indeed, from the mechanist standpoint the microscope is viewed as extending the sensory powers of the subject, and indeed her intellectual powers (notably in the Spinozist sense that it augments the causal power of the body and correspondingly "that part of the mind which is eternal"[21]). In contrast, the empiricist has a kind of first-person view of experience (*not, however,* reducing it to uniquely mental states as an animist would: for there is a difference between criticizing iatromechanism for its inability to capture features of living bodies, and criticizing it for missing the existence of an additional, spiritual substance,[22] a difference which also separates the vision of embodiment presented here from more phenomenologically inspired accounts of the *corps propre*

[17]Sydenham/Locke, *Anatomia* (1668) (the text is attributed both to Sydenham and to Locke, but its attribution does not affect the claims made here); Locke's 'version' is Locke ms., National Archives PRO 30/24/72/2 ff. 36v–37r., *cit.* Walmsley 2008, 70. It is also transcribed in Dewhurst 1966, 85–93. On this resistance to the microscope see Wolfe 1961.

[18]Sydenham/Locke, *Anatomia*, *op. cit.*, in Dewhurst 1966, 85 (also quoted in Dewhurst 1958, 7–8).

[19]Mandeville 1730/1976, 168.

[20]Van Speybroeck, et al. 2002, 18.

[21]Spinoza, *Ethics* V, prop. 39; Garrett 2003, 221.

[22]On medicine in this context as articulating a 'functional' rather than a 'substantival' dualism see Wright 2000; on how some vitalistic critiques of mechanism are better seen as 'complexifications' of mechanism see Wolfe & Terada 2008. "It is something of an 'ism' paradox that the eighteenth-century 'mechanists' generally described the body in non-quantitative terms whereas the 'animists' used mathematics to demonstrate the need of a soul to power the machine of the body" (French 1990, 103).

versus the body inhabiting Cartesian or Newtonian space[23]), a first-person state of experience in which "the investigator … [acquires] unmediated ocular evidence [*or tactile evidence, etc. CW*] of the way things stood in the body rather than testing hypotheses by means of artificial experiment."[24] That is, there is not a demarcation between an 'experiment' as something depersonalized and an 'experience' as something that is *my own*. Hence the emphasis on observation is not just a reflection of one particular 'way of knowing', or epistemological worry; for Sydenham and Locke, "observation without instruments is a moral imperative."[25]

We would of course want to ask here: what *kind* of knowledge is this embodied, first-person knowledge characteristic of the medical empiricist? Is it scientific knowledge? (Or, how does it fit in a more historically appropriate distinction between, e.g., natural philosophy, natural history, and the mathematical sciences? It clearly is close to natural history, from Bacon to Locke to Diderot; yet it is something else again.) Note in addition that by distinguishing the embodied, medical focus of this kind of empiricism from a more quantitative, experimentalist, 'Royal Society' kind of empiricism, I am not reiterating the older, fairly a-historical claim (which one finds in Charles Gillispie, but also in Koyré or Alistair Crombie) that medicine in the early modern period was more of an 'art' than a 'science' – Crombie goes as far as to declare that medicine was closer to religion than to science.[26] This does not mean that medicine was never opposed to 'science' as a body of formalized statements traceable back to first principles: one story tells how, in 1678, after the favourite of the viceroy of Naples was killed by a chemical remedy administered by a Galenist, the viceroy wished to protest to the profession, but was told that regulation and hence penalty was impossible, because medicine was not a *scientia*![27] Instead, one can consider that *personal* knowledge does not necessarily have to be 'art' rather than 'science'.[28] Witness Sydenham's efforts to not be identified with the 'empiricks' even though he also attacks 'learned medicine'.

If medical empiricism in its early modern form is to be distinguished from mechanistic medicine, and from a more 'rationalist' belief in the measurable transparency of Nature, it starts to resemble a slightly different creature, namely, medical vitalism.

---

[23]From Husserl to Hans Jonas and beyond, this anti-Cartesian tradition also completely misinterprets empiricism in 'disembodied' terms, as a theory of ideas condemned to skepticism and/or as a naïve, early psycho-physics. For an interesting attempt to enlist Hume in an embodied, 'enactive' phenomenological project, see Froese 2009.

[24]Dear 2006, 112. On the rich contextual sense of 'experimentum' and 'experimenta' see in addition Cynthia Klestinec's essay in this volume.

[25]Wolfe 1961, 209.

[26]"The effect aimed at by medicine is health. It shares this end rather with religion than with science" (Crombie, as quoted in Cook, 1990, 403).

[27]As described in French 2003, 188.

[28]As in Alan Salter's remark that Harvey put the phrase 'Per me' on the frontispiece of his Lumleian Lectures (an academic exercise if there was one). Of course this also has the effect of shifting the problem to another ground: if medical knowledge is a kind of personal knowledge, a kind of knowledge by acquaintance, akin to a 'craft knowledge', is it scientific knowledge? Cf. Cook 2007, 15, 34, 37.

The importance given to nosology – the history and taxonomy of diseases rather than the 'essentialist' concern with their internal causes, whether mechanistic or Helmontian – and the insistence on the legitimacy of (archaic) Hippocrates and Hippocratic medicine over and against 'modern' (e.g. mechanical) medicine, are elements common to the Locke/Sydenham view and to medical vitalism.[29] Sydenham had emphatically insisted on the need for a history and taxonomy of diseases rather than a theory of their internal causal composition. Illnesses needed to be repeatedly *observed* over their duration, not explained mechanistically, with the goal of being able to delineate them into *species*: "It is necessary that all diseases be reduced to definite and certain species …, with the same care which we see exhibited by botanists in their phytologies," he says in his *Medical Observations Concerning the History and Cure of Acute Diseases*.[30] Locke often praised Sydenham for his "practical" abilities, and many years after their collaboration, in an unpublished text entitled *On the Conduct of the Understanding* which was originally intended as a chapter of the *Essay*, Locke wrote, "were it my business to understand physick, would not the safer and readier way be to consult nature herself and informe my self in the *history of diseases* and their cures than espousing the principles of the Dogmatists, Methodists or Chymists"[31]? This "history of diseases" is, of course, what was also called "nosology." As such, it resonates with a later Hippocratic avatar in south-west France, the Montpellier vitalist school.

Montpellier vitalism (that is, the doctrine or cluster of positions associated with the Medical Faculty at Montpellier, from the early eighteenth to the early nineteenth centuries) consistently praises observation and disparages experiment, whether the latter consists of anatomical and pathological studies of corpses, or worse (in their opinion), of the vivisection of live animals. It also tends to include favourable references to empiricism, particularly to Condillac, but also to Locke, and in the case of Barthez – the 'leader' of and most influential spokesperson for the Montpellier vitalists in the late eighteenth century – to Hume, even if it is also true that these philosophical references tend to render Barthez's point murkier than it would have been otherwise. Henri Fouquet speaks of Haller's theory of the irritability of muscular fibres as dependent on a "horrific experimental set-up [*appareil d'expériences*]," ostensibly "guided by the desire to help humankind, but leaving out no painful instrument, no source of torment for … an infinite number of animals."[32] Jean-Joseph Ménuret de Chambaud (Ménuret) restricts his criticism to epistemology when he describes experimental phenomena in his programmatic

---

[29]On nosology and Montpellier vitalism see Martin 1990. Bichat in the early nineteenth century still has this specifically *medical* hostility to the microscope, as a source of potential errors (as noted in Grmek and Bernabeo 1997, 35).

[30]Sydenham, *Medical Observations* (orig. 1676), Preface to the 3rd edition, § 7, in Sydenham 1848, I, 13.

[31]Locke 1697, § 35 (emphasis mine).

[32]Fouquet 1765, 50b.

article "Observation" as having been arbitrarily "decomposed and combined," giving rise to conditions "far different from those present in nature."[33] When the physiologist dissolves blood freshly drawn from an animal, using other liquids, the knowledge he derives regarding the resulting mixture "is no longer the fruit of pure observation; ... knowledge acquired by this means is quite mediocre and imperfect" (*ibid.*, 314a). Louis de La Caze insists that 'observation' means "what can be observed on a healthy or sick body," whereas 'experiment' means "whatever can be observed of a dead body,"[34] and also targets Boerhaave's "useless experimentation" (*ibid.*, 47–48, 66).

Empiricism in its medical guise and vitalism in its Hippocratic, anti-substantialist mode thus inhabit a shared *Denkform* or rather discursive space. Common to them both, in addition to the above features, is an involvement with 'chymistry', particularly chimiatric medicine. From the empiricist standpoint, this was far removed from the more speculative, abstract character of iatromechanism; from the vitalist standpoint, chemical processes and particularly the relations and differences between 'aggregates' and 'mixts' were the best candidate for differentiating the organic from the inorganic. An additional feature in vitalist argumentation which is *not* present in the Sydenham–Locke critique of anatomy and experiment, is its usage of Newtonian analogies (see Hall 1968), which arguably complicates the distinction I suggested above between medical empiricism and Boyle–Newton, Royal Society empiricism. But in fact, contrary to the initial impression that empiricism should be a kind of 'epiphenomenalism' which disregards the essences of things, whereas vitalism would be an ontological claim about the unique nature of a particular class of beings, namely living beings,[35] the Newtonian motif in vitalism renders it equally 'agnostic' as to the ontological status of its entities. As Barthez declared, "I am as indifferent as could be regarding *Ontology* considered as the science of entities."[36]

The popularity of the analogy between vital force and Newtonian gravitation, in the eyes of the vitalists, lay in the combination this offered: explanatory power *and* the absence of obligations to provide an account of vitality in terms of micro-structure (given that iatromechanists, whether Descartes, Borelli, Baglivi or Boerhaave, consistently affirmed in widely known and quoted passages of their works that micro-structural explanations dispel ambiguities inherent in chimiatric language and subsume the variety of functions under a fixed number of mechanical, indeed mathematical laws[37]). This Newtonian feature of vitalism nicely complements the sustained references to Hippocrates and the Hippocratic tradition (as a non-reductive, non-essentialist type of medicine); taken together, they form a coherent anti-metaphysical position as we would expect from medical empiricism.

---

[33]Ménuret 1765, 313b.

[34]La Craz 1755, 9.

[35]Thanks to John Sutton for pointing out this issue.

[36]Barthez 1806, 96, note 17. See also French 1994, 315–316 for a parallel between Newtonian ontological agnosticism and Harvey's method.

[37]Baglivi 1696/1704, 135–136; Boerhaave 1708/1751, 81.

In sum: medical empiricism is different from the proof-and-validation experimentalism of Royal Society empiricism; it has a vitalistic flavour as distinct from mechanical medicine. This is not as shocking as might seem from a post-nineteenth century standpoint (that is, after the appearance of the 'neo-vitalisms' of Driesch or Bergson, which were less focused on medicine or the body than on the mechanisms of generation and, for better or worse, an ensuing metaphysics), as eighteenth century vitalism was a heavily pragmatic, heuristic sort of enterprise which, contrary to popular belief, has no concern with 'vital forces' or other extra-causal agents but rather with extending structural accounts of the organizational features of living bodies.

Now, in itself, the existence of a medicine of observation and nosology in contrast (indeed, in opposition) to a medicine of experiment and anatomy, is non-controversial – trivial, as it were – but its articulation *with*, or *as*, a kind of empiricism is less trivial. It implies, among other things, that 'biology', 'physiology' and other fields particularly concerned with living organisms do not have to wait for some of the 'classics' of Enlightenment popular science (Trembley's polyp, Bonnet's aphids, Haller's irritable muscle fibres, Maupertuis' African albino child and so on[38]) to suddenly appear, after centuries of so-called 'mechanism'. Not only is there a new, dynamic and active set of concerns grouped here under the heading 'medical empiricism' but even mechanism itself (a topic for a different paper) was far more concerned with the functional properties of life than is generally recognized, as recent work on Cartesian physiology has shown.

This is not tantamount to claiming that our uniquely medical, embodied empiricism is a 'paradigm' or brings about a 'paradigm shift'[39], nor that it was a specific research program; perhaps that it was a particular "historical form of mental activity," of experimental and experiential culture, of "cognitive practice."[40] And every attempt to distinguish between categories of a construct such as 'empiricism' will necessarily meet with disagreement since certain key figures (Harvey, Boyle and Locke come to mind) can be made to fit in several of these; other figures are hostile to the new program of experimental science for entirely different reasons (Hobbes). However, it remains the case that neither the history of science nor the history of philosophy have been able to account for this particularly 'embodied' context of

[38]See Barsanti 2000, 124 for this typical view (and a long enumeration of biologically relevant discoveries in the eighteenth century that on his view 'could not have been understood' a hundred years earlier). Similar arguments are made using the consolidation of chemistry as a basis for justifying the autonomy of 'biomedical' ontology and explanation, but they are *à double tranchant*.

[39]See Salomon-Bayet 1978, 15 and Hall 1968, 25 for interesting criticisms of the applicability of paradigms (or rather the very notion of paradigms) in early Enlightenment life science.

[40]I borrow these terms from the description of "historical cognitive science" in Sutton 1995. Of course, the idea of historically retrieving moments of 'tacit knowledge' is paradoxical by essence – but if we reject it outright then we also cut ourselves off from a variety of important interpretive moments, from the history of *mentalités* to recent discussions on 'experience' in the early modern period.

empiricism. It is clear that most, and perhaps all historians of the Scientific Revolution, be they internalist or externalist, inclined towards astronomy, alchemy or colonial voyages, miss this dimension of 'Life', and the specificity of living beings, which of course is much more loudly articulated in the subsequent hundred-odd years (from Leibniz's debate with Stahl, the so-called *Negotium otiosum*, to materialists such as Buffon and Diderot; from Haller to Bichat). Archaic medical invocations like Hippocratism and rather 'baroque' obsessions such as monsters are in fact signs of a dissatisfaction with Galilean, physico-mathematical explanations (it is eloquent in this sense that the *Encyclopédie* contains no entry on Galileo) faced with a kind of 'undecidability' of Nature.[41]

Should all of this revise our picture of the place of 'Life' within the Scientific Revolution (hitherto ignored or unacknowledged)?[42] Or is it sufficient to revise our vision of empiricism? It remains to be seen.

References

D'Alembert, Jean le Rond de. 1756. Expérimental (Philosophie nat.). *Encyclopédie* VI, 298–301. Paris: Briasson.

Baglivi, Giorgio. 1696/1704. *De Praxi Medica*, trans. G. Sewell & J.T. Desaguliers, *The Practice of Physick*. London: A. Bell et al., 2nd edition.

Barsanti, Giulio. 2000. Lamarck: Taxonomy and Theoretical Biology. *Asclepio* 52(2): 1191–131.

Barthez, Paul-Joseph. 1806. *Nouveaux Éléments de la Science de L'Homme*, 2 vols, 2nd edition. Paris: Goujon & Brunot.

Bloch, Olivier. 1992. L'Héritage Libertin dans la Pensée des Lumières. *Dix-Huitième Siècle* 24: 73–82.

Boerhaave, Hermann. 1708/1752. *Dr. Boerhaave's Academical Lectures on the Theory of Physic, Being a Translation of His Institutes and Explanatory Comments*, vol. 1 of 6. London: W. Innys (a translation of *Institutiones medicae*, 1708).

Boyle, Robert. 1999. *The Sceptical Chymist* (1661). In *Works*, eds. M. Hunter & E.B. Davis, vol. 2. London: Pickering & Chatto.

Canguilhem, Georges. 1955/1977. *La Formation du Concept de Réflexe aux XVIIe et XVIIIe Siècles*, 2nd revised edition. Paris: Vrin.

Chambers, Ephraim. 1728. *Cyclopaedia, or a Universal Dictionary of Arts and Sciences...*, 2 vols. London: J. Knapton.

Cook, Harold J. 1990. The New Philosophy and Medicine in Seventeenth-Century England. In *Reappraisals of the Scientific Revolution*, eds. D. Lindberg and R. Westman, 397–436. Cambridge: Cambridge University Press.

Cook, Harold J. 2007. *Matters of Exchange: Commerce, Medicine, and Science in the Dutch Golden Age*. New Haven: Yale University Press.

Cranefield, Paul. 1970. On the Origin of the Phrase *Nihil Est In Intellectu*. *Journal of the History of Medicine and the Allied Sciences* 25: 77–80.

---

[41]On monsters as a stand-in for the nature of living beings, see Wolfe 2005.

[42]On this question see Cook 1990 (especially 401–404), Duchesneau 1996, Rey 1994, Salomon-Bayet, Claire, 1978, 1981. However, it is also difficult to follow Cook's opposite claim, that medicine and natural history are "the big science of the early modern period" (Cook 2007, 410); compare the author's more measured statement (*ibid.*, 267).

Dear, Peter. 2006. The Meanings of Experience. In *The Cambridge History of Science*, vol. 3: *Early Modern Science*, eds. K. Park & L. Daston, 106–131. Cambridge: Cambridge University Press.
Dewhurst, Kenneth. 1958. Locke and Sydenham on the Teaching of Anatomy. *Medical History* 2: 1–12.
Dewhurst, Kenneth. 1966. *Dr. Thomas Sydenham (1624–1689), his Life and Original Writings*. London: Wellcome Historical Medical Library.
Duchesneau, François. 1996. Révolution Scientifique et Problématique de l'Étre Vivant. *Revue Philosophique de Louvain* 94: 568–598.
Fouquet, Henri. 1765. Sensibilité, Sentiment (Médecine). *Encyclopédie* XV, 38–52. Paris: Briasson.
French, Roger K. 1990. Sickness and the soul: Stahl, Hoffmann and Sauvages on Pathology. In *The Medical Enlightenment of the Eighteenth Century*, eds. A. Cunningham & R. French, 88–110. Cambridge: Cambridge University Press.
French, Roger K. 1994. *William Harvey's Natural Philosophy*. Cambridge: Cambridge University Press.
French, Roger K. 2003. Medicine Before Science. *The Business of Medicine from the Middle Ages to the Enlightenment*. Cambridge: Cambridge University Press.
Froese, Tom. 2009. Hume and the Enactive Approach to Mind. *Phenomenology and the Cognitive Sciences* 8(1): 95–133.
Garrett, Aaron V. 2003. *Meaning in Spinoza's Method*. Cambridge: Cambridge University Press.
Gaukroger, Stephen. 2005. The Autonomy of Natural Philosophy: From Truth to Impartiality. In *Natural Philosophy in the Seventeenth Century*, eds. P. Anstey & J. Schuster, 131–164. Dordrecht & New York: Springer.
Grmek, Mirko & Bernabeo, Raffaele. 1997. La machine du corps. In *Histoire de la pensée médicale en Occident*, ed. M. Grmek, vol. 2: *De la Renaissance aux Lumières*, 7–36. Paris: Éditions du Seuil.
Hall, T.S. 1968. On Biological Analogs of Newtonian paradigms. *Philosophy of Science* 35(1): 6–27.
Hankinson, R.J. 1995. The Growth of Medical Empiricism. In *Knowledge and the Scholarly Medical Traditions*, ed. D. Bates, 60–83. Cambridge: Cambridge University Press.
Kaplan, Barbara Beigun. 1993. *"Divulging of Useful Truths in Physic": The Medical Agenda of Robert Boyle*. Baltimore: John Hopkins Press.
de La Caze, Louis. 1755. *Idée de l'homme physique et moral pour servir d'introduction à un traité de médecine*. Paris: Guérin & Delatour.
Lamy, Guillaume. 1996. D*iscours Anatomiques & Explication Méchanique et Physique des Fonctions de l'Âme Sensitive*, ed. Anna Minerbi Belgrado. Oxford: Voltaire Foundation.
Locke, John. 1697. *On the Conduct of the Understanding*. Online at http://www.ilt.columbia.edu/publications/CESdigital/locke/conduct/toc.html.
Locke, John. 1975. *An Essay Concerning Human Understanding*, ed. P. Nidditch. Oxford: Oxford University Press.
Mandeville, Bernard. 1730/1976. *A Treatise of the Hypochondriack and Hysterick Diseases*, 2nd corrected edition. London: Tonson; reprint, Delmar, N.Y.: Scholars' Reprints.
Martensen, Robert L. 2004. *The Brain Takes Shape: An Early History*. New York: Oxford University Press.
Martin, Julian. 1990. Sauvages' Nosology. Medical Enlightenment in Montpellier. In *The Medical Enlightenment of the Eighteenth Century*, eds. A. Cunningham and R.K. French, 111–138. Cambridge: Cambridge University Press.
Ménuret de Chambaud, Jean-Joseph. 1765. Observation (Gram. Physiq. Méd.). *Encyclopédie* XI, 313–321. Paris: Briasson.
Milton, J.R. 2001. Locke, Medicine and the Mechanical Philosophy. *British Journal of the History of Philosophy* 9(2): 221–243.
Mothu, Alain. 1993. La mort de Guillaume Lamy. *La Lettre clandestine* 2. Online at http://www.univ-paris12.fr/scd/lc2-2d.htm.
Norton, David F. 1981. The Myth of British Empiricism. *History of European Ideas* 1(4): 331–344.

Pentzopoulou-Vallas, Teresa. 1990. Experience and Causal Explanation in Medical Empiricism. In *Greek Studies in the Philosophy and History of Science*, ed. P. Nicolacopoulos, BSPS, vol. 121, 91–107. Dordrecht: Kluwer.

Rey, Roselyne. 1994. Naissance de la Biologie et Redistribution des Savoirs. *Revue de Synthèse* 1(2): 167–197.

Salomon-Bayet, Claire. 1978. *L'Institution de la Science et l'Expérience du Vivant: Méthode et Expérience à l'Académie Royale des Sciences, 1666–1793*. Paris: Flammarion.

Salomon-Bayet, Claire. 1981. 1802, 'Biologie' et Médecine. In *Epistemological and Social Problems of the Sciences in the Early Nineteenth Century*, eds. H.N. Jahnke & M. Otte, 35–54. Dordrecht: Reidel.

Salter, Alan & Wolfe, Charles T. 2009. Empiricism contra Experiment: Harvey, Locke and the Revisionist View of Experimental Philosophy. *Bulletin de la Société d'Histoire et d'Épistémologie des Sciences de la Vie* 16(2): 113–140.

Sutton, John. 1995. Body, Mind, and Order: Local Memory and the Control of Mental Representations in Medieval and Renaissance Sciences of Self. In *1543 and All That: Word and Image in the Proto-Scientific Revolution*, eds. A. Corones & G. Freeland, 117–150. Dordrecht: Kluwer.

Sydenham, Thomas. 1848–1850. *Works of Thomas Sydenham*, trans. & ed. R.G. Latham, 2 vols. London: Sydenham Society.

Van Speybroeck, Linda, De Waele, Dani & Van De Vijver, Gertrudis. 2002. Theories in Early Embryology. Close Connections Between Epigenesis, Preformationism, and Self-Organization. *Annals of the New York Academy of Sciences* 981:7–49.

Waldow, Anik & Wolfe, Charles T. (ms. 2008). Empiricism as a Moral Project in Locke and Hume.

Walzer, Richard and Frede, Michael. 1985. *Galen. Three Treatises on the Nature of Science*. Indianapolis: Hackett.

Walmsley, J.C. 2008. Sydenham and the Development of Locke's Natural Philosophy. *British Journal for the History of Philosophy* 16(1): 65–83.

Wolfe, Charles T. 2005. The Materialist Denial of Monsters. In *Monsters and Philosophy*, ed. C.T. Wolfe, 187–204. London: College Publications.

Wolfe, Charles T. 2009a. 'Cabinet d'Histoire Naturelle', or: The Interplay of Nature and Artifice in Diderot's Naturalism. *Perspectives on Science* 17(1): 58–77.

Wolfe, Charles T., 2009b. A Happiness Fit for Organic Bodies: La Mettrie's Medical Epicureanism. In *Epicureanism in the Enlightenment*, eds. N. Leddy & A. Lifschitz, SVEC series, 69–83. Oxford: Voltaire Foundation.

Wolfe, Charles T. & Aury, Mathieu. 2008. Sommes-Nous les Héritiers des Lumières Matérialistes? *Phares* 8: 11–33 (Université de Laval, Québec), also online at http://www.ulaval.ca/phares/vol8-08/texte02.html.

Wolfe, Charles T. & Terada, Motoichi. 2008. The Animal Economy as Object and Program in Montpellier Vitalism. In *Vitalism without Metaphysics? A Reevaluation of Enlightenment Vitalism, Science in Context*, ed. C.T. Wolfe, vol. 21(4) special issue, 537–579.

Wolfe, D.E. 1961. Sydenham and Locke on the Limits of Anatomy. *Bulletin of the History of Medicine* 35: 193–220.

Wright, John P. 2000. Substance vs. Function Dualism in 18th Century Medicine. In *Psyche and Soma. Physicians and Metaphysicians on the Mind–Body Problem from Antiquity to the Enlightenment*, eds. J.P. Wright & P. Potter, 237–254. Oxford: Clarendon.

Index

FSC
www.fsc.org
MIX
Papier | Fördert
gute Waldnutzung
FSC® C083411

Kolibri 360 GmbH
Ferdinand-Jühlke-Straße 7
99095 Erfurt, Deutschland
produktsicherheit@kolibri360.de